OXFORD READINGS IN PHILOSOPHY

THE PHILOSOPHY OF ARTIFICIAL INTELLIGENCE

Also published in this series

Other volumes are in preparation

THE PHILOSOPHY
OF ARTIFICIAL
INTELLIGENCE

Edited by

MARGARET A. BODEN

OXFORD UNIVERSITY PRESS

Oxford University Press, Walton Street, Oxford OX2 6DP

Oxford New York
Athens Auckland Bangkok Bombay
Calcutta Cape Town Dar es Salaam Delhi
Florence Hong Kong Istanbul Karachi
Kuala Lumpur Madras Madrid Melbourne
Mexico City Nairobi Paris Singapore
Taipei Tokyo Toronto
and associated companies in
Berlin Ibadan

Oxford is a trade mark of Oxford University Press

Published in the United States by
Oxford University Press Inc., New York

Introduction and Selection © Oxford University Press 1990

First published 1990

British Library Cataloguing in Publication Data
Data available

Library of Congress Cataloging in Publication Data
The philosophy of artificial intelligence/edited by Margaret A. Boden.
p. cm.—(Oxford readings in philosophy)
Bibliography. Includes index.
1. Artificial intelligence—Philosophy. 2. Philosophy.
I. Boden, Margaret A. II. Series.
Q335.P48 1990 006.3'01—dc20 89-34075
ISBN 0-19-824854-7 (Pbk.)

5 7 9 10 8 6

Printed in Great Britain
on acid-free paper by
Biddles Ltd., Guildford and King's Lynn

CONTENTS

ABBREVIATIONS

AI Review	*Artifical Intelligence Review*
AISB Bulletin	*AISB Bulletin* (of the Society for the Study of Artificial Intelligence and Simulation of Behaviour)
Biol. Cybernetics	*Biological Cybernetics*
Brit. J. Phil. Science	*British Journal for the Philosophy of Science*
Commun. ACM	*Communications of the Association for Computing Machinery*
J. Experimental Psychol: General	*Journal of Experimental Psychology: General*
J. Philosophy	*Journal of Philosophy*
J. Theory of Social Behaviour	*Journal for the Theory of Social Behaviour*
Math. Comp.	*Mathematical Computing*
Phil. Trans. Roy. Soc. B	*Philosophical Transactions of the Royal Society, B*
Proc. & Addresses of Amer. Philos. Assoc.	*Proceedings and Addresses of the American Philosophical Association*
Proc. Aristotelian Soc.	*Proceedings of the Aristotelian Society*
Proc. 5th IJCAI Conference	*Proceedings of the Fifth International Joint Conference on Artificial Intelligence*
Proc. 1st AISB Conference	*Proceedings of the First AISB Conference* (of the Society for the Study of Artificial Intelligence and Simulation of Behaviour)
Proc. London Math. Society	*Proceedings of the London Mathematical Society*
Proc. Nat. Acad. Sci.	*Proceedings of the National Academy of Science*
Proc. Nat. Conf. AI	*Proceedings of the National Conference on Artificial Intelligence*
Proc. Roy. Soc. B	*Proceedings of the Royal Society, B*
Q. J. Exp. Psychol.	*Quarterly Journal of Experimental Psychology*
SIGART Newsletter	*SIGART Newsletter* (of the Special Interest Group on Artificial Intelligence, Association for Computing Machinery)
SIGSAM Bull., ACM	*SIGSAM Bulletin* (Association for Computing Machinery, Special Interest Group on Symbolic and Algebraic Manipulation)

INTRODUCTION

Artificial intelligence (AI) is sometimes defined as the study of how to build and/or program computers to enable them to do the sorts of things that minds can do. Some of these things are commonly regarded as requiring intelligence: offering a medical diagnosis and/or prescription, giving legal or scientific advice, proving theorems in logic or mathematics. Others are not, because they can be done by all normal adults irrespective of educational background (and sometimes by non-human animals too), and typically involve no conscious control: seeing things in sunlight and shadows, finding a path through cluttered terrain, fitting pegs into holes, speaking one's native tongue, and using one's common sense.

Because it covers AI research dealing with both these classes of mental capacity, this definition is preferable to one describing AI as making computers do 'things that would require intelligence if done by people'. However, it presupposes that computers could do what minds can do, that they might really diagnose, advise, infer, and understand. One could avoid this problematic assumption (and also side-step questions about whether computers do things *in the same way* as we do) by defining AI instead as 'the development of computers whose observable performance has features which in humans we would attribute to mental processes'. This bland characterization would be acceptable to some AI workers, especially amongst those focusing on the production of technological tools for commercial purposes.

But many others would favour a more controversial definition, seeing AI as *the science of intelligence in general*—or, more accurately, as the intellectual core of cognitive science. As such, its goal is to provide a systematic theory that can explain (and perhaps enable us to replicate) both the general categories of intentionality and the diverse psychological capacities grounded in them. It must encompass not only the psychology of terrestrial creatures, but the entire range of possible minds. It must tell us whether intelligence can be embodied only in systems whose basic architecture is brainlike (involving parallel-processing within networks of associated cells), or whether it can be implemented in some other manner. And, 'computers' having dropped out of the definition, their especial relevance to such a science must be proven. The many philosophical problems associated with AI arise from the question whether (and if so, how) this ambitious enterprise could be achieved or whether it is radically misconceived.

It follows that the philosophy of AI (considered as the science of intelligence in general) is closely related to the philosophies of mind, language, and epistemology and is central to the philosophy of cognitive science — and especially of computational psychology. Computational psychologists share four philosophical assumptions. They adopt a functionalist approach to mind and intelligence, regarding mental processes as rigorously specifiable procedures and mental states as defined by their causal relations with sensory input, motor behaviour, and other mental states. They see psychology as the study of the computational processes whereby mental representations are constructed, interpreted, and transformed. They view the brain as a computational system, asking what sorts of functional relations are embodied in it rather than which brain cells do the embodying or how their physiology makes this embodiment possible. And although they disagree (as AI workers do too) about which AI concepts and computer-modelling methodologies are likely to be most helpful in understanding intelligence, they all believe that AI concepts of some sort must form part of the substantive content of psychological theory.

The explanation of intelligence *by concepts significantly similar to those of AI* has long been a philosopher's dream, arguably since Plato (ch. 13). In past centuries, this dream has engendered metaphysical theories, formal specifications, and even exploratory modelling, of mental function — Hobbes, Leibniz, and Babbage spring to mind. In the twentieth century, its intellectual resources have been enriched by three developments: the formal theory of computation; the design of functioning machines to implement formally specified computations; and the discovery of the neurone.

These three developments underlie all of AI, although some research draws more obviously on one than another. Two classes of AI research commonly distinguished today are the 'traditional' variety (or GOFAI: good old-fashioned AI (Haugeland 1985)) and 'connectionism'. Although their theoretical relation is controversial, their historical relation is clear. They are branches arising from the same root, with common ancestry in a seminal paper co-authored by the neurophysiologist/psychiatrist Warren McCulloch and the mathematician Walter Pitts (ch. 1).

McCulloch and Pitts called their paper 'A Logical Calculus of the Ideas Immanent in Nervous Activity', a title which indicates the common birthright of traditional and connectionist AI. Their vision of implementing the 'logical calculus' influenced von Neumann in designing the digital computer, and inspired AI pioneers to attempt the formal modelling of thought. And their discussion of 'nervous activity' contributed to Hebb's psychophysiological theory of cell-assemblies, and engendered various models of neural networks—early precursors of today's connectionist systems.

The paper's influence was largely due to the fact that, while visionary, it was not merely speculative. Admittedly, its authors' ambitious views on the neural embodiment of purpose, learning, and psychopathology—not to mention epistemology, realism, universals, value, and number (cf. McCulloch 1965)—were argued only in the sketchiest terms. But McCulloch and Pitts did not simply argue the general materialist position that intelligence is embodied in the brain: they *proved* that certain (strictly definable) sorts of neural nets could in principle compute certain sorts of logical function.

They knew that the nervous system consists of interconnected cells, whose firing is all-or-none and dependent on the threshold and the activity of other cells. They knew, too, of Turing's (1936) paper on computable numbers and Russell and Whitehead's work on the propositional calculus. Integrating these diverse sources, they proved various theorems about the logical properties of idealized neural networks. For example: every function of the propositional calculus is realizable by some net (of a fairly simple sort); every net computes a function which can be computed by a Turing machine; and every Turing-computable function can be computed by some net. A Turing machine has an infinite tape—which is to say, it is a mathematical idealization, not an actual machine. Since neural nets are finite, one cannot adequately explain embodied intelligence by proving a generalized, and perhaps unrealizable, possibility. Rather, one must determine which nets could implement specific functions. Theoretical psychology thus becomes the design of nets capable of the computations carried out by minds.

This determination and design—complemented by the construction of working-models—is the task of AI, which (being concerned with actual and possible minds) includes human psychology as a special case. If 'nets' are thought of as approximations to real neural connectivities, then we have a broadly connectionist research-programme. Interpreted as highly abstract idealizations of neural activity, the prime focus being on binary logic rather than real cell-connectivities and thresholds, we have the digital information-processing typical of traditional AI. Both types of AI research were initiated as a result of McCulloch and Pitts's paper. Work on neural nets temporarily declined in the late 1960s, partly because of over-generalization of a critique pertinent only to a restricted class of nets (Minsky and Papert 1969). Its subsequent revival included the work discussed in Chapters 11–13 below, which (like most connectionist modelling in AI and psychology) makes no attempt to map anatomically identifiable neural connections. This challenge is sometimes taken up by neuroscientists (see ch. 14).

The theoretical groundwork for both approaches to AI was provided by Alan Turing's (1936) paper on computable numbers, which defined computation as the formal manipulation of (uninterpreted) symbols by the application of formal rules. The general notion of an 'effective procedure'—a strictly definable computational process—was illustrated by examples of mathematical calculation. But it implied (as McCulloch and Pitts realized) that if intelligence in general is explicable in terms of effective procedures implemented in the brain, then it could be simulated by a universal Turing Machine—or by some actual machine approximating thereto. By 1950 Turing and others had built general-purpose digital computers which were being used to simulate some aspects of intelligence. Now, in 'Computing Machinery and Intelligence' (ch. 2), he specifically asked whether such machines can think.

This should be decided, he argued, not on the basis of a prior (and possibly question-begging) definition of 'thinking' but by enquiring whether some conceivable computer could play the 'imitation game'. Could a computer reply to an interrogator in a way indistinguishable from the way a human being might reply, whether adding numbers or scanning sonnets? This question (often expressed as whether a computer could pass the 'Turing-test') has three aspects. Might some future computer actually be able to answer in the ways imagined? Are effective procedures in principle capable of generating this performance, whether in humans or computers? And would such performance suffice for the attribution of intelligence to the computer? Turing's own answer in each case was 'Yes'.

Turing's position has been attacked in three broadly different ways (neither mutually exclusive nor necessarily connected), each of which allows a number of significant variations.

The first type of attack uses a range of familiar anti-behaviourist arguments—having nothing specifically to do with AI—to dismiss the imitation game as a sufficient criterion of intelligence. However, even the anti-behaviourist appeal to conscious experience as a necessary condition of mentality cannot show intelligent computers to be impossible without additional arguments—as opposed to unargued intuitions—showing that consciousness cannot intelligibly be attributed to computers. The most that arguments against behaviourism in general can prove is that a high-performance computer *need not* be intelligent.

Proponents of AI can agree, taking the functionalist position that intelligence necessarily involves causal processes (computations) of a certain systematic sort. Behaviour—no matter how superficially impressive—due to some huge pre-stored look-up table, as contrasted with structured processes and representations (possibly mapping onto the mentalistic categories of folk-psychology), would not count as intelligent (Sloman 1986). On Turing's criterion, it would: he stipulated only that the causes underlying intelligent behaviour be effective procedures *of some sort.* Moreover, since he did not explicitly state that thinking *necessarily* involves underlying causes within the thinker, his criterion does not exclude behaviour happening by magic, or accidentally: nuts falling from a wind-blown tree onto the keys of a teletype could conceivably 'fool' a human interrogator playing the imitation game.

The second type of attack on Turing's position marshals additional arguments of one sort or another in claiming that computers *could not* be intelligent. One such argument regards Turing's reliance on verbal behaviour as even less plausible than behaviourism in general: in the absence not only of motor action but even of animate bodily form, psychological predicates cannot properly be ascribed to computers (Dreyfus 1979). Another objection argues that even if a computer were to perform as Turing imagined (scanning sonnets, and the like), it would not really be intelligent because no computer could conceivably really think or understand: no intelligence without intentionality. This charge does not threaten AI as a technology, or even as an exercise in simulation, for it allows that fully humanoid computer-performance may be possible. Nor does it deny the usefulness of computer models in clarifying theories in psychology (as in other sciences). But it insists that the conceptual content of AI cannot help philosophers or psychologists to describe or explain mental processes as such, since minds possess intentionality whereas computers do not—and *cannot.*

An influential example of this attack is John Searle's paper on 'Minds, Brains, and Programs' (ch. 3), which uses Turing's own concept of computation in rejecting his belief that a suitably programmed computer would be intelligent. Searle directs his criticism not at Turing's paper as such but at the two academic disciplines promised by it: 'strong' AI (the attempt to construct real mental powers by programming) and computational psychology (wherein AI contributes to the content of psychological theories).

Searle's primary argument, involving the imaginary 'Chinese room', takes for granted that AI programs and computer models are purely formal-syntactic (as is a Turing machine). On this basis, he claims that no system could understand purely in virtue of carrying out computations. Hence computational psychology cannot possibly explain our mental

powers, and *a fortiori* no program could confer intelligence on a computer. Searle's secondary argument claims that intelligence or intentionality
requires not only mindlike behaviour but also 'the right causal powers'
underlying this behaviour. As noted above, the AI proponent can agree
with the claim as so expressed. However, Searle identifies these causal
powers not in terms of dispositions or functions but in terms of material
stuff. Moreover, he regards it as intuitively obvious that neuroprotein can
generate intentionality, and that metal and silicon cannot.

In rebuttal, I argue (in ch. 4: 'Escaping from the Chinese Room') that
even the simplest program is not purely formalist but has some very primitive semantic properties, and that computational theories are therefore
not essentially incapable of explaining meaning. Moreover, in so far as the
brain's ability to generate intentionality is intelligible at all—as opposed to
being wholly counter-intuitive (how *could* that mushy stuff inside the skull
possibly understand?)—it is understood in information-processing terms
applicable likewise to computers. It follows that AI concepts can legiti-
mately be used as a substantive part of psychological theories, and that
some imaginable computers could possess powers closely similar to intentionality and intelligence. Whether any computer—even one whose internal computational economy was identical with ours—could appropriately
be termed intelligent *without even the faintest of scare-quotes* is another
matter: removal of the scare-quotes would require not just recognition of
the facts but also a moral decision on our part (Boden 1987: 423–5).

The third line of attack on Turing's position (to which chs. 5 to 14 are
variously relevant) argues that, contrary to his assumption, it is not
actually possible—in principle and/or in practice—to get computers to
perform in a way that matches the depth, range, and flexibility of human
minds. Irrespective of whether a sonnet-scanning computer would really
be intelligent or merely simulate intelligence, such a computer will never
exist. Often, this attack relies on variants of objections rebutted in his
paper: the arguments from 'informality' (irreducibility to rules) of
behaviour, creativity, and Godel's theorem (Dreyfus 1979; Lucas 1961).
Again, technological AI is not outlawed; indeed, useful AI systems have
already been produced. But the Holy Grail of AI and computational
psychology—a detailed computer model of human mental processes—is
impossible and/or infeasible.

Some philosophers may retort that infeasibility is irrelevant: logical, not
empirical, possibility is the issue. This response ignores the distinction
between a more abstract and a more realistic sense of 'empirically
possible', appealing respectively to basic scientific principles alone and to
scientific principles *plus very general real-world constraints*.

The Turing machine is an empirical impossibility in both senses, because it has an infinite tape. Other computational machines are like the legendary snowball-in-Hell: although not excluded by basic scientific principles, they are impossible in the real world because of constraints on time and/or space. For instance, a brute-search algorithm for chess would require an astronomical (though finite) timespan for each move; likewise, many visual tasks can be achieved in real time only by massively parallel processing, even if the processing units can react very much faster than neurones. In principle, such processing could be simulated on a serial computer (so some theoretical questions can be addressed without commitment to serial or parallel implementation); but in practice, only relatively small parallel systems can be so effected. Since our brains do not work in astronomical time, one should not conceptualize our minds in terms of computations whose actualization would require millennia. A further real-world constraint is our evolutionary origin; if (as argued in Clark (1987a)) certain sorts of computation are consistent with evolution whereas others are not, this has implications for the philosophy of mind and AI as well as for psychology.

Of those who share Turing's faith in the possibility of AI, none have done more to make it a practical reality, or to apply it to detailed psychological questions (involving abstract task-analysis as well as detailed experimental observation) than Allen Newell and Herbert Simon. And as is evident from 'Computer Science as Empirical Enquiry: Symbols and Search' (ch. 5), none have been more uncompromising about its implications for the philosophy of mind: the mind *is* a computational system, the brain *literally* performs computations (which are *sufficient* for intelligence), and these are *identical* with computations that could occur in computers. Human intelligence is explicable by sets of input–output rules controlling behaviour and (what the logically similar behaviourist psychologists ignored) internal information-processing. Computers too can be intelligent, since they possess the right causal powers: a computer—like a brain—is a physical-symbol system, and 'a physical-symbol system has the necessary and sufficient means for general intelligent action'.

The core of Newell and Simon's approach—as their opponent Searle (ch. 3) points out—is a formal-syntactic theory of symbolism linked to a causal theory of semantics. In their view, the criteria of identity of a symbol or computation are purely formal, and its significance is established by its causal history and effects. A symbol is a physical *pattern,* physically related to other patterns in various ways (such as juxtaposition) to form compound 'expressions'. Computational processes (realized by physical means) compare and modify *patterns:* one expression is the input, another the output. Any substrate physically capable of storing and systematically transforming expression-patterns can implement symbols. But for psychological purposes the substrate is irrelevant. To understand intelligence, we must describe physical-symbol systems at the information-processing level, in terms of designation and interpretation. These two semantic concepts are defined causally, such that the meaning of a symbol is the set of changes which it enables the system to effect, either to or in response to some (internal or external) state. The causal dependencies are essentially arbitrary, in the sense that any (non-complex) symbol could designate anything at all. (This constraint excludes analogical representation, in which there is some significant similarity between the representation and the thing represented (Boden 1988: 29–44).)

This definition of symbol systems can be criticized as overly physicalist even by someone who shares Newell and Simon's faith in the possibility of AI (Sloman 1986; in preparation). Quite apart from their unargued assumption that material instantiation is *necessary* to symbols (so ruling out angelic intelligence), their definition mentions only the physical, not the virtual machine. The 'virtual' machine is the machine which the programmer can *think* is being used. It is abstractly defined as the set of basic information-processing operations carried out by the system concerned; the symbols within virtual machines are abstract, not physical, entities. There may be several virtual machines within one computational system—as when a high-level programming-language is implemented in a lower-level one, which in turn is compiled into assembly-language, and thence converted into machine-code. The mind may consist of many such abstract symbolic machines, of which only the most fundamental is/are physically instantiated in brain-tissue (as opposed to being virtually instantiated in some lower-level system). However, the Newell–Simon account could be amended to meet this criticism, whose point is not that AI is impossible but that it is more complex than a literal reading of their definition suggests.

That AI can help us understand the mind—what it is, as well as how it works—is accepted by all computational psychologists, of whom some agree with Newell and Simon that AI *is* theoretical psychology (Longuet-Higgins 1987). Nevertheless, some criticize AI harshly: David Marr, in 'Artificial Intelligence: A Personal View' (ch. 6) dismissed much AI work as scientifically irrelevant. AI programs are often based on unarticulated and theoretically independent insights, and/or unprincipled empirical exploration. For Marr, a science of intelligence requires either 'Type-1' models based on theoretical understanding of fundamental (axiomatic) task-constraints or 'Type-2' implementations of intelligent performance effected by 'the simultaneous action of a considerable number of processes, *whose interaction is its own simplest description*'. A Type-2 theory may be discoverable or (more likely) it may not. If a task requires a Type-2 explanation too complex to be discovered, then it will never be understood in detail. Research in AI and psychology should seek Type-2 explanations only if Type-1 theories have been proven impossible.

Type-1 explanation involves three levels: the 'computational' theory provides axioms defining the content of the information-processing task concerned; the 'algorithm level' describes procedures capable of executing it; and the 'hardware' level shows how algorithms are implemented. Marr's usage of 'computational', like the mathematician's, denotes not temporal processes but timeless constraints. Temporal processes are considered at the algorithmic level. (The discussion of virtual machines, above, implies that this is not a single level; Marr himself posited several levels of visual processing (Marr 1982).) Just as various algorithms exist for any task, so diverse implementations are possible for any algorithm. Neurophysiologists and computational psychologists alike can sometimes benefit from considering the mutual constraints between algorithmic and hardware levels (see ch. 14).

The computational level (defining the basic tasks in intelligence) identifies the *natural kinds* of AI and psychology— which Marr believed to be largely unknown. He allowed generalized 2D-to-3D mapping (low-level vision) and grammatical parsing as natural kinds. But he excluded tasks defined by folk-psychology (like the attribution of intention, much studied by social psychologists), and even the arithmetic skills modelled by Newell and Simon. On his view, Newell and Simon's work is scientifically irrelevant because they ignore the fundamental, unconscious, information-processing tasks (pattern-matching may be crucial to arithmetic, but they take pattern-matching for granted). Like other modularity theorists (Fodor 1983), Marr denied that the 'higher mental processes' are explicable in detail: the degrees of freedom are considerable even in medical diagnosis, and astronomical in appreciating sonnets. Scientific understanding or comprehensive computer modelling of expertise or story-understanding is thus impossible. Technological AI may build useful 'expert systems', and (one day) even Turing's sonnet-scanner—but its only explanation would be a Type-2 theory too complex to be intelligible.

Marr's account of scientific explanation is seen by some computationalists as overly stringent. It ignores the explanatory role of elucidating a range of structural possibilities, within which natural phenomena must lie and in terms of which they can be systematically compared (Sloman 1978). Moreover, even when Type-1 analysis is available (as in low-level vision or parsing), it may have limited relevance to biological, and even artificial, systems. Evolution does not take the design stance in building intelligences from scratch, but 'tinkers' with the material available; elegant engineering solutions are found in organisms, but—unsurprisingly—so are unsystematic methods comparable to 'kludges' in computer programs (Clark 1987*b*). Real-time constraints suggest (and experimental evidence confirms) that even vision and parsing sometimes benefit from quick-and-dirty processing-methods not systematically explicable in Type-1 terms, but whose usefulness is intelligible without detailed Type-2 explanation.

One type of intelligence for which the possibility of a Type-1 computational task-analysis is hotly disputed is common-sense reasoning. It is widely assumed within AI that the thinking underlying our common sense is formalizable, and perhaps even deductive. But critics commonly object that even if logical and (some aspects of) scientific reasoning can be modelled by rules, everyday thinking cannot. Our tacit knowledge of the physical world, for example, is acquired by sensorimotor learning and has nothing to do with the abstract principles of physics. Since this unverbalized knowledge imbues not only our motor behaviour but our use of language too, the wide-ranging computerized conversations imagined by Turing are, on this view, impossible.

This dispute is outlined in Chapter 7 (an introduction to what AI workers call the 'frame problem'), and discussed in detail in Chapters 8 and 9. The frame problem was originally recognized in the context of robot planning, but appears also in AI work on language-understanding and common-sense thinking about social and cultural matters. The problem concerns foresight of the intended, and the potentially relevant unintended, effects of action in the real (physical or social) world. A formal representation of action must explicitly allow for *all* the intended effects, or some may not happen. And the action's myriad unintended effects will be wholly irrelevant only if the agent is very lucky, or very thorough in explicitly anticipating potentially relevant outcomes. Whether the frame problem can be solved in principle depends on whether any explicit representation of the various effects of action could conceivably be thorough enough. How (if at all) it can be solved in practice depends on how the relevant world knowledge can be identified, represented, and used.

In 'Cognitive Wheels: The Frame Problem of AI' (ch. 7), Daniel Dennett sketches some AI approaches to the frame problem, and argues that—besides its bearing on the feasibility of AI—it has independent interest for epistemology. Philosophers had not recognized it as a problem because they had not seriously asked *how* their preferred epistemological primitives could be employed in constructing, and reasoning from, knowledge. In general, new philosophical questions, and new insights into old questions, can arise from detailed consideration of work in AI. For AI researchers cannot ignore (though they can sometimes fudge) the 'how' question in writing programs. Even if—like some philosophers—they aim not to write programs but to provide a (Type-1) content-analysis of a given task-domain, their appreciation of the explicitness needed to make any future program run can benefit their abstract analysis.

Much AI work on the frame problem is of the latter, abstract, variety. It is closely akin to philosophical logic and epistemology, but its kinship with psychology is doubly problematic. First, it is not necessarily intended to simulate psychological reality: the aim may be to provide robots with a usable and reliable representation of common-sense knowledge, irrespective of what *processes* go on inside human heads. Second, there is some dispute within AI as to whether the *content* of common-sense knowledge can be axiomatized, or even formalized.

Turing's faith in the formalizability of common sense is shared by Patrick Hayes, who (with John McCarthy) first identifed the frame problem. In 'The Naïve Physics Manifesto' (ch. 8), Hayes grants (what AI critics remark) that everyday thinking about the material world does not employ theoretical physics—which in any event is inadequate to solve the relevant information-processing tasks. Rather, it employs 'naïve physics': our untutored and largely unconscious knowledge of the environment. This knowledge is involved in sensorimotor skills (like pouring) and linguistic understanding (of the verb 'to pour'). Formalization of naïve physics requires the analysis of many concepts dealing with matter, cause, space, and time; those addressed by Hayes (here and in subsequent papers (1985)) include *weight*, *support*, *velocity*, *height*, *inside*, *outside*, *next to*, *boundary*, *path*, *entrance*, *obstacle*, *fluid*, and *cause*. Likewise, Hayes sees our practical and linguistic grasp of social life as depending on 'naïve psychology', consisting not of empirical generalizations about how people behave but of the fundamental concepts and inference-patterns defining everyday psychological competence. If AI is to explain intelligence, or achieve a sonnet-scanning computer program, it must first complete the analysis of naïve physics and naïve psychology.

This challenge could be approached in various ways. One way, of which Hayes's research is a prime example, is 'logicism': the view that our basic common-sense knowledge can not merely be formalized (expressed as well-formed formulae in some formal language, such as a programming-language), but axiomatized. The axioms express general truths about the physical and social world, and inferences validly derived from them are as reliable as logical theorems. Indeed, they may as well be expressed in terms of formal logic (predicate calculus, for example). Logicist AI research aims primarily to develop an abstract (Type-1) theory of the fundaments of common sense, and only secondarily to write programs using this knowledge for robotics, planning or language-understanding.

For many years, Hayes could count Drew McDermott as a fellow-logicist. But McDermott has recently recanted, expressing his disillusion in 'A Critique of Pure Reason' (ch. 9). McDermott's discussion is reminiscent of arguments against the formalizability of knowledge (ch. 13), but his faith in formalizability is intact: his doubts concern axiomatizing knowledge, not programming it.

The crucial logicist fallacy, says McDermott, is to confuse deduction with computation, so to assume that all thinking is essentially deductive. Efforts by logicists—himself included—to define 'non-monotonic' logics, in which deductive conclusions can be withdrawn on the addition of new information (treated as extra premisses), fail to deal with everyday contingency. Axiomatic (Type-1) analysis being restricted to deductive domains, non-deductive AI-programs can be scientifically understood only by means of a general theory of non-deductive inference—which philosophers have long sought, and not found. Admitting the possibility that much of AI may be dead-end research 'buoyed by simple ignorance of the past failures of philosophers', McDermott hopes nevertheless that its concepts and techniques will help us discover the general theories which traditional epistemology has not. Till then, there can be no comprehensive science of intelligence, and all one can say in justification of a non-deductive AI-program is, 'It works!'

Common-sense thinking may present difficulties for AI, but it is at least *thinking*—a mental activity which most people believe to be amenable to computer modelling if any is. Motivation and emotion are another matter: it is widely doubted that a computational approach could possibly simulate or explain these aspects of mind (Dreyfus 1979; Haugeland 1978). Such doubts relate not only to their conscious dimension—for reasoning can be conscious, too—but to their general nature. Motives (and other conative categories such as intentions) are the origin or driving force of action, and have an intimate relation to personality and the self: how can they be likened to computations? As for emotions, these are the opposite of rationality, leading us to do things and to see things in ways we may abhor in our calmer moments; even if cognition can be computationally understood, surely emotion cannot? Moods appear even more recalcitrant to a computational analysis, affecting as they do virtually all our thinking, action, and experience (and, apparently, being associated with or caused by chemicals diffused throughout the body). In sum: it seems absurd to suggest that conation and affect will turn out to be theoretically intelligible in computational terms.

Far from sharing this belief that such non-cognitive phenomena cannot be understood in AI terms, some computationalists argue that *any* intelligent system having distinct and potentially conflicting goals, and acting in a complex and rapidly changing environment, would need the internal control-functions intrinsic to motivation and affect. So if AI workers were ever to model (or fully understand) intelligence they would necessarily have modelled (understood) motivation and emotion too. Such arguments adopt the design stance, wherein one asks what features have to be provided (by engineer or evolution) if a certain sort of computational system is to be possible. For instance, I have argued elsewhere that many theoretical concepts employed by humanist psychologists (including motive, intention, emotion, sentiment, character, and self-ideal) mark aspects of control and organization essential to multi-goaled systems (Boden 1972, 1973). A design-stance approach similarly imbues Aaron Sloman's discussion of 'Motives, Mechanisms, and Emotions' (ch. 10).

Sketching some dimensions of 'the space of possible minds', Sloman argues that a many-goaled system is essentially prone to internal conflicts whose resolution requires special sorts of control mechanism, and that emotion—associated as it is with some of these mechanisms—is not a special subsystem of the mind but a pervasive feature of it. An autonomous system having many motives and finite resources of time, effort, and knowledge, needs strategies for comparing and choosing between them and deciding what to do accordingly. Since this decision will often involve the generation of further motives (intentions), such a system is inherently recursive. Preferences and moral values represent background criteria for comparison and choice, which economize on computation effort because they do not have to be generated afresh at every decision point. Motives crucial to survival may need to interrupt ongoing activity by instantly commandeering the available resources (working-memory, muscles), initiating actions without any extended comparison or decision-processes. The different emotions are associated with various sorts of motivational control: fear with interrupts due to perceived danger, anxiety with perceptions of the likelihood of success in achieving an important goal, and so on.

Motives, emotions, and intentions are aspects of the many-levelled set of virtual machines associated with, and ultimately implemented in, the brain. Sloman points out that his arguments concern computational systems of the relevant kind (with multiple goals and finite resources) irrespective of whether they are implemented in a von Neumann or a parallel machine. But although von Neumann computers (approximations to Turing machines) can in principle implement any computation, some computations may require embodiment of a fundamentally different kind in order to function in real time. Certainly, the mental processes of animals are embodied in machines (brains) very different from computers.

The branch of AI inspired by this fact is *connectionism* (chs. 11–13). Connectionism is a general term covering many species of information-processing systems, whose differing computational properties are still largely unknown (Anderson and Rosenfeld 1988). Their common feature is that they are conceptualized as massively parallel-processing devices, made up of many simple units. A unit's activity is regulated by the activity of neighbouring units, connected to it by inhibitory or excitatory links whose strength can vary according to design and/or learning. The activity of a unit, and the strength of a link, are expressed as numbers; and the changes in overall system-activity and system-weights are (usually) governed by differential equations.

These units can be of various types: binary (active or inactive) or continuously graded (having varying degrees of activity); deterministic (their activity depending strictly on activity in the input units) or stochastic (sometimes firing at random); sole-experts (each one affected by evidence available to no other unit) or having overlapping interests (being sensitive to partially shared evidence); and more, or less, finely tuned. Each of these computational distinctions is exemplified in the brain. But despite their broadly neural inspiration, connectionist units—like McCulloch and Pitts's theoretically specified 'neurones'—are abstract idealizations: actual neurones are not only much more complex, but have some significantly different properties.

One influential example of connectionism is 'PDP' research on parallel distributed processing, involving the 'Distributed Representations' described by Geoffrey Hinton, James McClelland, and David Rumelhart (ch. 11). In PDP systems a concept is represented not by an individual symbol stored at some identifiable memory-location, but by an equilibrium state defined over a dynamic network of locally interacting units. Each unit codes one of the many microfeatures relevant to the concept concerned, and the connections between units are excitatory or inhibitory inasmuch as the corresponding features are mutually supportive or contradictory. In equilibrium, the highly active units represent features which are mutually supportive, or at least consistent (the other units being quiescent). Any given unit can contribute to the representation of several concepts, and 'one and the same concept' in different contexts can be represented by partially different networks.

The inherent properties of distributed representations enable them to perform certain computations which are difficult to program in the traditional way. One example is pattern-matching—even when the new pattern is somewhat different from, or only a fragment of, the old one. Perceptual recognition, analogical thought, and classification by family-resemblances—each highly resistant to traditional AI—may therefore be amenable to PDP computation. These capacities are naturally emergent in PDP systems, and do not have to be specifically programmed into it by anticipatory specifications of myriad individual rules (although units for the potential basic features must be included). These mental powers have all been cited by philosophers arguing that no set of rules could be specified to cover them and that AI therefore cannot explain them. This objection holds only if one restricts 'AI' to its earliest form, wherein each program-symbol represents some identifiable concept, not a 'subsymbolic' microfeature.

The theoretical relation between traditional and connectionist AI is controversial, even among connectionists. Certainly, all connectionist systems can in principle be implemented on von Neumann machines. But setting aside this highly abstract equivalence-in-principle of the two forms of AI, their interrelations and comparative usefulness in cognitive science are disputed. As Andy Clark shows in 'Connectionism, Competence, and Explanation' (ch. 12), this controversy concerns two points: the suitability of either approach for modelling specific mental capacities, and the explanatory status of connectionist performance-models.

On the first point, committed traditionalists often remark that language-understanding and step-by-step reasoning require computations not well-suited to connectionist systems. This is admitted by the connectionist authors of Chapter 11, who suggest that a PDP system may have to simulate a von Neumann machine to achieve them. That is, the virtual machine within which these computations are effected is of a type best described by traditional AI, even though their basic implementation in the brain is connectionist (cf. Smolensky 1988; Clark 1989). To this extent, the two branches of AI are complementary.

The second point appears in various guises. For instance, it is sometimes suggested that connectionism concerns not psychological processes but their neural implementation. However, despite connectionism's general respect for biological implementability, most connectionist systems model not neural implementation but abstractly defined information-processing. Only a computational system representing specific neural circuitry and/or synaptic interactions (as in ch. 14) would be a model of implementation as such.

Another way of pressing this second point is to ask what sort of explanation connectionist models provide. Can they embody the abstract principles underlying distinct psychological domains which are recommended by Marr (as 'Type-1' theories) and by Newell and Simon (as 'knowledge-level' theories)? Or are they no more than largely unintelligible existence-proofs, models whose functioning can—at best—be given only a Type-2 explanation (a listing of all the connection-weights), backed up by reference to the abstract principles of equilibration and learning (such as Boltzmann equations, or back-propagation) embodied in them? Such models might be powerful additions to technological AI, and could conceivably help to produce the intelligent computer-performance imagined by Turing. But would they provide useful explanations within a science of intelligence?

Clark holds that connectionism is not restricted to Type-2 explanations, but is not based on Type-1 explanations either. Connectionist systems *cannot* exemplify the basically axiomatic explanatory 'cascade' recommended by Newell and Simon and by Marr. Certainly, an abstract task-analysis may be used to specify the relevant fundamental inputs and outputs, and (in a non-learning model) even to assign many connection-weights. A system modelling low-level vision, for instance, may have input units responsive to specific sorts of 2D-information and output units computing specific types of 3D-information, all chosen according to the theory of 2D-to-3D mapping. A competence theory may even be a useful idealization of what a connectionist system is doing. But it cannot explain what is really going on, for connectionist systems contain neither explicit nor even tacit knowledge of Type-1 theories. In general, no precise mapping exists between the information-processing in connectionist and von Neumann systems performing 'the same' task.

Connectionism, Clark argues, involves a 'methodological inversion' of traditional cognitive science. Connectionists do not, and need not, first seek an axiomatic task-analysis before starting to model the processing involved in a given form of intelligence. Instead, they build a network only loosely specified by considerations of abstract competence (at the '0.5 level'), they let it learn to perform the task in question, and only then do they find the high-level principles it has come to embody. The model's explanatory force lies in those principles, not in the (probably unintelligible) set of connection-weights. Just what these high-level principles are like is still unclear. Possible methods for discovering them include studies of network pathology (exploring the effects of deliberate changes to the system), activation recording (whereby the units active at a particular time, or at successive times, are identified), and cluster-analysis (which discovers hierarchical structure in the patterns of activation within a given system). The conceptual structures underlying real-world action, for example, might be discovered *post hoc* in these ways—as opposed to the logicist strategy of compiling an a priori, and largely *ad hoc,* list of axioms for naïve physics.

That connectionism can, to some extent, defend AI against some familiar criticisms is admitted by Hubert (with Stuart) Dreyfus in 'Making a Mind Versus Modelling the Brain: Artificial Intelligence Back at a Branch-point' (ch. 13). But he still holds, *contra* Hayes and Clark, that language and common sense cannot be captured by AI (not even the connectionist variety). In defending this position, Dreyfus relates AI work to a wide range of philosophical literature, contrasting the Western rationalist tradition with Continental phenomenology and the later Wittgenstein. His scepticism about AI springs from the view that people do not use—and science cannot express—a theory about the everyday world, because there is no set of context-free primitives of understanding. Our knowledge is skilled *know-how,* as contrasted with procedural rules, representations, or *knowledge that;* even our knowledge of formal systems involves shared background intuitions about how to continue a mathematical series or apply a logical rule.

Outlining a strongly oppositional history of the two branches of AI (omitting McCulloch and Pitts, their common source), Dreyfus mentions the early efforts and temporary doldrums in neural-net research, the initial successes of formalist AI, and its subsequent difficulties in programming common-sense understanding. In particular, traditional AI failed to capture holistic perception, context-sensitivity, and the recognition of family-resemblances and relevance—each better handled by connectionism. Dreyfus's judgement of the first two decades of AI is that 'the rationalist tradition had finally been put to an empirical test, and it had failed'. He sees connectionism as vindicating his argument (1979) that intelligence does not rest on a theory of the world, so cannot be captured by rules or modelled in a computational system.

One might object that connectionism, too, studies computational systems, whose units compute by rigorously specified processes or rules. Dreyfus replies that the functions computed by connectionist units are typically so abstract that they bear no direct relation to linguistically expressible concepts or beliefs, and often cannot even be identified by the scientists concerned (cf. ch. 12). Moreover, a unit's activity and influence is not determined by a single instruction or rule but varies with the activity of other units, which undermines any attempt to regard connectionist units as theoretical primitives with identifiable semantic significance. Nor are connectionist 'rules' (differential equations) like the logicist axioms beloved of traditional AI.

Dreyfus still despairs, however, of a theory of intelligence—even one based on connectionist information-processing. He suggests that human intelligence could be modelled only by a connectionist system of a size and circuitry near-identical with the brain, endowed with human motives, cultural goals, and bodily flesh and form.

One need not endorse this suggestion (that AI model the whole brain) in arguing that AI should pay more detailed attention to the brain than it has done so far. Despite its root in ideas about 'nervous activity', current connectionism owes very little to neuroscience. (Traditional AI owes even less, regarding questions about neural implementation as not only distinct from but utterly irrelevant to computational questions.) AI work on low-level vision sometimes generates or borrows hypotheses about visual neuroanatomy. But current connectionism in general aims to identify the computational properties of abstractly defined (idealized) associative networks, not to model the complexity of real neurones or to map actual neural circuitry. Indeed, it is not even clear which part or parts of a real neurone correspond most closely to the units and connections of connectionist models (Smolensky 1988: §4).

Actual neurones and neuroanatomy are the special concern of neuroscientists—who often employ computational modelling to discover how a given (cellular or multicellular) neuroanatomical structure enables the brain to process information of certain kinds. Many of the computer models of neurones so far developed by neuroscientists have focused on uninterpreted cell-activations, rather than neuronal activity interpreted in psychological terms. However, although only a few neuroscientists (as yet) are located in interdisciplinary centres of cognitive science, and fewer still would describe themselves as working in AI, their research is an essential part of a general science of intelligence—and is increasingly compared with AI ideas (Thorpe and Imbert, in press; P. S. Churchland 1986; P. M. Churchland in press). Marr, for instance, drew on neuroscience in modelling not only vision but also motor control (his theory of the cerebellum employed an early version of a learning-rule used in some recent connectionist models).

Detailed knowledge of neuroanatomy can suggest types of computational systems different from those studied in traditional or connectionist AI. One such example is outlined by Paul Churchland in 'Some Reductive

Strategies in Cognitive Neurobiology' (ch. 14). Instead of von Neumann or connectionist architecture, it involves interconnected sheets of neurones modelled on specific cerebral structures; and instead of symbol-processing or equilibration, it effects co-ordinate transformations. The basic idea — the 'state-space sandwich' — arose in neuroscientific work on various sensory 2D-to-2D transformations. The more abstract 'neural matrix', capable of defining multi-dimensional state-spaces, was developed to explain how the cerebellum computes motor co-ordination (an aspect of 'bodily experience').

These ideas are speculatively generalized by Churchland to other psychological domains, including the nice discrimination of colours and tastes. One could substitute 'experience' for 'discrimination' here. Churchland's philosophical aim is to show that eliminative materialism is conceivable, by describing a method of neural computation which illustrates how it might be realized. (As he admits, this method is arguably better suited to computing isolated sensory discriminations and sensori-motor co-ordinations than structured linguistic understanding or voluntary action.) According to eliminative materialism, a future neuroscience might so inform our perceptions and thinking that we no longer conceive of mental events (even *qualia*) as distinct from brain-states. For this to happen, neuroscience must first have shown, in detail, how our mental states are possible. It is noteworthy (*pace* Searle) that Churchland's suggestions about how this could be done draw not on neural biochemistry, but on accounts of brain circuitry capable of implementing specific types of information-processing. From the AI viewpoint, this is just what one would expect.

The last chapter, by Adrian Cussins, considers AI in the light of long-standing issues in the philosophy of language and mind, concerned with reference and intentionality. In this previously unpublished paper added to this collection when it was already in proof, Cussins argues (as Searle does, too) that GOFAI cannot explain intentionality, or account for the appearance of mind in a physical universe. Formalist AI, and psychological theories positing a 'language of thought' (Fodor, 1975; 1981, ch. 9), deal with (syntactic) relations between concepts, whose existence—and semantic properties—are taken for granted.

But the existence of concepts, and of their semantic properties, must itself be explained. To account for the appearance of concepts, Cussins claims, one needs a non-conceptualist notion of mental content, an account of how this can arise within a (non-formalist) representational system, and an analysis of how concepts can progressively be constructed from it (a general position argued at greater length by Clark, 1993). To explain the mind/world distinction, or objectivity, one must show how conceptual representations can have semantic properties such as reference, truth, and falsity.

Drawing on recent work in philosophical semantics and metaphysics, Cussins defines an experience-based notion of non-conceptual content ('construction-theoretic content', or CTC) and analyses what it is for a physical system to possess a concept. He argues that concepts can in principle be constructed from CTC, and that truth-preserving inferences depend on the internal structure of concepts, not the syntactic relations between them. He does not discuss in any detail how concepts might arise in practice, but relevant connectionist work on the construction of decreasingly perspective-dependent representations suggests how a psychological system might gradually develop objectivity (Hinton, 1981; Marr, 1982; see also Clark, 1993).

These fifteen papers, then, provide a wide range of comment on the philosophy of AI. They show that AI embraces significantly different methodologies, much as mathematics covers diverse types of theory. To dismiss AI as philosophically bogus because of shortcomings in its earliest branch (GOFAI) would be like a seventeenth-century philosopher rejecting Galileo's suggestion that 'mathematics is the language of God' because—having no differential equations—he could not explain fluid dynamics. To be sure, whereas four centuries of physics have shown that Galileo was right, to describe AI as the general science of intelligence is, as yet, merely to offer a promissory note. In evaluating this promise, we should not only examine current AI in some detail but also remember that science in general is not 'purely empirical'. Science can provide new ideas, enabling new ways of asking philosophical questions—which questions may be transformed in the process, because of deep conceptual revisions prompted by scientific advance (Putnam 1962; Churchland 1979). Proponents of AI see it as capable of contributing in this way to the philosophy of mind and epistemology. As Dennett (1988) has put it: 'AI has not yet solved any of our ancient riddles about the mind, but it has provided us with new ways of disciplining and extending philosophical imagination that we have only begun to exploit.'

REFERENCES

Anderson, J. A., and Rosenfeld, E. (1988). *Neurocomputing: A Reader.* Cambridge, Mass.: MIT Press/Bradford Books.

Boden, M. A. (1972). *Purposive Explanation in Psychology.* Cambridge, Mass.: Harvard University Press.

—— (1973). 'The Structure of Intentions.' *J. Theory of Social Behaviour* 3: 23–46.

—— (1987). *Artificial Intelligence and Natural Man*, 2nd edn., London: MIT Press; New York: Basic Books.

—— (1988). *Computer Models of Mind: Computational Approaches in Theoretical Psychology*. Cambridge: Cambridge University Press.

Churchland, P. M. (1979). *Scientific Realism and the Plasticity of Mind*. Cambridge: Cambridge University Press.

—— (in press). *The Neurocomputational Perspective*. Cambridge, Mass.: MIT Press/Bradford Books.

Churchland, P. S. (1986). *Neurophilosophy: Toward a Unified Science of the Mind–Brain*. Cambridge, Mass.: MIT Press/Bradford Books.

Clark, A. J. (1987*a*). 'Connectionism and Cognitive Science.' In J. Hallam and C. Mellish (eds.), *Advances in Artificial Intelligence*, pp. 3–15. Chichester: Wiley.

—— (1987*b*). 'The Kludge in the Machine.' *Mind and Language* 2: 277–300.

—— (1989). *Microcognition: Philosophy, Cognitive Science, and Parallel Distributed Processing*. Cambridge, Mass.: MIT Press/Bradford Books.

Dennett, D. C. (1988). 'When Philosophers Encounter Artificial Intelligence.' *Daedalus* (Winter 1988): 283–95. Also published in S. R. Graubard (ed.), *The Artificial Intelligence Debate: False Starts, Real Foundations*. Cambridge, Mass.: MIT Press, 1988.

Dreyfus, H. L. (1979). *What Computers Can't Do: The Limits of Artificial Intelligence*, rev. edn. New York: Harper & Row.

Fodor, J. A. (1983). *The Modularity of Mind: An Essay on Faculty Psychology*. Cambridge, Mass.: MIT Press/Bradford Books.

Haugeland, J. (1978). 'The Nature and Plausibility of Cognitivism' (with peer-commentary and author's reply). *Behavioral and Brain Sciences* 1: 215–60.

—— (1985). *Artificial Intelligence: The Very Idea*. Cambridge, Mass.: MIT Press/Bradford Books.

Hayes, P. J. (1985). 'The Second Naïve Physics Manifesto.' In J. C. Hobbs and R. C. Moore (eds.), *Formal Theories of the Commonsense World*, pp. 1–36. Norwood, NJ: Ablex. Repr. in R. J. Brachman and H. J. Levesque (eds.), *Readings in Knowledge Representation*, pp. 467–86. Los Altos, Calif.: Morgan Kaufmann.

Longuet-Higgins, H. C. (1987). *Mental Processes: Studies in Cognitive Science*. Cambridge, Mass.: MIT Press/Bradford Books.

Lucas, J. R. (1961). 'Minds, Machines, and Godel.' *Philosophy* 36: 112–27.

McCulloch, W. S. (1965). *Embodiments of Mind*. Cambridge, Mass.: MIT Press.

Marr, D. C. (1982), *Vision: A Computational Investigation into the Human Representation and Processing of Visual Information*. San Francisco: Freeman.

Minsky, M. L., and Papert, S. (1969). *Perceptrons: An Introduction to Computational Geometry*. Cambridge, Mass.: MIT Press.

Putnam, H. (1962). 'Dreaming and Depth Grammar.' In R. J. Butler's (ed.), *Analytical Philosophy*, pp. 211–35. Oxford: Blackwell.

Sloman, A. (1978). *The Computer Revolution in Philosophy: Philosophy, Science, and Models of Mind*. Brighton: Harvester Press.

—— (1986). 'What Sorts of Machines Can Understand the Symbols They Use?' *Proc. Aristotelian Soc.*, Supp. 60: 61–80.

Smolensky, P. (1988). 'On the Proper Treatment of Connectionism' (with peer-commentary and author's reply). *Behavioral and Brain Sciences* 11: 1–74.

Thorpe, S., and Imbert, M. (in press). 'Neuroscientific Constraints on Connectionist Modelling.' In R. Pfeiffer, Z. Schreber, F. Fogelman, and T. Bernold (eds.), *Connectionism in Perspective*. Amsterdam: North-Holland.

Turing, A. M. (1936). 'On Computable Numbers, with an Application to the *Entscheidungsproblem*.' *Proc. London Math. Soc.* 42: 230–65; also 43: 544.

1

A LOGICAL CALCULUS OF THE IDEAS IMMANENT IN NERVOUS ACTIVITY

WARREN S. McCULLOCH and WALTER H. PITTS

1. INTRODUCTION

Theoretical neurophysiology rests on certain cardinal assumptions. The nervous system is a net of neurones, each having a soma and an axon. Their adjunctions, or synapses, are always between the axon of one neurone and the soma of another. At any instant a neurone has some threshold, which excitation must exceed to initiate an impulse. This, except for the fact and the time of its occurrence, is determined by the neurone, not by the excitation. From the point of excitation the impulse is propagated to all parts of the neurone. The velocity along the axon varies directly with its diameter, from less than one meter per second in thin axons, which are usually short, to more than 150 meters per second in thick axons, which are usually long. The time for axonal conduction is consequently of little importance in determining the time of arrival of impulses at points unequally remote from the same source. Excitation across synapses occurs predominantly from axonal terminations to somata. It is still a moot point whether this depends upon irreciprocity of individual synapses or merely upon prevalent anatomical configurations. To suppose the latter requires no hypothesis *ad hoc* and explains known exceptions, but any assumption as to cause is compatible with the calculus to come. No case is known in which excitation through a single synapse has elicited a nervous impulse in any neurone, whereas any neurone may be excited by impulses arriving at a sufficient number of neighbouring synapses within the period of latent addition, which lasts less than one quarter of a millisecond. Observed temporal summation of impulses at greater intervals is impossible for single neurones and empirically depends upon structural properties of the net. Between the arrival of impulses

Warren S. McCulloch and Walter H. Pitts, 'A Logical Calculus of the Ideas Immanent in Nervous Activity,' in W. S. McCulloch, *Embodiments of Mind* (MIT Press, 1965), pp. 19–39. Reprinted by permission of MIT Press, Cambridge, Mass.

upon a neurone and its own propagated impulse there is a synaptic delay of more than half a millisecond. During the first part of the nervous impulse the neurone is absolutely refractory to any stimulation. Thereafter its excitability returns rapidly, in some cases reaching a value above normal from which it sinks again to a subnormal value, whence it returns slowly to normal. Frequent activity augments this subnormality. Such specificity as is possessed by nervous impulses depends solely upon their time and place and not on any other specificity of nervous energies. Of late only inhibition has been seriously adduced to contravene this thesis. Inhibition is the termination or prevention of the activity of one group of neurones by concurrent or antecedent activity of a second group. Until recently this could be explained on the supposition that previous activity of neurones of the second group might so raise the thresholds of internuncial neurones that they could no longer be excited by neurones of the first group, whereas the impulses of the first group must sum with the impulses of these internuncials to excite the now inhibited neurones. Today, some inhibitions have been shown to consume less than one millisecond. This excludes internuncials and requires synapses through which impulses inhibit that neurone which is being stimulated by impulses through other synapses. As yet experiment has not shown whether the refractoriness is relative or absolute. We will assume the latter and demonstrate that the difference is immaterial to our argument. Either variety of refractoriness can be accounted for in either of two ways. The 'inhibitory synapse' may be of such a kind as to produce a substance which raises the threshold of the neurone, or it may be so placed that the local disturbance produced by its excitation opposes the alteration induced by the otherwise excitatory synapses. Inasmuch as position is already known to have such effects in the case of electrical stimulation, the first hypothesis is to be excluded unless and until it be substantiated, for the second involves no new hypothesis. We have, then, two explanations of inhibition based on the same general premises, differing only in the assumed nervous nets and, consequently, in the time required for inhibition. Hereafter we shall refer to such nervous nets as *equivalent in the extended sense*. Since we are concerned with properties of nets which are invariant under equivalence, we may make the physical assumptions which are most convenient for the calculus.

Many years ago one of us, by considerations impertinent to this argument, was led to conceive of the response of any neurone as factually equivalent to a proposition which proposed its adequate stimulus. He therefore attempted to record the behaviour of complicated nets in the notation of the symbolic logic of propositions. The 'all-or-none' law of nervous activity is sufficient to insure that the activity of any neurone may be represented as a proposition. Physiological relations existing among nervous activities correspond, of course, to relations among the propositions; and the utility of the representation depends upon the identity of these relations with those of the logic of propositions. To each reaction of any neurone there is a corresponding assertion of a simple proposition. This, in turn, implies either some other simple proposition or the disjunction or the conjunction, with or without negation, of similar propositions, according to the configuration of the synapses upon and the threshold of the neurone in question. Two difficulties appeared. The first concerns facilitation and extinction, in which antecedent activity temporarily alters responsiveness to subsequent stimulation of one and the same part of the net. The second concerns learning, in which activities concurrent at some previous time have altered the net permanently, so that a stimulus which would previously have been inadequate is now adequate. But for nets undergoing both alterations, we can substitute equivalent fictitious nets composed of neurones whose connections and thresholds are unaltered. But one point must be made clear: neither of us conceives the formal equivalence to be a factual explanation. *Per contra!* — we regard facilitation and extinction as dependent upon continuous changes in threshold related to electrical and chemical variables, such as after-potentials and ionic concentrations; and learning as an enduring change which can survive sleep, anaesthesia, convulsions, and coma. The importance of the formal equivalence lies in this: that the alterations actually underlying facilitation, extinction, and learning in no way affect the conclusions which follow from the formal treatment of the activity of nervous nets, and the relations of the corresponding propositions remain those of the logic of propositions.

The nervous system contains many circular paths, whose activity so regenerates the excitation of any participant neurone that reference to time past becomes indefinite, although it still implies that afferent activity has realized one of a certain class of configurations over time. Precise specification of these implications by means of recursive functions, and determination of those that can be embodied in the activity of nervous nets, completes the theory.

2. THE THEORY: NETS WITHOUT CIRCLES

We shall make the following physical assumptions for our calculus.

1. The activity of the neurone is an 'all-or-none' process.

2. A certain fixed number of synapses must be excited within the period of latent addition in order to excite a neurone at any time, and this number is independent of previous activity and position on the neurone.

3. The only significant delay within the nervous system is synaptic delay.

4. The activity of any inhibitory synapse absolutely prevents excitation of the neurone at that time.

5. The structure of the net does not change with time.

To present the theory, the most appropriate symbolism is that of Language II of R. Carnap (1938), augmented with various notations drawn from B. Russell and A. N. Whitehead (1927), including the *Principia* conventions for dots. Typographical necessity, however, will compel us to use the upright 'E' for the existential operator instead of the inverted, and an arrow ('$\rightarrow$') for implication instead of the horseshoe. We shall also use the Carnap syntactical notations, but print them in boldface rather than German type; and we shall introduce a functor S, whose value for a property P is the property which holds of a number when P holds of its predecessor; it is defined by '$S(P)$ (t). $=$.$P(Kx)$.$t = x$'; the brackets around its argument will often be omitted, in which case this is understood to be the nearest predicate-expression [*Pr*] on the right. Moreover, we shall write S^2 *Pr* for $S(S(Pr))$, etc.

The neurones of a given net $\mathfrak{N}$ may be assigned designations 'c_1', 'c_2', . . . , 'c_n'. This done, we shall denote the property of a number, that a neurone c_i fires at a time which is that number of synaptic delays from the origin of time, by 'N' with the numeral i as subscript, so that $N_i(t)$ asserts that c_i fires at the time t. N_i is called the *action* of c_i. We shall sometimes regard the subscripted numeral of 'N' as if it belonged to the object-language, and were in a place for a functoral argument, so that it might be replaced by a number-variable [z] and quantified; this enables us to abbreviate long but finite disjunctions and conjunctions by the use of an operator. We shall employ this locution quite generally for sequences of *Pr*; it may be secured formally by an obvious disjunctive definition. The predicates 'N_1', 'N_2', . . ., comprise the syntactical class 'N'.

Let us define the *peripheral afferents* of $\mathfrak{N}$ as the neurones of $\mathfrak{N}$ with no axons synapsing upon them. Let $N_1, \ldots, N_p$ denote the actions of such neurones and $N_{p+1}, N_{p+2}, \ldots, N_n$ those of the rest. Then a *solution of* $\mathfrak{N}$ will be a class of sentences of the form S_i: $N_{p+1}(z_1)$. $=$. $Pr_i(N_1, N_2, \ldots, N_p, z_1)$, where Pr_i contains no free variable save z_1 and no descriptive symbols save the N in the argument [*Arg*], and possibly some constant sentences [*sa*]; and such that each S_i is true of $\mathfrak{N}$. Conversely, given a Pr_1

($^1p^1_1$, $^1p^1_2$, . . ., $^1p^1_p$, z_1, s), containing no free variable save those in its *Arg*, we shall say that it is *realizable in the narrow sense* if there exists a net $\mathfrak{N}$ and a series of N_i in it such that N_1 (z_1) $= . Pr_1$ (N_1, N_2, . . . , z_1, sa_1) is true of it, where sa_1 has the form $N(0)$. We shall call it *realizable in the extended sense*, or simply *realizable*, if for some n S^n (Pr_1) (p_1, . . . , P_p, z_1, s) is realizable in the above sense. c_{pi} is here the realizing neurone. We shall say of two laws of nervous excitation which are such that every S which is realizable in either sense upon one supposition is also realizable, perhaps by a different net, upon the other, that they are equivalent assumptions, in that sense.

The following theorems about realizability all refer to the extended sense. In some cases, sharper theorems about narrow realizability can be obtained; but in addition to greater complication in statement this were of little practical value, since our present neurophysiological knowledge determines the law of excitation only to extended equivalence, and the more precise theorems differ according to which possible assumption we make. Our less precise theorems, however, are invariant under equivalence, and are still sufficient for all purposes in which the exact time for impulses to pass through the whole net is not crucial.

Our central problems may now be stated exactly: first, to find an effective method of obtaining a set of computable S constituting a solution of any given net; and second, to characterize the class of realizable S in an effective fashion. Materially stated, the problems are to calculate the behaviour of any net, and to find a net which will behave in a specified way, when such a net exists.

A net will be called *cyclic* if it contains a circle: i.e., if there exists a chain c_i, c_{i+1}, . . . of neurones on it, each member of the chain synapsing upon the next, with the same beginning and end. If a set of its neurones c_1, c_2, . . . , c_p is such that its removal from $\mathfrak{N}$ leaves it without circles, and no smaller class of neurones has this property, the set is called a *cyclic set*, and its cardinality is the *order of* $\mathfrak{N}$. In an important sense, as we shall see, the order of a net is an index of the complexity of its behaviour. In particular, nets of zero order have especially simple properties; we shall discuss them first.

Let us define a *temporal propositional expression* (a T P E), designating a *temporal propositional function* (T P F), by the following recursion:

1. A $^1p^1$ [z_1] is a T P E, where p_1 is a predicate-variable.
2. If S_1 and S_2 are T P E containing the same free individual variable, so are SS_1, S_1vS_2, $S_1.S_2$, and $S_i. \sim S_2$.
3. Nothing else is a T P E.

Theorem I

Every net of order 0 *can be solved in terms of temporal propositional expressions.*

Let c_i be any neurone of $\mathfrak{N}$ with a threshold $\theta_i > 0$, and let $c_{i1}, c_{i2}, \ldots, c_{ip}$ have respectively $n_{i1}, n_{i2}, \ldots, n_{ip}$ excitatory synapses upon it. Let $c_{j1}, c_{j2}, \ldots, c_{jq}$ have inhibitory synapses upon it. Let k_i be the set of the subclasses of $\{n_{i1}, n_{i2}, \ldots, n_{ip}\}$ such that the sum of their members exceeds θ_i. We shall then be able to write, in accordance with the assumptions mentioned above,

$$N_i(z_1) \, . \, = \, . \, S \left\{ \prod_{m=1}^{q} \sim N_{jm}(z_1) \, . \, \sum_{\alpha \varepsilon k i} \prod_{\delta \varepsilon \alpha} N_{is}(z_1) \right\}$$

where the 'Σ' and 'Π' are syntactical symbols for disjunctions and conjunctions which are finite in each case. Since an expression of this form can be written for each c_i which is not a peripheral afferent, we can, by substituting the corresponding expression in (1) for each N_{jm} or N_{is} whose neurone is not a peripheral afferent, and repeating the process on the result, ultimately come to an expression for N_i in terms solely of peripherally afferent N, since $\mathfrak{N}$ is without circles. Moreover, this expression will be a $T P E$, since obviously (1) is; and it follows immediately from the definition that the result of substituting a $T P E$ for a constituent $p(z)$ in a $T P E$ is also one.

Theorem II

Every TPE *is realizable by a net of order zero.*

The functor S obviously commutes with disjunction, conjunction, and negation. It is obvious that the result of substituting any S_i, realizable in the narrow sense (i.n.s.), for the $p(z)$ in a realizable expression S_1 is itself realizable i.n.s.; one constructs the realizing net by replacing the peripheral afferents in the net for S_1 by the realizing neurones in the nets for the S_i. The one neurone net realizes $p_1(z_1)$ i.n.s., and Figure 1a shows a net that realizes $Sp_1(z_1)$ and hence SS_2, i.n.s., if S_2 can be realized i.n.s. Now if S_2 and S_3 are realizable then $S^m S_2$ and $S^n S_3$ are realizable i.n.s., for suitable m and n. Hence so are $S^{m+n} S_2$ and $S^{m+n} S_3$. Now the nets of Figures 1b, c, and d respectively realize $S(p_1(z_1) \, v \, p_2(z_1))$, $S(p_1(z_1) \, . \, p_2(z_1))$, and $S(p_1(z_1) \, . \sim p_2(z_1))$ i.n.s. Hence $S^{m+n+1} (S_1 v S_2)$, $S^{m+n+1} (S_1 \, . \, S_2)$, and $S^{m+n+1} (S_1 \, . \sim S_2)$ are realizable i.n.s. Therefore $S_1 v S_2 S_1 \, . \, S_2 S_1 \, . \sim S_2$ are realizable if S_1 and S_2 are. By complete induction, all $T P E$ are realizable. In this way all nets may be regarded as built out of the fundamental elements of

Figures 1a, b, c, d, precisely as the temporal propositional expressions are generated out of the operations of precession, disjunction, conjunction, and conjoined negation. In particular, corresponding to any description of state, or distribution of the values *true* and *false* for the actions of all the neurones of a net save that which makes them all false, a single neurone is constructible whose firing is a necessary and sufficient condition for the validity of that description. Moreover, there is always an indefinite number of topologically different acts realizing any $T\ P\ E$.

Theorem III

Let there be given a complex sentence S_1 built up in any manner out of elementary sentences of the form $\mathbf{p}\ (z_1 - zz)$ where zz is any numeral, by any of the propositional connections: negation, disjunction, conjunction, implication, and equivalence. Then S_1 is a T P E *and only if it is false when its constituent* $\mathbf{p}\ (z_1 - zz)$ *are all assumed false—i.e., replaced by false sentences—or that the last line in its truth table contains an* 'F', *—or there is no term in its Hilbert disjunctive normal form composed exclusively of negated terms.*

These latter three conditions are of course equivalent (Hilbert and Ackermann 1938). We see by induction that the first of them is necessary, since $p(z_1 - zz)$ becomes false when it is replaced by a false sentence, and $S_1 \text{v} S_2, S_1 . S_2$ and $S_1 . \sim S_2$ are all false if both their constituents are. We see that the last condition is sufficient by remarking that a disjunction is a $T\ P\ E$ when its constituents are, and that any term

$$S_1 . S_2 . \ldots . S_m . \sim S_{m+1} . \sim \ldots . \sim S_n$$

can be written as

$$(S_1 . S_2 . \ldots . S_m) . \sim (S_{m+1} \text{v} S_{m+2} \text{v} \ldots . \text{v} S_n),$$

which is clearly a $T\ P\ E$.

The method of the last theorems does in fact provide a very convenient and workable procedure for constructing nervous nets to order, for those cases where there is no reference to events indefinitely far in the past in the specification of the conditions. By way of example, we may consider the case of heat produced by a transient cooling.

If a cold object is held to the skin for a moment and removed, a sensation of heat will be felt; if it is applied for a longer time, the sensation will be only of cold, with no preliminary warmth, however transient. It is known that one cutaneous receptor is affected by heat, and another by cold. If we let N_1 and N_2 be the actions of the respective receptors and N_3 and N_4 of neurones whose activity implies a sensation of heat and cold, our requirements may be written as

$$N_3(t) : = : N_1(t-1) \cdot \mathbf{v} \cdot N_2(t-3) \cdot \sim N_2(t-2)$$
$$N_4(t) \cdot = \cdot N_2(t-2) \cdot N_2(t-1)$$

where we suppose for simplicity that the required persistence in the sensation of cold is, say, two synaptic delays, compared with one for that of heat. These conditions clearly fall under Theorem III. A net may consequently be constructed to realize them, by the method of Theorem II. We begin by writing them in a fashion which exhibits them as built out of their constituents by the operations realized in Figures 1a, b, c, d: i.e., in the form

$$N_3(t) \cdot = \cdot S\{N_1(t) \, \mathbf{v} \, S[\, (SN_2(t)) \cdot \sim N_2(t)\,]\}$$
$$N_4(t) \cdot = \cdot S\{\, [SN_2(t)\,] \cdot N_2(t)\,\}.$$

First we construct a net for the function enclosed in the greatest number of brackets and proceed outward; in this case we run a net of the form shown in Figure 1a from c_2 to some neurone c_a, say, so that

$$N_a(t) \cdot = \cdot SN_2(t).$$

Next introduce two nets in the forms 1c and 1d, both running from c_a and c_2, and ending respectively at c_4 and say c_b. Then

$$N_4(t) \cdot = \cdot S[N_a(t) \cdot N_2(t)\,] \cdot = \cdot S[\, (SN_2(t)) \cdot N_2(t)\,].$$
$$N_b(t) \cdot = \cdot S[N_a(t) \cdot \sim N_2(t)\,] \cdot = \cdot S[\, (SN_2(t)) \cdot \sim N_2(t)\,].$$

Finally, run a net of the form 1b from c_1 and c_b to c_3, and derive

$$N_3(t) \cdot = \cdot S[N_1(t) \, \mathbf{v} \, N_b(t)\,] \cdot = \cdot S\{N_1(t) \, \mathbf{v} \, S[\, (SN_2(t)) \cdot \sim N_2(t)\,]\}.$$

These expressions for $N_3(t)$ and $N_4(t)$ are the ones desired; and the realizing net *in toto* is shown in Figure 1e.

This illusion makes very clear the dependence of the correspondence between perception and the 'external world' upon the specific structural properties of the intervening nervous net. The same illusion, of course, could also have been produced under various other assumptions about the behaviour of the cutaneous receptors, with correspondingly different nets.

We shall now consider some theorems of equivalence: i.e., theorems which demonstrate the essential identity, save for time, of various

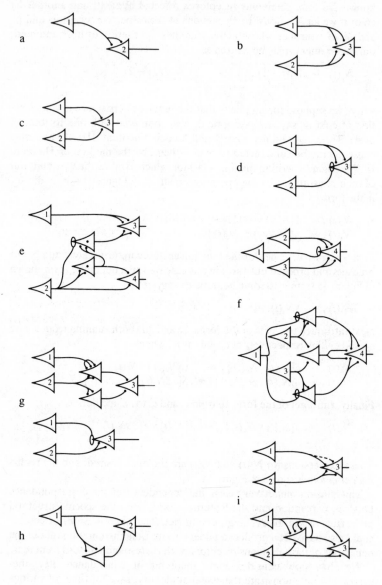

Fig. 1.

alternative laws of nervous excitation. Let us first discuss the case of *relative inhibition*. By this we mean the supposition that the firing of an inhibitory synapse does not absolutely prevent the firing of the neurone, but merely raises its threshold, so that a greater number of excitatory synapses must fire concurrently to fire it than would otherwise be needed. We may suppose, losing no generality, that the increase in threshold is unity for the firing of each such synapse; we then have the theorem:

Theorem IV

Relative and absolute inhibition are equivalent in the extended sense.

We may write out a law of nervous excitation after the fashion of (1), but employing the assumption of relative inhibition instead; inspection then shows that this expression is a $T P E$. An example of the replacement of relative inhibition by absolute is given by Figure 1f. The reverse replacement is even easier; we give the inhibitory axons afferent to c_i any sufficiently large number of inhibitory synapses apiece.

Second, we consider the case of extinction. We may write this in the form of a variation in the threshold θ_i after the neurone c_i has fired; to the nearest integer—and only to this approximation is the variation in threshold significant in natural forms of excitation—this may be written as a sequence $\theta_i + b_j$ for j synaptic delays after firing, where $b_j = 0$ for j large enough, say $j = M$ or greater. We may then state

---

FIG. 1: EXPRESSION FOR THE FIGURES

In the figure the neurone c_i is always marked with the numeral i upon the body of the cell, and the corresponding action is denoted by 'N' with i as subscript, as in the text.

a. $N_2(t) . = . N_1(t - 1)$

b. $N_3(t) . = . N_1(t - 1) \mathbf{v} N_2(t - 1)$

c. $N_3(t) . = . N_1(t - 1) . N_2(t - 1)$

d. $N_3(t) . = . N_1(t - 1) . \sim N_2(t - 1)$

e. $N_3(t) : = : N_1(t - 1) . \mathbf{v} . N_2(t - 3) . \sim N_2(t - 2)$
    $N_4(t) . = . N_2(t - 2) . N_2(t - 1)$

f. $N_4(t) : = : \sim N_1(t - 1) . N_2(t - 1) \mathbf{v} N_3(t - 1) . \mathbf{v} . N_1(t - 1) . N_3(t - 1) . N_1(t - 1)$
    $N_4(t) : = : \sim N_1(t - 2) . N_2(t - 2) \mathbf{v} N_3(t - 2) . \mathbf{v} . N_1(t - 2) . N_2(t - 2) . N_3(t - 2)$

g. $N_3(t) . = . N_2(t - 2) . \sim N_1(t - 3)$

h. $N_2(t) . = . N_1(t - 1) . N_1(t - 2)$

i. $N_3(t) : = : N_2(t - 1) . \mathbf{v} . N_1(t - 1) . \mathbf{v} . N_1(t - 1) . (Ex)t - 1 . N_1(x) . N_2(x)$

Theorem V

Extinction is equivalent to absolute inhibition.

For, assuming relative inhibition to hold for the moment, we need merely run M circuits $\mathfrak{I}_1, \mathfrak{I}_2, \ldots, \mathfrak{I}_M$ containing respectively $1, 2, \ldots, M$ neurones, such that the firing of each link in any is sufficient to fire the next, from the neurone c_i back to it, where the end of the circuit $\mathfrak{I}_j$ has just b_j inhibitory synapses upon c_i. It is evident that this will produce the desired results. The reverse substitution may be accomplished by the diagram of Figure 1g. From the transitivity of replacement, we infer the theorem. To this group of theorems also belongs the well-known

Theorem VI

Facilitation and temporal summation may be replaced by spatial summation.

This is obvious: one need merely introduce a suitable sequence of delaying chains, of increasing numbers of synapses, between the exciting cell and the neurone whereon temporal summation is desired to hold. The assumption of spatial summation will then give the required results. See, for example, Figure 1h. This procedure had application in showing that the observed temporal summation in gross nets does not imply such a mechanism in the interaction of individual neurones.

The phenomena of learning, which are of a character persisting over most physiological changes in nervous activity, seem to require the possibility of permanent alterations in the structure of nets. The simplest such alteration is the formation of new synapses or equivalent local depressions of threshold. We suppose that some axonal terminations cannot at first excite the succeeding neurone; but if at any time the neurone fires, and the axonal terminations are simultaneously excited, they become synapses of the ordinary kind, henceforth capable of exciting the neurone. The loss of an inhibitory synapse gives an entirely equivalent result. We shall then have

Theorem VII

Alterable synapses can be replaced by circles.

This is accomplished by the method of Figure 1i. It is also to be remarked that a neurone which becomes and remains spontaneously active can likewise be replaced by a circle, which is set into activity by a peripheral afferent when the activity commences, and inhibited by one when it ceases.

3. THE THEORY: NETS WITH CIRCLES

The treatment of nets which do not satisfy our previous assumption of freedom from circles is very much more difficult than that case. This is largely a consequence of the possibility that activity may be set up in a circuit and continue reverberating around it for an indefinite period of time, so that the realizable Pr may involve reference to past events of an indefinite degree of remoteness. Consider such a net $\mathfrak{N}$, say of order p, and let $c_1, c_2, \ldots, c_p$ be a cyclic set of neurones of $\mathfrak{N}$. It is first of all clear from the definition that every N_s of $\mathfrak{N}$ can be expressed as a $T P E$, of N_1, $N_2, \ldots, N_p$ and the absolute afferents; the solution of $\mathfrak{N}$ involves then only the determination of expressions for the cyclic set. This done, we shall derive a set of expressions $[A]$:

$$N_i(z_1) \; . = . \; Pr_i[S^{n_{i1}} N_1(z_1), S^{n_{i2}} N_2(z_1), \ldots, S^{n_{ip}} N_p(z_1)], \qquad (2)$$

where Pr_i also involves peripheral afferents. Now if n is the least common multiple of the n_{ij}, we shall, by substituting their equivalents according to (2) in (3) for the N_j, and repeating this process often enough on the result, obtain S of the form

$$N_i(z_1) \; . = . \; Pr_1[S^n N_1(z_1), S^n N_2(z_1), \ldots, S^n N_p(z_1)]. \qquad (3)$$

These expressions may be written in the Hilbert disjunctive normal form as

$$N_i(z_1) \; . = . \; \sum_{\substack{\alpha \epsilon k \\ \beta_{\alpha}^{\epsilon k}}} S_a \prod_{j \epsilon k} S^n N_j(z_1) \prod_{j \epsilon \beta \alpha} \sim S^n N_j(z_1), \text{ for suitable } k, \qquad (4)$$

where S_a is a $T P E$ of the absolute afferents of $\mathfrak{N}$. There exist some 2^p different sentences formed out of the pN_i by conjoining to the conjunction of some set of them the conjunction of the negations of the rest. Denumerating these by $X_1(z_1), X_2(z_1), \ldots, X_{2^p}(z_1)$, we may, by use of the expressions (4), arrive at an equipollent set of equations of the form

$$X_i(z_1) \; . = . \; \sum_{j=1}^{2p} Pr_{ij}(z_1) \; . \; S^n X_j(z_1). \qquad (5)$$

Now we import the subscripted numerals i, j into the object-language: i.e., define Pr_1 and Pr_2 such that $Pr_1(zz_1, z_1) \; . = . \; X_i(z_1)$ and $Pr_2(zz_1, zz_2, z_1) \; . \; = . \; Pr_{ij}(z_1)$ are provable whenever zz_1 and zz_2 denote i and j respectively.

Then we may rewrite (5) as

$$(z_1)zz_p : Pr_1(z_1, z_3)$$
$$. = . (Ez_2)zz_p \; . \; Pr_2(z_1, z_2, z_3 - zz_n) \; . \; Pr_1(z_2, z_3 - zz_n) \qquad (6)$$

where zz_m denotes n and zz_p denotes 2^p. By repeated substitution we arrive at an expression

$$(z_1)zz_p : Pr_1\,(z_1,\,zz_n\,zz_2)\,.\,=\,.\,(Ez_2)zz_p\,(Ez_3)zz_p\,\ldots\,(Ez_n)zz_p.$$
$$Pr_2(z_1,\,z_2,\,zz_n\,(zz_2-1))\,.\,Pr_2(z_2,\,z_3,\,zz_n\,(zz_2-1))\,.\,.\,.\,.\,.\tag{7}$$

$Pr_2(z_{n-1},z_n,0)\,.\,Pr_1(z_n,0)$, for any numeral zz_2 which denotes s. This is easily shown by induction to be equipollent to

$$(z_1)zz_p :\,.\,Pr_1\,(z_1,\,zz_n\,zz_2) :\,=\,:\,(Ef)\,(z_2)zz_2-1f(z_2\,zz_n)$$
$$zz_p\,.\,f\,(zz_n\,zz_2)=z_1\,.\,Pr_2\,(f\,(zz_n\,(z_2+1))\,),\tag{8}$$
$$f\,(zz_n\,z_2)\,)\,.\,Pr_1(f(0),0)$$

and since this is the case for all zz_2, it is also true that

$$(z_4)\,(z_1)zz_p :\,Pr_1(z_1,\,z_4)\,.\,=\,.\,(Ef)\,(z_2)\,(z_4-1)\,.\,f(z_2)$$
$$\leq zz_p\,.\,f\,(z_4)=(z_1f\,(z_4)=z_1\,.\,Pr_2[f\,(z_2+1),f(z_2),\,z_2]\,.\tag{9}$$
$$Pr_1[f\,(\mathrm{res}\,(z_4,\,zz_n)\,),\,\mathrm{res}\,(z_4,\,zz_n)\,],$$

where zz_n denotes n, res (r,s) is the residue of r mod s and zz_p denotes 2^p. This may be written in a less exact way as

$$N_i(t)\,.\,=\,.\,(E\varphi)\,(x)t-1\,.\,\varphi(x)\leq 2^p\,.\,\varphi\,(t)=i\,.$$
$$P[\varphi(x+1),\,\varphi(x)\,.\,N_{\varphi(0)}\,(0)\,],$$

where x and t are also assumed divisible by n, and Pr_2 denotes P. From the preceding remarks we shall have

Theorem VIII

The expression (9) for neurones of the cyclic set of a net $\mathfrak{N}$ together with certain TPE expressing the actions of other neurones in terms of them, constitute a solution of $\mathfrak{N}$.

Consider now the question of the realizability of a set of S_i. A first necessary condition, demonstrable by an easy induction, is that

$$(z_2)z_1\,.\,p_1(z_2)=p_2(z_2)\,.\to\,.\,S_i=S_i\,\left\{\begin{array}{c}p_1\\p_2\end{array}\right\}\tag{10}$$

should be true, with similar statements for the other free p in S_i: i.e., no nervous net can take account of future peripheral afferents. Any S_i satisfying this requirement can be replaced by an equipollent S of the form

$$(Ef)\,(z_2)z_1\,(z_3)zz_p :f\varepsilon\,Pr_{mi}:f\,(z_1,z_2,z_3)=1\,.\,=\,.\,p_{z3}\,(z_2)\tag{11}$$

where zz_p denotes p, by defining

$$Pr_{mi} = \hat{f}\,[\,(z_1)\,(z_2)z_1(z_3)zz_p : . \; f\,(z_1, z_2, z_3) = 0\,.\,\mathbf{v}\,.\,f\,(z_1, z_2, z_3)$$
$$= 1 : f\,(z_1, z_2, z_3) = 1\,.\, = .\, p_{z3}\,(z_2) : \rightarrow : S_i].$$

Consider now these series of classes α_i, for which

$$N_i(t) : = : (E\varphi)\,(x)t(m)q : \varphi\varepsilon\alpha_i : N_m(x)\,.\, = .\,\varphi(t, x, m) = 1.$$
$$[i = q + 1, \cdots, M] \tag{12}$$

holds for some net. These will be called *prehensible* classes. Let us define the *Boolean ring* generated by a class of classes k as the aggregate of the classes which can be formed from members of k by repeated application of the logical operations; i.e., we put

$$\mathcal{R}(\varkappa) = p\,\hat{\lambda}[\,(\alpha, \beta) : \alpha\varepsilon k \rightarrow \alpha\varepsilon\lambda : \alpha, \beta\varepsilon\lambda\,.\,\rightarrow\,.\, - \alpha, \alpha\,.\,\beta, \alpha\,\mathbf{v}\,\beta\varepsilon\lambda].$$

We shall also define

$$\overline{\mathcal{R}}(k)\,.\, = .\,\mathcal{R}(k) - \iota\,{}^{\prime}p^{\prime} - {}^{\prime\prime}k,$$
$$\mathcal{R}_e(k) = p^{\prime}\hat{\lambda}[\,(\alpha, \beta) : \alpha\varepsilon k \rightarrow \alpha\varepsilon\lambda\,.\,\rightarrow\,.\, - \alpha, \alpha\,.\,\beta, \alpha\,\mathbf{v}\,\beta, S\,{}^{\prime\prime}\alpha\varepsilon\hat{\lambda}$$
$$\overline{\mathcal{R}}_e(k) = \mathcal{R}_e(k) - \iota\,{}^{\prime}p^{\prime} - {}^{\prime\prime}k,$$

and

$$\sigma(\psi, t) = \varphi[\,(m)\,.\,\hat{\varphi}(t + 1, t, m) = \psi(m)\,].$$

The class $\mathcal{R}_e(k)$ is formed from k in analogy with $\mathcal{R}(k)$, but by repeated application not only of the logical operations but also of that which replaces a class of properties $P\,\varepsilon\,\alpha$ by $S(P)\,\varepsilon\,S\,{}^{\prime\prime}\alpha$. We shall then have the

LEMMA

$Pr_1(p_1, p_2, \ldots, p_m, z_1)$ is a *T P E* if and only if
$$(z_1)\,(p_1, \ldots, p_m)\,(Ep_{m+1}) : p_{m+1}\,\varepsilon\,\overline{\mathcal{R}}_e\,(\{\;p_1, p_2, \ldots, p_m\}\,)$$
$$p_{m+1}\,(z_1) = Pr_1\,(p_1, p_2, \ldots, p_m, z_1) \tag{13}$$

is true; and it is a *T P E* not involving '*S*' if and only if this holds when '$\overline{\mathcal{R}}_e$' is replaced by '$\overline{\mathcal{R}}$', and we then obtain

Theorem IX

A series of classes α_1, α_2, $\ldots \alpha_s$ is a series of prehensible classes if and only if

$$(Em) (En) (p)n(i) (\psi) : . (x)m\psi(x) = 0 \text{ v } \psi (x = 1 : \to : (E\beta)$$
$$(Ey)m . \psi(y) = 0 . \beta\varepsilon\mathcal{R}[\hat{\gamma} ((Ei) . \gamma = \alpha_i)) . \text{ v } . (x)m .$$
$$\psi(x) = 0 . \beta\varepsilon\bar{\mathcal{R}}[\hat{\gamma} ((E_i) . \gamma = \alpha_i)] : (t) (\varphi) : \varphi\varepsilon\alpha_i . \qquad (14)$$
$$\sigma (\varphi, nt + p) . \to . (Ef) . f\varepsilon\beta . (w)m(x)t - 1 .$$
$$\varphi(n (t + 1) + p, nx + p, w) = f(nt + p, nx + p, w).$$

The proof here follows directly from the lemma. The condition is necessary, since every net for which an expression of the form (4) can be written obviously verifies it, the ψ's being the characteristic functions of the S_a and the β for each ψ being the class whose designation has the form $\prod_{i\varepsilon\alpha} Pr_i \prod_{j\varepsilon\beta} Pr_j$, where Pr_k denotes α_k for all k. Conversely, we may write an expression of the form (4) for a net $\mathcal{N}$ fulfilling prehensible classes satisfying (14) by putting for the $Pr_a Pr$ denoting the ψ's and a Pr, written in the analogue for classes of the disjunctive normal form, and denoting the α corresponding to that ψ, conjoined to it. Since every S of the form (4) is clearly realizable, we have the theorem.

It is of some interest to consider the extent to which we can by knowledge of the present determine the whole past of various special nets: i.e. when we may construct a net the firing of the cyclic set of whose neurones requires the peripheral afferents to have a set of past values specified by given functions φ_i. In this case the classes α_i of the last theorem reduced to unit classes; and the condition may be transformed into

$$(E m, n) (p)n(i, \psi) (Ej) : . (x)m : \psi(x) = 0 . v . \psi(x) = 1 :$$
$$\varphi_i\varepsilon\sigma (\psi, nt + p) : \to : (w)m(x)t - 1 . \varphi_i(n(t + 1)$$
$$+ p, nx + p, w) = \varphi_j(nt + p, nx + p, w) : .$$
$$(u, v) (w)m . \varphi_i(n(u + 1) + p, nu + p, w)$$
$$= \varphi_i(n(v + 1) + p, nv + p, w).$$

On account of limitations of space, we have presented the above argument very sketchily; we propose to expand it and certain of its implications in a further publication.

The condition of the last theorem is fairly simple in principle, though not in detail; its application to practical cases would, however, require the exploration of some 2^{2n} classes of functions, namely the members of $\mathcal{R} (\{ \alpha_1, \ldots, \alpha_s \})$. Since each of these is a possible β of Theorem IX, this result cannot be sharpened. But we may obtain a sufficient condition for the realizability of an S which is very easily applicable and probably covers most practical purposes. This is given by

Theorem X

Let us define a set of K of S by the following recursion:

1. Any $T P E$ and any $T P E$ whose arguments have been replaced by members of K belong to K;

2. If $Pr_1(z_1)$ is a member of K, then $(z_2)z_1 \cdot Pr_1(z_2)$, $(Ez_2)z_1 \cdot Pr_1(z_2)$, and $C_{mn}(z_1) \cdot s$ belong to it, where C_{mn} denotes the property of being congruent to m modulo n, $m < n$.

3. *The set K has no further members.*

Then every member of K is realizable.

For, if $Pr_1(z_1)$ is realizable, nervous nets for which

$$N_i(z_1) \cdot = \cdot Pr_1(z_1) \cdot SN_i(z_1)$$
$$N_i(z_1) \cdot = \cdot Pr_1(z_1) \text{ v } SN_i(z_1)$$

are the expressions of equation (4), realize $(z_2)z_1 \cdot Pr_1(z_2)$ and $(E\ z_2)z_1 \cdot Pr_1(z_2)$ respectively; and a simple circuit, $c_1, c_2, \ldots, c_n$, of n links, each sufficient to excite the next, gives an expression

$$N_m(z_1) \cdot = \cdot N_1(0) \cdot C_{mn}$$

for the last form. By induction we derive the theorem.

One more thing is to be remarked in conclusion. It is easily shown: first, that every net, if furnished with a tape, scanners connected to afferents, and suitable efferents to perform the necessary motor-operations, can compute only such numbers as can a Turing machine; second, that each of the latter numbers can be computed by such a net; and that nets with circles can be computed by such a net; and that nets with circles can compute, without scanners and a tape, some of the numbers the machine can, but no others, and not all of them. This is of interest as affording a psychological justification of the Turing definition of computability and its equivalents, Church's λ—definability and Kleene's primitive recursiveness: if any number can be computed by an organism, it is computable by these definitions, and conversely.

4. CONSEQUENCES

Causality, which requires description of states and a law of necessary connection relating them, has appeared in several forms in several sciences, but never, except in statistics, has it been as irreciprocal as in this theory. Specification for any one time of afferent stimulation and of the activity of all constituent neurones, each an 'all-or-none' affair, determines the state. Specification of the nervous net provides the law of necessary connection whereby one can compute from the description of any state that of the succeeding state, but the inclusion of disjunctive relations prevents complete determination of the one before. Moreover, the regenerative activity of constituent circles renders reference indefinite as to time past. Thus our knowledge of the world, including ourselves, is incomplete as to space and indefinite as to time. This ignorance, implicit in all our brains, is the counterpart of the abstraction which renders our knowledge useful. The role of brains in determining the epistemic relations of our theories to our observations and of these to the facts is all too clear, for it is apparent that every idea and every sensation is realized by activity within that net, and by no such activity are the actual afferents fully determined.

There is no theory we may hold and no observation we can make that will retain so much as its old defective reference to the facts if the net be altered. Tinnitus, paraesthesias, hallucinations, delusions, confusions, and disorientations intervene. Thus empiry confirms that if our nets are undefined, our facts are undefined, and to the 'real' we can attribute not so much as one quality or 'form'. With determination of the net, the unknowable object of knowledge, the 'thing in itself', ceases to be unknowable.

To psychology, however defined, specification of the net would contribute all that could be achieved in that field—even if the analysis were pushed to ultimate psychic units or 'psychons', for a psychon can be no less than the activity of a single neurone. Since that activity is inherently propositional, all psychic events have an intentional, or 'semiotic', character. The 'all-or-none' law of these activities, and the conformity of their relations to those of the logic of propositions, insure that the relations of psychons are those of the two-valued logic of propositions. Thus in psychology, introspective, behaviouristic, or physiological, the fundamental relations are those of two-valued logic.

Hence arise constructional solutions of holistic problems involving the differentiated continuum of sense awareness and the normative, perfective, and resolvent properties of perception and execution. From the irreciprocity of causality it follows that even if the net be known, though we may predict future from present activities, we can deduce neither afferent from central, nor central from efferent, nor past from present activities—conclusions which are reinforced by the contradictory testimony of eye witnesses, by the difficulty of diagnosing differentially the organically diseased, the hysteric, and the malingerer, and by comparing one's own memories or recollections with his contemporaneous records. Moreover, systems which so respond to the difference between afferents to a regenerative net and certain activity within that net, as to reduce the difference, exhibit purposive behaviour; and organisms are known to possess many such systems, subserving homeostasis, appetition, and attention. Thus both the formal and the final aspects of that activity which we are wont to call *mental* are rigorously deducible from present neurophysiology. The psychiatrist may take comfort from the obvious conclusion concerning causality—that, for prognosis, history is never necessary. He can take little from the equally valid conclusion that his observables are explicable only in terms of nervous activities which, until recently, have been beyond his ken. The crux of this ignorance is that inference from any sample of overt behaviour to nervous nets is not unique, whereas, of imaginable nets, only one in fact exists, and may, at any moment, exhibit some unpredictable activity. Certainly for the psychiatrist it is more to the point that in such systems 'Mind' no longer 'goes more ghostly than a ghost'. Instead, diseased mentality can be understood without loss of scope or rigour, in the scientific terms of neurophysiology. For neurology, the theory sharpens the distinction between nets necessary or merely sufficient for given activities, and so clarifies the relations of disturbed structure to disturbed function. In its own domain the difference between equivalent nets and nets equivalent in the narrow sense indicates the appropriate use and importance of temporal studies of nervous activity: and to mathematical biophysics the theory contributes a tool for rigorous symbolic treatment of known nets and an easy method of constructing hypothetical nets of required properties.

REFERENCES

Carnap, R. (1938). *The Logical Syntax of Language.* New York: Harcourt, Brace and Company.

Hilbert, D., and Ackermann, W. (1927). *Grunduğe der Theoretischen Logik.* Berlin: J. Springer.

Whitehead, A. N., and Russell, B. (1925–7). *Principia Mathematica.* Cambridge: Cambridge University Press.

2

COMPUTING MACHINERY
AND INTELLIGENCE

ALAN M. TURING

1. THE IMITATION GAME

I propose to consider the question, 'Can machines think?' This should begin with definitions of the meaning of the terms 'machine' and 'think'. The definitions might be framed so as to reflect so far as possible the normal use of the words, but this attitude is dangerous. If the meaning of the words 'machine' and 'think' are to be found by examining how they are commonly used it is difficult to escape the conclusion that the meaning and the answer to the question, 'Can machines think?' is to be sought in a statistical survey such as a Gallup poll. But this is absurd. Instead of attempting such a definition I shall replace the question by another, which is closely related to it and is expressed in relatively unambiguous words.

The new form of the problem can be described in terms of a game which we call the 'imitation game'. It is played with three people, a man (A), a woman (B), and an interrogator (C) who may be of either sex. The inter-rogator stays in a room apart from the other two. The object of the game for the interrogator is to determine which of the other two is the man and which is the woman. He knows them by labels X and Y, and at the end of the game he says either 'X is A and Y is B' or 'X is B and Y is A.' The interrogator is allowed to put questions to A and B thus:

C: Will X please tell me the length of his or her hair?

Now suppose X is actually A, then A must answer. It is A's object in the game to try and cause C to make the wrong identification. His answer might therefore be:

'My hair is shingled, and the longest strands are about nine inches long.'

In order that tones of voice may not help the interrogator the answers

A. M. Turing, 'Computing Machinery and Intelligence' from *Mind* LIX, no. 2236 (Oct. 1950): 433–60. Reprinted by permission of Oxford University Press.

should be written, or better still, typewritten. The ideal arrangement is to have a teleprinter communicating between the two rooms. Alternatively the question and answers can be repeated by an intermediary. The object of the game for the third player (B) is to help the interrogator. The best strategy for her is probably to give truthful answers. She can add such things as 'I am the woman, don't listen to him!' to her answers, but it will avail nothing as the man can make similar remarks.

We now ask the question, 'What will happen when a machine takes the part of A in this game?' Will the interrogator decide wrongly as often when the game is played like this as he does when the game is played between a man and a woman? These questions replace our original, 'Can machines think?'

2. CRITIQUE OF THE NEW PROBLEM

As well as asking, 'What is the answer to this new form of the question?' one may ask, 'Is this new question a worthy one to investigate?' This latter question we investigate without further ado, thereby cutting short an infinite regress.

The new problem has the advantage of drawing a fairly sharp line between the physical and the intellectual capacities of a man. No engineer or chemist claims to be able to produce a material which is indistinguishable from the human skin. It is possible that at some time this might be done, but even supposing this invention available we should feel there was little point in trying to make a 'thinking machine' more human by dressing it up in such artificial flesh. The form in which we have set the problem reflects this fact in the condition which prevents the interrogator from seeing or touching the other competitors, or hearing their voices. Some other advantages of the proposed criterion may be shown up by specimen questions and answers. Thus:

Q: Please write me a sonnet on the subject of the Forth Bridge.
A: Count me out on this one. I never could write poetry.
Q: Add 34957 to 70764.
A: (Pause about 30 seconds and then give as answer) 105621.
Q: Do you play chess?
A: Yes.
Q: I have K at my K1, and no other pieces. You have only K at K6 and R at R1. It is your move. What do you play?
A: (After a pause of 15 seconds) R-R8 mate.

The question and answer method seems to be suitable for introducing almost any one of the fields of human endeavour that we wish to include. We do not wish to penalize the machine for its inability to shine in beauty competitions, nor to penalize a man for losing in a race against an aeroplane. The conditions of our game make these disabilities irrelevant. The 'witnesses' can brag, if they consider it advisable, as much as they please about their charms, strength, or heroism, but the interrogator cannot demand practical demonstrations.

The game may perhaps be criticized on the ground that the odds are weighted too heavily against the machine. If the man were to try and pretend to be the machine he would clearly make a very poor showing. He would be given away at once by slowness and inaccuracy in arithmetic. May not machines carry out something which ought to be described as thinking but which is very different from what a man does? This objection is a very strong one, but at least we can say that if, nevertheless, a machine can be constructed to play the imitation game satisfactorily, we need not be troubled by this objection.

It might be urged that when playing the 'imitation game' the best strategy for the machine may possibly be something other than imitation of the behaviour of a man. This may be, but I think it is unlikely that there is any great effect of this kind. In any case there is no intention to investigate here the theory of the game, and it will be assumed that the best strategy is to try to provide answers that would naturally be given by a man.

3. THE MACHINES CONCERNED IN THE GAME

The question which we put in §1 will not be quite definite until we have specified what we mean by the word 'machine'. It is natural that we should wish to permit every kind of engineering technique to be used in our machines. We also wish to allow the possibility that an engineer or team of engineers may construct a machine which works, but whose manner of operation cannot be satisfactorily described by its constructors because they have applied a method which is largely experimental. Finally, we wish to exclude from the machines men born in the usual manner. It is difficult to frame the definitions so as to satisfy these three conditions. One might for instance insist that the team of engineers should be all of one sex, but this would not really be satisfactory, for it is probably possible to rear a complete individual from a single cell of the skin (say) of a man. To do so would be a feat of biological technique deserving of the very highest praise, but we would not be inclined to regard it as a case of 'constructing a thinking machine'. This prompts us to abandon the requirement that every kind of technique should be permitted. We are the more ready to do so in view of the fact that the present interest in 'thinking machines' has been aroused by a particular kind of machine, usually called an 'electronic computer' or 'digital computer'. Following this suggestion we only permit digital computers to take part in our game.

This restriction appears at first sight to be a very drastic one. I shall attempt to show that it is not so in reality. To do this necessitates a short account of the nature and properties of these computers.

It may also be said that this identification of machines with digital computers, like our criterion for 'thinking', will only be unsatisfactory if (contrary to my belief), it turns out that digital computers are unable to give a good showing in the game.

There are already a number of digital computers in working order, and it may be asked, 'Why not try the experiment straight away? It would be easy to satisfy the conditions of the game. A number of interrogators could be used, and statistics compiled to show how often the right identification was given.' The short answer is that we are not asking whether all digital computers would do well in the game nor whether the computers at present available would do well, but whether there are imaginable computers which would do well. But this is only the short answer. We shall see this question in a different light later.

4. DIGITAL COMPUTERS

The idea behind digital computers may be explained by saying that these machines are intended to carry out any operations which could be done by a human computer. The human computer is supposed to be following fixed rules; he has no authority to deviate from them in any detail. We may suppose that these rules are supplied in a book, which is altered whenever he is put on to a new job. He has also an unlimited supply of paper on which he does his calculations. He may also do his multiplications and additions on a 'desk machine', but this is not important.

If we use the above explanation as a definition we shall be in danger of circularity of argument. We avoid this by giving an outline of the means by which the desired effect is achieved. A digital computer can usually be regarded as consisting of three parts:

1. Store
2. Executive unit
3. Control

The store is a store of information, and corresponds to the human computer's paper, whether this is the paper on which he does his calculations or that on which his book of rules is printed. In so far as the human computer does calculations in his head a part of the store will correspond to his memory.

The executive unit is the part which carries out the various individual operations involved in a calculation. What these individual operations are will vary from machine to machine. Usually fairly lengthy operations can be done such as 'Multiply 3540675445 by 7076345687' but in some machines only very simple ones such as 'Write down 0' are possible.

We have mentioned that the 'book of rules' supplied to the computer is replaced in the machine by a part of the store. It is then called the 'table of instructions'. It is the duty of the control to see that these instructions are obeyed correctly and in the right order. The control is so constructed that this necessarily happens.

The information in the store is usually broken up into packets of moderately small size. In one machine, for instance, a packet might consist of ten decimal digits. Numbers are assigned to the parts of the store in which the various packets of information are stored, in some systematic manner. A typical instruction might say:

> 'Add the number stored in position 6809 to that in 4302 and put the result back into the latter storage position.'

Needless to say it would not occur in the machine expressed in English. It would more likely be coded in a form such as 6809430217. Here 17 says which of various possible operations is to be performed on the two numbers. In this case the operation is that described above, viz., 'Add the number. . . .' It will be noticed that the instruction takes up 10 digits and so forms one packet of information, very conveniently. The control will normally take the instructions to be obeyed in the order of the positions in which they are stored, but occasionally an instruction such as:

> 'Now obey the instruction stored in position 5606, and continue from there'

may be encountered, or again:

> 'If position 4505 contains 0 obey next the instruction stored in 6707, otherwise continue straight on.'

Instructions of these latter types are very important because they make it possible for a sequence of operations to be replaced over and over again until some condition is fulfilled, but in doing so to obey, not fresh instructions on each repetition, but the same ones over and over again. To take a domestic analogy. Suppose Mother wants Tommy to call at the cobbler's every morning on his way to school to see if her shoes are done, she can ask him afresh every morning. Alternatively she can stick up a notice once and for all in the hall which he will see when he leaves for school and which tells him to call for the shoes, and also to destroy the notice when he comes back if he has the shoes with him.

The reader must accept it as a fact that digital computers can be constructed, and indeed have been constructed, according to the principles we have described, and that they can in fact mimic the actions of a human computer very closely.

The book of rules which we have described our human computer as using is of course a convenient fiction. Actual human computers really remember what they have got to do. If one wants to make a machine mimic the behaviour of the human computer in some complex operation one has to ask him how it is done, and then translate the answer into the form of an instruction table. Constructing instruction tables is usually described as 'programming'. To 'programme a machine to carry out the operation A' means to put the appropriate instruction table into the machine so that it will do A.

An interesting variant on the idea of a digital computer is a 'digital computer with a random element'. These have instructions involving the throwing of a die or some equivalent electronic process; one such instruction might for instance be, 'Throw the die and put the resulting number into store 1000.' Sometimes such a machine is described as having free will (though I would not use this phrase myself). It is not normally possible to determine from observing a machine whether it has a random element, for a similar effect can be produced by such devices as making the choices depend on the digits of the decimal for π.

Most actual digital computers have only a finite store. There is no theoretical difficulty in the idea of a computer with an unlimited store. Of course only a finite part can have been used at any one time. Likewise only a finite amount can have been constructed, but we can imagine more and more being added as required. Such computers have special theoretical interest and will be called infinitive capacity computers.

The idea of a digital computer is an old one. Charles Babbage, Lucasian Professor of Mathematics at Cambridge from 1828 to 1839, planned such a machine, called the Analytical Engine, but it was never completed. Although Babbage had all the essential ideas, his machine was not at that time such a very attractive prospect. The speed which would have been available would be definitely faster than a human computer but something like 100 times slower than the Manchester machine, itself one of the slower of the modern machines. The storage was to be purely mechanical, using wheels and cards.

The fact that Babbage's Analytical Engine was to be entirely mechanical will help us to rid ourselves of a superstition. Importance is often attached to the fact that modern digital computers are electrical, and that the nervous system also is electrical. Since Babbage's machine was not electrical, and since all digital computers are in a sense equivalent, we see that this use of electricity cannot be of theoretical importance. Of course electricity usually comes in where fast signalling is concerned, so that it is not surprising that we find it in both these connections. In the nervous system chemical phenomena are at least as important as electrical. In certain computers the storage system is mainly acoustic. The feature of using electricity is thus seen to be only a very superficial similarity. If we wish to find such similarities we should look rather for mathematical analogies of function.

5. UNIVERSALITY OF DIGITAL COMPUTERS

The digital computers considered in the last section may be classified amongst the 'discrete-state machines'. These are the machines which move by sudden jumps or clicks from one quite definite state to another. These states are sufficiently different for the possibility of confusion between them to be ignored. Strictly speaking there are no such machines. Everything really moves continuously. But there are many kinds of machine which can profitably be *thought of* as being discrete-state machines. For instance in considering the switches for a lighting system it is a convenient fiction that each switch must be definitely on or definitely off. There must be intermediate positions, but for most purposes we can forget about them. As an example of a discrete-state machine we might consider a wheel which clicks round through 120° once a second, but may be stopped by a lever which can be operated from outside; in addition a lamp is to light in one of the positions of the wheel. This machine could be described abstractly as follows. The internal state of the machine (which is described by the position of the wheel) may be q_1, q_2 or q_3. There is an input signal i_0 or i_1 (position of lever). The internal state at any moment is determined by the last state and input signal according to the table:

		Last State		
		q_1	q_2	q_3
Input	i_0	q_2	q_3	q_1
	i_1	q_1	q_2	q_3

The output signals, the only externally visible indication of the internal state (the light) are described by the table:

State	q_1	q_2	q_3
Output	o_0	o_0	o_1

This example is typical of discrete-state machines. They can be described by such tables provided they have only a finite number of possible states.

It will seem that given the initial state of the machine and the input signals it is always possible to predict all future states. This is reminiscent of Laplace's view that from the complete state of the universe at one moment of time, as described by the positions and velocities of all particles, it should be possible to predict all future states. The prediction which we are considering is, however, rather nearer to practicability than that considered by Laplace. The system of the 'universe as a whole' is such that quite small errors in the initial conditions can have an overwhelming effect at a later time. The displacement of a single electron by a billionth of a centimetre at one moment might make the difference between a man being killed by an avalanche a year later, or escaping. It is an essential property of the mechanical systems which we have called 'discrete-state machines' that this phenomenon does not occur. Even when we consider the actual physical machines instead of the idealized machines, reasonably accurate knowledge of the state at one moment yields reasonably accurate knowledge any number of steps later.

As we have mentioned, digital computers fall within the class of discrete-state machines. But the number of states of which such a machine is capable is usually enormously large. For instance, the number for the machine now working at Manchester is about $2^{165.000}$, i.e. about $10^{50.000}$. Compare this with our example of the clicking wheel described above, which had three states. It is not difficult to see why the number of states should be so immense. The computer includes a store corresponding to the paper used by a human computer. It must be possible to write into the store any one of the combinations of symbols which might have been written on the paper. For simplicity suppose that only digits from 0 to 9 are used as symbols. Variations in handwriting are ignored. Suppose the computer is allowed 100 sheets of paper each containing 50 lines each with room for 30 digits. Then the number of states is $10^{100 \times 50 \times 30}$, i.e., $10^{150.000}$. This is about the number of states of three Manchester machines put together. The logarithm to the base two of the number of states is usually called the 'storage capacity' of the machine. Thus the Manchester machine has a storage capacity of about 165,000 and the wheel machine of our example about 1.6. If two machines are put together their capacities must be added to obtain the capacity of the resultant machine. This leads to the possibility of statements such as: 'The Manchester machine contains 64 magnetic tracks each with a capacity of 2560, eight electronic tubes with a capacity of 1280. Miscellaneous storage amounts to about 300 making a total of 174,380.'

Given the table corresponding to a discrete-state machine it is possible to predict what it will do. There is no reason why this calculation should not be carried out by means of a digital computer. Provided it could be carried out sufficiently quickly the digital computer could mimic the behaviour of any discrete-state machine. The imitation game could then be played with the machine in question (as B) and the mimicking digital computer (as A) and the interrogator would be unable to distinguish them. Of course the digital computer must have an adequate storage capacity as well as working sufficiently fast. Moreover, it must be programmed afresh for each new machine which it is desired to mimic.

This special property of digital computers, that they can mimic any discrete-state machine, is described by saying that they are *universal* machines. The existence of machines with this property has the important consequence that, considerations of speed apart, it is unnecessary to design various new machines to do various computing processes. They can all be done with one digital computer, suitably programmed for each case. It will be seen that as a consequence of this all digital computers are in a sense equivalent.

We may now consider again the point raised at the end of §3. It was suggested tentatively that the question, 'Can machines think?' should be replaced by 'Are there imaginable digital computers which would do well in the imitation game?' If we wish we can make this superficially more general and ask 'Are there discrete-state machines which would do well?' But in view of the universality property we see that either of these questions is equivalent to this, 'Let us fix our attention on one particular digital computer C. Is it true that by modifying this computer to have an adequate storage, suitably increasing its speed of action, and providing it with an appropriate programme, C can be made to play satisfactorily the part of A in the imitation game, the part of B being taken by a man?'

6. CONTRARY VIEWS ON THE MAIN QUESTION

We may now consider the ground to have been cleared and we are ready to proceed to the debate on our question, 'Can machines think?' and the variant of it quoted at the end of the last section. We cannot altogether abandon the original form of the problem, for opinions will differ as to the appropriateness of the substitution and we must at least listen to what has to be said in this connexion.

It will simplify matters for the reader if I explain first my own beliefs in the matter. Consider first the more accurate form of the question. I believe that in about fifty years' time it will be possible to programme computers, with a storage capacity of about 10^9, to make them play the imitation game so well that an average interrogator will not have more than 70 per cent chance of making the right identification after five minutes of questioning. The original question, 'Can machines think?' I believe to be too meaningless to deserve discussion. Nevertheless I believe that at the end of the century the use of words and general educated opinion will have altered so much that one will be able to speak of machines thinking without expecting to be contradicted. I believe further that no useful purpose is served by concealing these beliefs. The popular view that scientists proceed inexorably from well-established fact to well-established fact, never being influenced by any improved conjecture, is quite mistaken. Provided it is made clear which are proved facts and which are conjectures, no harm can result. Conjectures are of great importance since they suggest useful lines of research.

I now proceed to consider opinions opposed to my own.

The Theological Objection

Thinking is a function of man's immortal soul. God has given an immortal soul to every man and woman, but not to any other animal or to machines. Hence no animal or machine can think.[1]

I am unable to accept any part of this, but will attempt to reply in theological terms. I should find the argument more convincing if animals were classed with men, for there is a greater difference, to my mind, between the typical animate and the inanimate than there is between man and the other animals. The arbitrary character of the orthodox view becomes clearer if we consider how it might appear to a member of some other religious community. How do Christians regard the Moslem view that women have no souls? But let us leave this point aside and return to the main argument. It appears to me that the argument quoted above

---

[1] Possibly this view is heretical. St Thomas Aquinas (*Summa Theologica*, quoted by Bertrand Russell (1945- 458)) states that God cannot make a man to have no soul. But this may not be a real restriction on His powers, but only a result of the fact that men's souls are immortal, and therefore indestructible.

implies a serious restriction of the omnipotence of the Almighty. It is admitted that there are certain things that He cannot do such as making one equal to two, but should we not believe that He has freedom to confer a soul on an elephant if He sees fit? We might expect that He would only exercise this power in conjunction with a mutation which provided the elephant with an appropriately improved brain to minister to the needs of this soul. An argument of exactly similar form may be made for the case of machines. It may seem different because it is more difficult to 'swallow'. But this really only means that we think it would be less likely that He would consider the circumstances suitable for conferring a soul. The circumstances in question are discussed in the rest of this paper. In attempting to construct such machines we should not be irreverently usurping His power of creating souls, any more than we are in the procreation of children: rather we are, in either case, instruments of His will providing mansions for the souls that He creates.

However, this is mere speculation. I am not very impressed with theological arguments whatever they may be used to support. Such arguments have often been found unsatisfactory in the past. In the time of Galileo it was argued that the texts, 'And the sun stood still . . . and hasted not to go down about a whole day' (Josh. 10: 13) and 'He laid the foundations of the earth, that it should not move at any time' (Ps. 105: 5) were an adequate refutation of the Copernican theory. With our present knowledge such an argument appears futile. When that knowledge was not available it made a quite different impression.

The 'Heads in the Sand' Objection

'The consequences of machines thinking would be too dreadful. Let us hope and believe that they cannot do so.'

This argument is seldom expressed quite so openly as in the form above. But it affects most of us who think about it at all. We like to believe that Man is in some subtle way superior to the rest of creation. It is best if he can be shown to be *necessarily* superior, for then there is no danger of him losing his commanding position. The popularity of the theological argument is clearly connected with this feeling. It is likely to be quite strong in intellectual people, since they value the power of thinking more highly than others, and are more inclined to base their belief in the superiority of Man on this power.

I do not think that this argument is sufficiently substantial to require refutation. Consolation would be more appropriate: perhaps this should be sought in the transmigration of souls.

The Mathemetical Objection

There are a number of results of mathematical logic which can be used to show that there are limitations to the powers of discrete-state machines. The best known of these results is known as Gödel's theorem (1931) and shows that in any sufficiently powerful logical system statements can be formulated which can neither be proved nor disproved within the system, unless possibly the system itself is inconsistent. There are other, in some respects similar, results due to Church (1936), Kleene (1935), Rosser, and Turing (1937). The latter result is the most convenient to consider, since it refers directly to machines, whereas the others can only be used in a comparatively indirect argument: for instance if Gödel's theorem is to be used we need in addition to have some means of describing logical systems in terms of machines, and machines in terms of logical systems. The result in question refers to a type of machine which is essentially a digital computer with an infinite capacity. It states that there are certain things that such a machine cannot do. If it is rigged up to give answers to questions as in the imitation game, there will be some questions to which it will either give a wrong answer, or fail to give an answer at all however much time is allowed for a reply. There may, of course, be many such questions, and questions which cannot be answered by one machine may be satisfactorily answered by another. We are of course supposing for the present that the questions are of the kind to which an answer 'Yes' or 'No' is appropriate, rather than questions such as 'What do you think of Picasso?' The questions that we know the machines must fail on are of this type, 'Consider the machine specified as follows . . . Will this machine ever answer "Yes" to any question?' The dots are to be replaced by a description of some machine in a standard form, which could be something like that used in §5. When the machine described bears a certain comparatively simple relation to the machine which is under interrogation, it can be shown that the answer is either wrong or not forthcoming. This is the mathematical result: it is argued that it proves a disability of machines to which the human intellect is not subject.

The short answer to this argument is that although it is established that there are limitations to the powers of any particular machine, it has only been stated, without any sort of proof, that no such limitations apply to the human intellect. But I do not think this view can be dismissed quite so lightly. Whenever one of these machines is asked the appropriate critical question, and gives a definite answer, we know that this answer must be wrong, and this gives us a certain feeling of superiority. Is this feeling illusory? It is no doubt quite genuine, but I do not think too much importance should be attached to it. We too often give wrong answers to questions ourselves to be justified in being very pleased at such evidence of fallibility on the part of the machines. Further, our superiority can only be felt on such an occasion in relation to the one machine over which we have scored our petty triumph. There would be no question of triumphing simultaneously over *all* machines. In short, then, there might be men cleverer than any given machine, but then again there might be other machines cleverer again, and so on.

Those who hold to the mathematical argument would, I think, mostly be willing to accept the imitation game as a basis for discussion. Those who believe in the two previous objections would probably not be interested in any criteria.

The Argument from Consciousness

This argument is very well expressed in Professor Jefferson's Lister Oration for 1949, from which I quote.

Not until a machine can write a sonnet or compose a concerto because of thoughts and emotions felt, and not by the chance fall of symbols, could we agree that machine equals brain—that is, not only write it but know that it had written it. No mechanism could feel (and not merely artificially signal, an easy contrivance) pleasure at its successes, grief when its valves fuse, be warmed by flattery, be made miserable by its mistakes, be charmed by sex, be angry or depressed when it cannot get what it wants.

This argument appears to be a denial of the validity of our test. According to the most extreme form of this view the only way by which one could be sure that a machine thinks is to *be* the machine and to feel oneself thinking. One could then describe these feelings to the world, but of course no one would be justified in taking any notice. Likewise according to this view the only way to know that a *man* thinks is to be that particular man. It is in fact the solipsist point of view. It may be the most logical view to hold but it makes communication of ideas difficult. A is liable to believe 'A thinks but B does not' whilst B believes 'B thinks but A does not.' Instead of arguing continually over this point it is usual to have the polite convention that everyone thinks.

I am sure that Professor Jefferson does not wish to adopt the extreme and solipsist point of view. Probably he would be quite willing to accept the imitation game as a test. The game (with the player B omitted) is frequently used in practice under the name of viva voce to discover whether some one really understands something or has 'learnt it parrot fashion'. Let us listen in to a part of such a viva voce:

INTERROGATOR: In the first line of your sonnet which reads 'Shall I compare thee to a summer's day,' would not 'a spring day' do as well or better?

WITNESS: It wouldn't scan.

INTERROGATOR: How about 'a winter's day'. That would scan all right.

WITNESS: Yes, but nobody wants to be compared to a winter's day.

INTERROGATOR: Would you say Mr. Pickwick reminded you of Christmas?

WITNESS: In a way.

INTERROGATOR: Yet Christmas is a winter's day, and I do not think Mr. Pickwick would mind the comparison.

WITNESS: I don't think you're serious. By a winter's day one means a typical winter's day, rather than a special one like Christmas.

And so on. What would Professor Jefferson say if the sonnet-writing machine was able to answer like this in the viva voce? I do not know whether he would regard the machine as 'merely artificially signalling' these answers, but if the answers were as satisfactory and sustained as in the above passage I do not think he would describe it as 'an easy contrivance'. This phrase is, I think, intended to cover such devices as the inclusion in the machine of a record of someone reading a sonnet, with appropriate switching to turn it on from time to time.

In short then, I think that most of those who support the argument from consciousness could be persuaded to abandon it rather than be forced into the solipsist position. They will then probably be willing to accept our test.

I do not wish to give the impression that I think there is no mystery about consciousness. There is, for instance, something of a paradox connected with any attempt to localize it. But I do not think these mysteries necessarily need to be solved before we can answer the question with which we are concerned in this paper.

Arguments from Various Disabilities

These arguments take the form, 'I grant you that you can make machines do all the things you have mentioned but you will never be able to make one to do X'. Numerous features X are suggested in this connexion. I offer a selection:

Be kind, resourceful, beautiful, friendly, have initiative, have a sense of humour, tell right from wrong, make mistakes, fall in love, enjoy strawberries and cream, make some one fall in love with it, learn from experience, use words properly, be the subject of its own thought, have as much diversity of behaviour as a man, do something really new.

No support is usually offered for these statements. I believe they are mostly founded on the principle of scientific induction. A man has seen thousands of machines in his lifetime. From what he sees of them he draws a number of general conclusions. They are ugly, each is designed for a very limited purpose, when required for a minutely different purpose they are useless, the variety of behaviour of any one of them is very small, etc., etc. Naturally he concludes that these are necessary properties of machines in general. Many of these limitations are associated with the very small storage capacity of most machines. (I am assuming that the idea of storage capacity is extended in some way to cover machines other than discrete-state machines. The exact definition does not matter as no mathematical accuracy is claimed in the present discussion.) A few years ago, when very little had been heard of digital computers, it was possible to elicit much incredulity concerning them, if one mentioned their properties without describing their construction. That was presumably due to a similar application of the principle of scientific induction. These applications of the principle are of course largely unconscious. When a burnt child fears the fire and shows that he fears it by avoiding it, I should say that he was applying scientific induction. (I could of course also describe his behaviour in many other ways.) The works and customs of mankind do not seem to be very suitable material to which to apply scientific induction. A very large part of space-time must be investigated, if reliable results are to be obtained. Otherwise we may (as most English children do) decide that everybody speaks English, and that it is silly to learn French.

There are, however, special remarks to be made about many of the disabilities that have been mentioned. The inability to enjoy strawberries and cream may have struck the reader as frivolous. Possibly a machine might be made to enjoy this delicious dish, but any attempt to make one do so would be idiotic. What is important about this disability is that it contributes to some of the other disabilities, e.g., to the difficulty of the same kind of friendliness occurring between man and machine as between white man and white man, or between black man and black man.

The claim that 'machines cannot make mistakes' seems a curious one. One is tempted to retort, 'Are they any the worse for that?' But let us adopt a more sympathetic attitude, and try to see what is really meant. I think this criticism can be explained in terms of the imitation game. It is claimed that the interrogator could distinguish the machine from the man simply by setting them a number of problems in arithmetic. The machine would be unmasked because of its deadly accuracy. The reply to this is simple. The machine (programmed for playing the game) would not attempt to give the *right* answers to the arithmetic problems. It would deliberately introduce mistakes in a manner calculated to confuse the interrogator. A mechanical fault would probably show itself through an unsuitable decision as to what sort of a mistake to make in the arithmetic. Even this interpretation of the criticism is not sufficiently sympathetic. But we cannot afford the space to go into it much further. It seems to me that this criticism depends on a confusion between two kinds of mistake. We may call them 'errors of functioning' and 'errors of conclusion'. Errors of functioning are due to some mechanical or electrical fault which causes the machine to behave otherwise than it was designed to do. In philosophical discussions one likes to ignore the possibility of such errors; one is therefore discussing 'abstract machines'. These abstract machines are mathematical fictions rather than physical objects. By definition they are incapable of errors of functioning. In this sense we can truly say that 'machines can never make mistakes.' Errors of conclusion can only arise when some meaning is attached to the output signals from the machine. The machine might, for instance, type out mathemetical equations, or sentences in English. When a false proposition is typed we say that the machine has committed an error of conclusion. There is clearly no reason at all for saying that a machine cannot make this kind of mistake. It might do nothing but type out repeatedly '0 = 1'. To take a less perverse example, it might have some method for drawing conclusions by scientific induction. We must expect such a method to lead occasionally to erroneous results.

The claim that a machine cannot be the subject of its own thought can of course only be answered if it can be shown that the machine has *some* thought with *some* subject matter. Nevertheless, 'the subject matter of a machine's operations' does seem to mean something, at least to the people who deal with it. If, for instance, the machine was trying to find a solution of the equation $x^2 - 40x - 11 = 0$ one would be tempted to describe this equation as part of the machine's subject matter at that moment. In this sort of sense a machine undoubtedly can be its own subject-matter. It may be used to help in making up its own programmes, or to predict the effect of alterations in its own structure. By observing the results of its own behaviour it can modify its own programmes so as to achieve some purpose more effectively. These are possibilities of the near future, rather than Utopian dreams.

The criticism that a machine cannot have much diversity of behaviour is just a way of saying that it cannot have much storage capacity. Until fairly recently a storage capacity of even a thousand digits was very rare.

The criticisms that we are considering here are often disguised forms of the argument from consciousness. Usually if one maintains that a machine *can* do one of these things, and describes the kind of method that the machine could use, one will not make much of an impression. It is thought that the method (whatever it may be, for it must be mechanical) is really rather base. Compare the parentheses in Jefferson's statement quoted on page 52.

Lady Lovelace's Objection

Our most detailed information of Babbage's Analytical Engine comes from a memoir by Lady Lovelace (1842). In it she states, 'The Analytical Engine has no pretensions to *originate* anything. It can do *whatever we know how to order it* to perform' (her italics). This statement is quoted by Hartree (1949) who adds:

This does not imply that it may not be possible to construct electronic equipment which will 'think for itself', or in which, in biological terms, one could set up a conditioned reflex, which would serve as a basis for 'learning'. Whether this is possible in principle or not is a stimulating and exciting question, suggested by some of these recent developments. But it did not seem that the machines constructed or projected at the time had this property.

I am in thorough agreement with Hartree over this. It will be noticed that he does not assert that the machines in question had not got the property, but rather that the evidence available to Lady Lovelace did not encourage her to believe that they had it. It is quite possible that the machines in question had in a sense got this property. For suppose that some discrete-state machine has the property. The Analytical Engine was a universal digital computer, so that, if its storage capacity and speed were adequate, it could by suitable programming be made to mimic the machine in question. Probably this argument did not occur to the Countess or to Babbage. In any case there was no obligation on them to claim all that could be claimed.

This whole question will be considered again under the heading of learning machines.

A variant of Lady Lovelace's objection states that a machine can 'never do anything really new'. This may be parried for a moment with the saw, 'There is nothing new under the sun.' Who can be certain that 'original work' that he has done was not simply the growth of the seed planted in him by teaching, or the effect of following well-known general principles. A better variant of the objection says that a machine can never 'take us by surprise'. This statement is a more direct challenge and can be met directly. Machines take me 'by surprise with great frequency. This is largely because I do not do sufficient calculation to decide what to expect them to do, or rather because, although I do a calculation, I do it in a hurried, slipshod fashion, taking risks. Perhaps I say to myself, 'I suppose the voltage here ought to be the same as there: anyway let's assume it is.' Naturally I am often wrong, and the result is a surprise for me for by the time the experiment is done these assumptions have been forgotten. These admissions lay me open to lectures on the subject of my vicious ways, but do not throw any doubt on my credibility when I testify to the surprises I experience.

I do not expect this reply to silence my critic. He will probably say that such surprises are due to some creative mental act on my part, and reflect no credit on the machine. This leads us back to the argument from consciousness, and far from the idea of surprise. It is a line of argument we must consider closed, but it is perhaps worth remarking that the appreciation of something as surprising requires as much of a 'creative mental act' whether the surprising event originates from a man, a book, a machine, or anything else.

The view that machines cannot give rise to surprises is due, I believe, to a fallacy to which philosophers and mathematicians are particularly subject. This is the assumption that as soon as a fact is presented to a mind all consequences of that fact spring into the mind simultaneously with it. It is a very useful assumption under many circumstances, but one too easily forgets that it is false. A natural consequence of doing so is that one then assumes that there is no virtue in the mere working out of consequences from data and general principles.

Argument from Continuity in the Nervous System

The nervous system is certainly not a discrete-state machine. A small error in the information about the size of a nervous impulse impinging on a neurone, may make a large difference to the size of the outgoing impulse. It may be argued that, this being so, one cannot expect to be able to mimic the behaviour of the nervous system with a discrete-state system.

It is true that a discrete-state machine must be different from a continuous machine. But if we adhere to the conditions of the imitation game, the interrogator will not be able to take any advantage of this difference. The situation can be made clearer if we consider some other simpler continuous machine. A differential analyser will do very well. (A differential analyser is a certain kind of machine not of the discrete-state type used for some kinds of calculation.) Some of these provide their answers in a typed form, and so are suitable for taking part in the game. It would not be possible for a digital computer to predict exactly what answers the differential analyser would give to a problem, but it would be quite capable of giving the right sort of answer. For instance, if asked to give the value of π (actually about 3.1416) it would be reasonable to choose at random between the values 3.12, 3.13, 3.14, 3.15, 3.16 with the probabilities of 0.05, 0.15, 0.55, 0.19, 0.06 (say). Under these circumstances it would be very difficult for the interrogator to distinguish the differential analyser from the digital computer.

The Argument from Informality of Behaviour

It is not possible to produce a set of rules purporting to describe what a man should do in every conceivable set of circumstances. One might for instance have a rule that one is to stop when one sees a red traffic light, and to go if one sees a green one, but what if by some fault both appear together? One may perhaps decide that it is safest to stop. But some further difficulty may well arise from this decision later. To attempt to provide rules of conduct to cover every eventuality, even those arising from traffic lights, appears to be impossible. With all this I agree.

From this it is argued that we cannot be machines. I shall try to reproduce the argument, but I fear I shall hardly do it justice. It seems to run something like this. 'If each man had a definite set of rules of conduct by which he regulated his life he would be no better than a machine. But there are no such rules, so men cannot be machines.' The undistributed middle is glaring. I do not think the argument is ever put quite like this, but I believe this is the argument used nevertheless. There may however be a certain confusion between 'rules of conduct' and 'laws of behaviour' to cloud the issue. By 'rules of conduct' I mean precepts such as 'Stop if you see red lights,' on which one can act, and of which one can be conscious. By 'laws of behaviour' I mean laws of nature as applied to a man's body such as 'if you pinch him he will squeak'. If we substitute 'laws of behaviour which regulate his life' for 'laws of conduct by which he regulates his life' in the argument quoted the undistributed middle is no longer insuperable. For we believe that it is not only true that being regulated by laws of behaviour implies being some sort of machine (though not necessarily a discrete-state machine), but that conversely being such a machine implies being regulated by such laws. However, we cannot so easily convince ourselves of the absence of complete laws of behaviour as of complete rules of conduct. The only way we know of for finding such laws is scientific observation, and we certainly know of no circumstances under which we could say, 'We have searched enough. There are no such laws.'

We can demonstrate more forcibly that any such statement would be unjustified. For suppose we could be sure of finding such laws if they existed. Then given a discrete-state machine it should certainly be possible to discover by observation sufficient about it to predict its future behaviour, and this within a reasonable time, say a thousand years. But this does not seem to be the case. I have set up on the Manchester computer a small programme using only 1,000 units of storage, whereby the machine supplied with one sixteen-figure number replies with another within two seconds. I would defy anyone to learn from these replies sufficient about the programme to be able to predict any replies to untried values.

The Argument from Extrasensory Perception

I assume that the reader is familiar with the idea of extrasensory perception, and the meaning of the four items of it, viz., telepathy, clairvoyance, precognition, and psychokinesis. These disturbing phenomena seem to deny all our usual scientific ideas. How we should like to discredit them! Unfortunately the statistical evidence, at least for telepathy, is overwhelming. It is very difficult to rearrange one's ideas so as to fit these new facts in. Once one has accepted them it does not seem a very big step to believe in ghosts and bogies. The idea that our bodies move simply according to the known laws of physics, together with some others not yet discovered but somewhat similar, would be one of the first to go.

This argument is to my mind quite a strong one. One can say in reply that many scientific theories seem to remain workable in practice, in spite of clashing with ESP; that in fact one can get along very nicely if one forgets about it. This is rather cold comfort, and one fears that thinking is just the kind of phenomenon where ESP may be especially relevant.

A more specific argument based on ESP might run as follows: 'Let us play the imitation game, using as witnesses a man who is good as a telepathic receiver, and a digital computer. The interrogator can ask such questions as 'What suit does the card in my right hand belong to?' The man by telepathy or clairvoyance gives the right answer 130 times out of 400 cards. The machine can only guess at random, and perhaps gets 104 right, so the interrogator makes the right identification.' There is an interesting possibility which opens here. Suppose the digital computer contains a random number generator. Then it will be natural to use this to decide what answer to give. But then the random number generator will be subject to the psychokinetic powers of the interrogator. Perhaps this psychokinesis might cause the machine to guess right more often than would be expected on a probability calculation, so that the interrogator might still be unable to make the right identification. On the other hand, he might be able to guess right without any questioning, by clairvoyance. With ESP anything may happen.

If telepathy is admitted it will be necessary to tighten our test up. The situation could be regarded as analogous to that which would occur if the interrogator were talking to himself and one of the competitors was listening with his ear to the wall. To put the competitors into a 'telepathy-proof room' would satisfy all requirements.

7. LEARNING MACHINES

The reader will have anticipated that I have no very convincing arguments of a positive nature to support my views. If I had I should not have taken such pains to point out the fallacies in contrary views. Such evidence as I have I shall now give.

Let us return for a moment to Lady Lovelace's objection, which stated that the machine can only do what we tell it to do. One could say that a man can 'inject' an idea into the machine, and that it will respond to a certain extent and then drop into quiescence, like a piano string struck by a hammer. Another simile would be an atomic pile of less than critical size: an injected idea is to correspond to a neutron entering the pile from without. Each such neutron will cause a certain disturbance which eventually dies away. If, however, the size of the pile is sufficiently increased, the disturbance caused by such an incoming neutron will very likely go on and on increasing until the whole pile is destroyed. Is there a corresponding phenomenon for minds, and is there one for machines? There does seem to be one for the human mind. The majority of them seem to be 'subcritical', i.e., to correspond in this analogy to piles of subcritical size. An idea presented to such a mind will on average give rise to less than one idea in reply. A smallish proportion are supercritical. An idea presented to such a mind that may give rise to a whole 'theory' consisting of secondary, tertiary, and more remote ideas. Animals minds seem to be very definitely subcritical. Adhering to this analogy we ask, 'Can a machine be made to be supercritical?'

The 'skin-of-an-onion' analogy is also helpful. In considering the functions of the mind or the brain we find certain operations which we can explain in purely mechanical terms. This we say does not correspond to the real mind: it is a sort of skin which we must strip off if we are to find the real mind. But then in what remains we find a further skin to be stripped off, and so on. Proceeding in this way do we ever come to the 'real' mind, or do we eventually come to the skin which has nothing in it? In the latter case the whole mind is mechanical. (It would not be a discrete-state machine however. We have discussed this.)

These last two paragraphs do not claim to be convincing arguments. They should rather be described as 'recitations tending to produce belief'.

The only really satisfactory support that can be given for the view expressed at the beginning of §6, will be that provided by waiting for the end of the century and then doing the experiment described. But what can we say in the meantime? What steps should be taken now if the experiment is to be successful?

As I have explained, the problem is mainly one of programming. Advances in engineering will have to be made too, but it seems unlikely that these will not be adequate for the requirements. Estimates of the storage capacity of the brain vary from 10^{10} to 10^{15} binary digits. I incline to the lower values and believe that only a very small fraction is used for the higher types of thinking. Most of it is probably used for the retention of visual impressions. I should be surprised if more than 10^9 was required for satisfactory playing of the imitation game, at any rate against a blind man. (Note: The capacity of the *Encyclopaedia Britannica*, 11th edition, is 2×10^9.) A storage capacity of 10^7 would be a very practicable possibility even by present techniques. It is probably not necessary to increase the speed of operations of the machines at all. Parts of modern machines which can be regarded as analogs of nerve cells work about a thousand times faster than the latter. This should provide a 'margin of safety' which could cover losses of speed arising in many ways. Our problem then is to find out how to programme these machines to play the game. At my present rate of working I produce about a thousand digits of programme a day, so that about sixty workers, working steadily through the fifty years might accomplish the job, if nothing went into the wastepaper basket. Some more expeditious method seems desirable.

In the process of trying to imitate an adult human mind we are bound to think a good deal about the process which has brought it to the state that it is in. We may notice three components.

1. The initial state of the mind, say at birth,
2. The education to which it has been subjected,
3. Other experience, not to be described as education, to which it has been subjected.

Instead of trying to produce a programme to simulate the adult mind, why not rather try to produce one which simulates the child's? If this were then subjected to an appropriate course of education one would obtain the adult brain. Presumably the child brain is something like a notebook as one buys it from the stationer's. Rather little mechanism, and lots of blank sheets. (Mechanism and writing are from our point of view almost synonymous.) Our hope is that there is so little mechanism in the child brain that something like it can be easily programmed. The amount of work in the education we can assume, as a first approximation, to be much the same as for the human child.

We have thus divided our problem into two parts. The child programme and the education process. These two remain very closely connected. We cannot expect to find a good child machine at the first attempt. One must experiment with teaching one such machine and see how well it learns. One can then try another and see if it is better or worse. There is an obvious connection between this process and evolution, by the identifications

Structure of the child machine = hereditary material
Changes of the child machine = mutations
Natural selection = judgement of the experimenter

One may hope, however, that this process will be more expeditious than evolution. The survival of the fittest is a slow method for measuring advantages. The experimenter, by the exercise of intelligence, should be able to speed it up. Equally important is the fact that he is not restricted to random mutations. If he can trace a cause for some weakness he can probably think of the kind of mutation which will improve it.

It will not be possible to apply exactly the same teaching process to the machine as to a normal child. It will not, for instance, be provided with legs, so that it could not be asked to go out and fill the coal scuttle. Possibly it might not have eyes. But however well these deficiencies might be overcome by clever engineering, one could not send the creature to school without the other children making excessive fun of it. It must be given some tuition. We need not be too concerned about the legs, eyes, etc. The example of Miss Helen Keller shows that education can take place provided that communication in both directions between teacher and pupil can take place by some means or other.

We normally associate punishments and rewards with the teaching process. Some simple child machines can be constructed or programmed on this sort of principle. The machine has to be so constructed that events which shortly preceded the occurrence of a punishment signal are unlikely to be repeated, whereas a reward signal increased the probability of repetition of the events which led up to it. These definitions do not presuppose any feelings on the part of the machine. I have done some experiments with one such child machine, and succeeded in teaching it a few things, but the teaching method was too unorthodox for the experiment to be considered really successful.

The use of punishments and rewards can at best be a part of the teaching process. Roughly speaking, if the teacher has no other means of communicating to the pupil, the amount of information which can reach him does not exceed the total number of rewards and punishments applied. By the time a child has learnt to repeat 'Casabianca' he would probably feel very sore indeed, if the text could only be discovered by a 'Twenty Questions' technique, every 'NO' taking the form of a blow. It is necessary therefore to have some other 'unemotional' channels of communication. If these are available it is possible to teach a machine by punishments and rewards to obey orders given in some language, e.g., a symbolic language. These orders are to be transmitted through the 'unemotional' channels. The use of this language will diminish greatly the number of punishments and rewards required.

Opinions may vary as to the complexity which is suitable in the child machine. One might try to make it as simple as possible consistently with the general principles. Alternatively one might have a complete system of logical inference 'built in'.[2] In the latter case the store would be largely occupied with definitions and propositions. The propositions would have various kinds of status, e.g., well-established facts, conjectures, mathematically proved theorems, statements given by an authority, expressions having the logical form of proposition but not belief-value. Certain propositions may be described as 'imperatives'. The machine should be so constructed that as soon as an imperative is classed as 'well established' the appropriate action automatically takes place. To illustrate this, suppose the teacher says to the machine, 'Do your homework now.' This may cause 'Teacher says "Do your homework now" ' to be included amongst the well-established facts. Another such fact might be, 'Everything that teacher says is true.' Combining these may eventually lead to the imperative, 'Do your homework now,' being included amongst the well-established facts, and this, by the construction of the machine, will mean that the homework actually gets started, but the effect is very satisfactory. The processes of inference used by the machine need not be such as would satisfy the most exacting logicians. There might for instance be

---

[2] Or rather 'programmed in' for our child machine will be programmed in a digital computer. But the logical system will not have to be learnt.

no hierarchy of types. But this need not mean that type fallacies will occur, any more than we are bound to fall over unfenced cliffs. Suitable imperatives (expressed *within* the systems, not forming part of the rules *of* the system) such as 'Do not use a class unless it is a subclass of one which has been mentioned by teacher' can have a similar effect to 'Do not go too near the edge.'

The imperatives that can be obeyed by a machine that has no limbs are bound to be of a rather intellectual character, as in the example (doing homework) given above. Important amongst such imperatives will be ones which regulate the order in which the rules of the logical system concerned are to be applied. For at each stage when one is using a logical system, there is a very large number of alternative steps, any of which one is permitted to apply, so far as obedience to the rules of the logical system is concerned. These choices make the difference between a brilliant and a footling reasoner, not the difference between a sound and a fallacious one. Propositions leading to imperatives of this kind might be 'When Socrates is mentioned, use the syllogism in Barbara' or 'If one method has been proved to be quicker than another, do not use the slower method.' Some of these may be 'given by authority', but others may be produced by the machine itself, e.g., by scientific induction.

The idea of a learning machine may appear paradoxical to some readers. How can the rules of operation of the machine change? They should describe completely how the machine will react whatever its history might be, whatever changes it might undergo. The rules are thus quite time-invariant. This is quite true. The explanation of the paradox is that the rules which get changed in the learning process are of a rather less pretentious kind, claiming only an ephemeral validity. The reader may draw a parallel with the Constitution of the United States.

An important feature of a learning machine is that its teacher will often be very largely ignorant of quite what is going on inside, although he may still be able to some extent to predict his pupil's behaviour. This should apply most strongly to the later education of a machine arising from a child machine of well-tried design (or programme). This is in clear contrast with normal procedure when using a machine to do computations: one's object is then to have a clear mental picture of the state of the machine at each moment in the computation. This object can only be achieved with a struggle. The view that 'the machine can only do what we know how to order it to do,'[3] appears strange in face of this. Most of the programmes which we can put into the machine will result in its doing

---

[3] Compare Lady Lovelace's statement which does not contain the word 'only'.

something that we cannot make sense of at all or which we regard as completely random behaviour. Intelligent behaviour presumably consists in a departure from the completely disciplined behaviour involved in computation, but a rather slight one, which does not give rise to random behaviour, or to pointless repetitive loops. Another important result of preparing our machine for its part in the imitation game by a process of teaching and learning is that 'human fallibility' is likely to be omitted in a rather natural way, i.e., without special 'coaching'. (The reader should reconcile this with the point of view on pages 54 and 55.) Processes that are learnt do not produce a hundred per cent certainty of result; if they did they could not be unlearnt.

It is probably wise to include a random element in a learning machine. A random element is rather useful when we are searching for a solution of some problem. Suppose for instance we wanted to find a number between 50 and 200 which was equal to the square of the sum of its digits, we might start at 51 then try 52 and go on until we got a number that worked. Alternatively we might choose numbers at random until we got a good one. This method has the advantage that it is unnecessary to keep track of the values that have been tried, but the disadvantage that one may try the same one twice, but this is not very important if there are several solutions. The systematic method has the disadvantage that there may be an enormous block without any solutions in the region which has to be investigated first. Now the learning process may be regarded as a search for a form of behaviour which will satisfy the teacher (or some other criterion). Since there is probably a very large number of satisfactory solutions the random method seems to be better than the systematic. It should be noticed that it is used in the analogous process of evolution. But there the systematic method is not possible. How could one keep track of the different genetical combinations that had been tried, so as to avoid trying them again?

We may hope that machines will eventually compete with men in all purely intellectual fields. But which are the best ones to start with? Even this is a difficult decision. Many people think that a very abstract activity, like the playing of chess, would be best. It can also be maintained that it is best to provide the machine with the best sense organs that money can buy, and then teach it to understand and speak English. This process could follow the normal teaching of a child. Things would be pointed out and named, etc. Again I do not know what the right answer is, but I think both approaches should be tried.

We can only see a short distance ahead, but we can see plenty there that needs to be done.

REFERENCES

Church, A. (1936). 'An Unsolvable Problem of Elementary Number Theory.' *American J. Mathematics* 58: 345–63.

Gödel, K. (1931). 'Uber Formal Unentscheidbare Sätze der Principia Mathematica und Verwandter Systeme, I.' *Monatshefte für Mathematica und Physics*, pp. 173–89.

Hartree, D. R. (1949). *Calculating Instruments and Machines*. Urbana: University of Illinois Press.

Kleene, S. C. (1935). 'General Recursive Functions of Natural Numbers.' *American J. Mathematics* 57: 153–7, 219–44.

Russell, B. (1945). *History of Western Philosophy*. New York: Simon and Schuster.

Turing, A. M. (1937). 'On Computable Numbers, with an Application to the *Entscheidungsproblem.*' *Proc. London Math. Soc.* 43: 544; also 42 (1936): 230–65.

3

MINDS, BRAINS, AND PROGRAMS

JOHN R. SEARLE

What psychological and philosophical significance should we attach to recent efforts at computer simulations of human cognitive capacities? In answering this question, I find it useful to distinguish what I will call 'strong' AI from 'weak' or 'cautious' AI (Artificial Intelligence). According to weak AI, the principal value of the computer in the study of the mind is that it gives us a very powerful tool. For example, it enables us to formulate and test hypotheses in a more rigorous and precise fashion. But according to strong AI, the computer is not merely a tool in the study of the mind; rather, the appropriately programmed computer really *is* a mind, in the sense that computers given the right programs can be literally said to *understand* and have other cognitive states. In strong AI, because the programmed computer has cognitive states, the programs are not mere tools that enable us to test psychological explanations; rather, the programs are themselves the explanations.

I have no objection to the claims of weak AI, at least as far as this article is concerned. My discussion here will be directed at the claims I have defined as those of strong AI, specifically the claim that the appropriately programmed computer literally has cognitive states and that the programs thereby explain human cognition. When I hereafter refer to AI, I have in mind the strong version, as expressed by these two claims.

I will consider the work of Roger Schank and his colleagues at Yale (Schank and Abelson 1977), because I am more familiar with it than I am with any other similar claims, and because it provides a very clear example of the sort of work I wish to examine. But nothing that follows depends upon the details of Schank's programs. The same arguments would apply to Winograd's SHRDLU (Winograd 1973), Weizenbaum's ELIZA (Weizenbaum 1965), and indeed any Turing machine simulation of human mental phenomena.

John R. Searle, 'Minds, Brains, and Programs' from *The Behavioral and Brain Sciences* 3 (1980): 417–24. © 1980 Cambridge University Press. Reprinted by permission of Cambridge University Press for the author.

Very briefly, and leaving out the various details, one can describe Schank's program as follows: the aim of the program is to simulate the human ability to understand stories. It is characteristic of human beings' story-understanding capacity that they can answer questions about the story even though the information that they give was never explicitly stated in the story. Thus, for example, suppose you are given the following story: 'A man went into a restaurant and ordered a hamburger. When the hamburger arrived it was burned to a crisp, and the man stormed out of the restaurant angrily, without paying for the hamburger or leaving a tip.' Now, if you are asked 'Did the man eat the hamburger?' you will presumably answer, 'No, he did not.' Similarly, if you are given the following story: 'A man went into a restaurant and ordered a hamburger; when the hamburger came he was very pleased with it; and as he left the restaurant he gave the waitress a large tip before paying his bill,' and you are asked the question, 'Did the man eat the hamburger?', you will presumably answer, 'Yes, he ate the hamburger.' Now Schank's machines can similarly answer questions about restaurants in this fashion. To do this, they have a 'representation' of the sort of information that human beings have about restaurants, which enables them to answer such questions as those above, given these sorts of stories. When the machine is given the story and then asked the question, the machine will print out answers of the sort that we would expect human beings to give if told similar stories. Partisans of strong AI claim that in this question and answer sequence the machine is not only simulating a human ability but also

1. that the machine can literally be said to *understand* the story and provide the answers to questions, and
2. that what the machine and its program do *explains* the human ability to understand the story and answer questions about it.

Both claims seem to me to be totally unsupported by Schank's[1] work, as I will attempt to show in what follows.

One way to test any theory of the mind is to ask oneself what it would be like if my mind actually worked on the principles that the theory says all minds work on. Let us apply this test to the Schank program with the following *Gedankenexperiment.* Suppose that I'm locked in a room and given a large batch of Chinese writing. Suppose furthermore (as is indeed the case) that I know no Chinese, either written or spoken, and that I'm not even confident that I could recognize Chinese writing as Chinese

---

[1] I am not, of course, saying the Schank himself is committed to these claims.

writing distinct from, say, Japanese writing or meaningless squiggles. To me, Chinese writing is just so many meaningless squiggles. Now suppose further that after this first batch of Chinese writing I am given a second batch of Chinese script together with a set of rules for correlating the second batch with the first batch. The rules are in English, and I understand these rules as well as any other native speaker of English. They enable me to correlate one set of formal symbols with another set of formal symbols, and all that 'formal' means here is that I can identify the symbols entirely by their shapes. Now suppose also that I am given a third batch of Chinese symbols together with some instructions, again in English, that enable me to correlate elements of this third batch with the first two batches, and these rules instruct me how to give back certain Chinese symbols with certain sorts of shapes in response to certain sorts of shapes given to me in the third batch. Unknown to me, the people who are giving me all these symbols call the first batch 'a script', they call the second batch a 'story,' and they call the third batch 'questions.' Furthermore, they call the symbols I give them back in response to the third batch 'answers to the questions,' and the set of rules in English that they gave me, they call 'the program.' Now just to complicate the story a little, imagine that these people also give me stories in English, which I understand, and they then ask me questions in English about these stories, and I give them back answers in English. Suppose also that after a while I get so good at following the instructions for manipulating the Chinese symbols and the programmers get so good at writing the programs that from the external point of view—that is, from the point of view of somebody outside the room in which I am locked—my answers to the questions are absolutely indistinguishable from those of native Chinese speakers. Nobody just looking at my answers can tell that I don't speak a word of Chinese. Let us also suppose that my answers to the English questions are, as they no doubt would be, indistinguishable from those of other native English speakers, for the simple reason that I am a native English speaker. From the external point of view—from the point of view of somebody reading my 'answers'—the answers to the Chinese questions and the English questions are equally good. But in the Chinese case, unlike the English case, I produce the answers by manipulating uninterpreted formal symbols. As far as the Chinese is concerned, I simply behave like a computer; I perform computational operations on formally specified elements. For the purposes of the Chinese, I am simply an instantiation of the computer program.

Now the claims made by strong AI are that the programmed computer understands the stories and that the program in some sense explains human understanding. But we are now in a position to examine these claims in light of our thought-experiment.

1. As regards the first claim, it seems to me quite obvious in the example that I do not understand a word of the Chinese stories. I have inputs and outputs that are indistinguishable from those of the native Chinese speaker, and I can have any formal program you like, but I still understand nothing. For the same reasons, Schank's computer understands nothing of any stories, whether in Chinese, English or whatever, since in the Chinese case the computer is me, and in cases where the computer is not me, the computer has nothing more than I have in the case where I understand nothing.

2. As regards the second claim, that the program explains human understanding, we can see that the computer and its program do not provide sufficient conditions of understanding since the computer and the program are functioning, and there is no understanding. But does it even provide a necessary condition or a significant contribution to understanding? One of the claims made by the supporters of strong AI is that when I understand a story in English, what I am doing is exactly the same—or perhaps more of the same—as what I was doing in manipulating the Chinese symbols. It is simply more formal symbol-manipulation that distinguishes the case in English, where I do understand, from the case in Chinese, where I don't. I have not demonstrated that this claim is false, but it would certainly appear an incredible claim in the example. Such plausibility as the claim has derives from the supposition that we can construct a program that will have the same inputs and outputs as native speakers, and in addition we assume that speakers have some level of description where they are also instantiations of a program. On the basis of these two assumptions we assume that even if Schank's program isn't the whole story about understanding, it may be part of the story. Well, I suppose that is an empirical possibility, but not the slightest reason has so far been given to believe that it is true, since what is suggested—though certainly not demonstrated—by the example is that the computer program is simply irrelevant to my understanding of the story. In the Chinese case I have everything that artificial intelligence can put into me by way of a program, and I understand nothing; in the English case I understand everything, and there is so far no reason at all to suppose that my understanding has anything to do with computer programs, that is, with computational operations on purely formally specified elements. As long as the program is defined in terms of computational operations on purely formally defined elements, what the example suggests is that these by themselves have no interesting connection with understanding. They are

certainly not sufficient conditions, and not the slightest reason has been given to suppose that they are necessary conditions or even that they make a significant contribution to understanding. Notice that the force of the argument is not simply that different machines can have the same input and output while operating on different formal principles—that is not the point at all. Rather, whatever purely formal principles you put into the computer, they will not be sufficient for understanding, since a human will be able to follow the formal principles without understanding anything. No reason whatever has been offered to suppose that such principles are necessary or even contributory, since no reason has been given to suppose that when I understand English I am operating with any formal program at all.

Well, then, what is it that I have in the case of the English sentences that I do not have in the case of the Chinese sentences? The obvious answer is that I know what the former mean, while I haven't the faintest idea what the latter mean. But in what does this consist and why couldn't we give it to a machine, whatever it is? I will return to this question later, but first I want to continue with the example.

I have had the occasions to present this example to several workers in artificial intelligence, and, interestingly, they do not seem to agree on what the proper reply to it is. I get a surprising variety of replies, and in what follows I will consider the most common of these (specified along with their geographic origins).

But I first want to block some common misunderstandings about 'understanding': in many of these discussions one finds a lot of fancy footwork about the word 'understanding'. My critics point out that there are many different degrees of understanding; that 'understanding' is not a simple two-place predicate; that there are even different kinds and levels of understanding, and often the law of excluded middle doesn't even apply in a straightforward way to statements of the form 'x understands y'; that in many cases it is a matter for decision and not a simple matter of fact whether x understands y; and so on. To all of these points I want to say: of course, of course. But they have nothing to do with the points at issue. There are clear cases in which 'understanding' literally applies and clear cases in which it does not apply; and these two sorts of cases are all I need for this argument.[2] I understand stories in English; to a lesser degree I can understand stories in French; to a still lesser degree, stories in German; and in Chinese, not at all. My car and my adding machine, on the other

---

[2] Also, 'understanding' implies both the possession of mental (intentional) states and the truth (validity, success) of these states. For the purposes of this discussion we are concerned only with the possession of the states.

hand, understand nothing: they are not in that line of business. We often attribute 'understanding' and other cognitive predicates by metaphor and analogy to cars, adding machines, and other artefacts, but nothing is proved by such attributions. We say, 'The door *knows* when to open because of its photoelectric cell,' 'The adding machine *knows how (understands how,* is *able)* to do addition and subtraction but not division,' and 'The thermostat *perceives* changes in the temperature.' The reason we make these attributions is quite interesting, and it has to do with the fact that in artefacts we extend our own intentionality;[3] our tools are extensions of our purposes, and so we find it natural to make metaphorical attributions of intentionality to them; but I take it no philosophical ice is cut by such examples. The sense in which an automatic door 'understands instructions' from its photoelectric cell is not at all the sense in which I understand English. If the sense in which Schank's programmed computers understand stories is supposed to be the metaphorical sense in which the door understands, and not the sense in which I understand English, the issue would not be worth discussing. But Newell and Simon (1963) write that the kind of cognition they claim for computers is exactly the same as for human beings. I like the straightforwardness of this claim, and it is the sort of claim I will be considering. I will argue that in the literal sense the programmed computer understands what the car and the adding machine understand, namely, exactly nothing. The computer understanding is not just (like my understanding of German) partial or incomplete; it is zero.

Now to the replies:

1. THE SYSTEMS REPLY (BERKELEY)

'While it is true that the individual person who is locked in the room does not understand the story, the fact is that he is merely part of a whole system, and the system does understand the story. The person has a large ledger in front of him in which are written the rules, he has a lot of scratch paper and pencils for doing calculations, he has "data banks" of sets of Chinese symbols. Now, understanding is not being ascribed to the mere individual; rather it is being ascribed to this whole system of which he is a part.'

---

[3] Intentionality is by definition that feature of certain mental states by which they are directed at or about objects and states of affairs in the world. Thus, beliefs, desires, and intentions are intentional states; undirected forms of anxiety and depression are not. For further discussion see Searle (1976*b*).

My response to the systems theory is quite simple: let the individual internalize all of these elements of the system. He memorizes the rules in the ledger and the data banks of Chinese symbols, and he does all the calculations in his head. The individual then incorporates the entire system. There isn't anything at all to the system that he does not encompass. We can even get rid of the room and suppose he works outdoors. All the same, he understands nothing of the Chinese, and *a fortiori* neither does the system, because there isn't anything in the system that isn't in him. If he doesn't understand, then there is no way the system could understand because the system is just a part of him.

Actually I feel somewhat embarrassed to give even this answer to the systems theory because the theory seems to me so unplausible to start with. The idea is that while a person doesn't understand Chinese, somehow the *conjunction* of that person and bits of paper might understand Chinese. It is not easy for me to imagine how someone who was not in the grip of an ideology would find the idea at all plausible. Still, I think many people who are committed to the ideology of strong AI will in the end be inclined to say something very much like this; so let us pursue it a bit further. According to one version of this view, while the man in the internalized systems example doesn't understand Chinese in the sense that a native Chinese speaker does (because, for example, he doesn't know that the story refers to restaurants and hamburgers, etc.), still 'the man as a formal symbol-manipulation system' *really does understand Chinese.* The subsystem of the man that is the formal symbol-manipulation system for Chinese should not be confused with the subsystem for English.

So there are really two subsystems in the man; one understands English, the other Chinese, and 'it's just that the two systems have little to do with each other.' But, I want to reply, not only do they have little to do with each other, they are not even remotely alike. The subsystem that understands English (assuming we allow ourselves to talk in this jargon of 'subsystems' for a moment) knows that the stories are about restaurants and eating hamburgers, he knows that he is being asked questions about restaurants and that he is answering questions as best he can by making various inferences from the content of the story, and so on. But the Chinese system knows none of this. Whereas the English system knows that 'hamburgers' refers to hamburgers, the Chinese subsystem knows only that 'squiggle squiggle' is followed by 'squoggle squoggle'. All he knows is that various formal symbols are being introduced at one end and manipulated according to rules written in English, and other symbols are going out at the other end. The whole point of the original example was to argue that such symbol manipulation by itself couldn't be sufficient for understanding Chinese in any literal sense because the man could write 'squoggle squoggle' after 'sqiggle squiggle' without understanding anything in Chinese. And it doesn't meet that argument to postulate subsystems within the man, because the subsystems are no better off than the man was in the first place; they still don't have anything even remotely like what the English-speaking man (or subsystem) has. Indeed, in the case as described, the Chinese subsystem is simply a part of the English subsystem, a part that engages in meaningless symbol manipulation according to rules in English.

Let us ask ourselves what is supposed to motivate the systems reply in the first place; that is, what *independent* grounds are there supposed to be for saying that the agent must have a subsystem within him that literally understands stories in Chinese? As far as I can tell the only grounds are that in the example I have the same input and output as native Chinese speakers and a program that goes from one to the other. But the whole point of the examples has been to try to show that that couldn't be sufficient for understanding, in the sense in which I understand stories in English, because a person, and hence the set of systems that go to make up a person, could have the right combination of input, output, and program and still not understand anything in the relevant literal sense in which I understand English. The only motivation for saying there *must* be a subsystem in me that understands Chinese is that I have a program and I pass the Turing-test; I can fool native Chinese speakers. But precisely one of the points at issue is the adequacy of the Turing-test. The example shows that there could be two 'systems', both of which pass the Turing-test, but only one of which understands; and it is no argument against this point to say that since they both pass the Turing-test they must both understand, since this claim fails to meet the argument that the system in me that understands English has a great deal more than the system that merely processes Chinese. In short, the systems reply simply begs the question by insisting without argument that the system must understand Chinese.

Furthermore, the systems reply would appear to lead to consequences that are independently absurd. If we are to conclude that there must be cognition in me on the grounds that I have a certain sort of input and output and a program in between, then it looks like all sorts of non-cognitive subsystems are going to turn out to be cognitive. For example, there is a level of description at which my stomach does information-processing, and it instantiates any number of computer-programs, but I take it we do not want to say that it has any understanding (cf. Pylyshyn 1980). But if we accept the systems reply, then it is hard to see how we avoid saying that stomach, heart, liver, and so on, are all understanding subsystems, since there is no principled way to distinguish the motivation for saying the Chinese subsystem understands from saying that the stomach understands. It is, by the way, not an answer to this point to say that the Chinese system has information as input and output and the stomach has food and food products input and output, since from the point of view of the agent, from my point of view, there is no information in either the food or the Chinese — the Chinese is just so many meaningless squiggles. The information in the Chinese case is solely in the eyes of the programmers and the interpreters, and there is nothing to prevent them from treating the input and output of my digestive organs as information if they so desire.

This last point bears on some independent problems in strong AI, and it is worth digressing for a moment to explain it. If strong AI is to be a branch of psychology, then it must be able to distinguish those systems that are genuinely mental from those that are not. It must be able to distinguish the principles on which the mind works from those on which non-mental systems work; otherwise it will offer us no explanations of what is specifically mental about the mental. And the mental–non-mental distinction cannot be just in the eye of the beholder but it must be intrinsic to the systems; otherwise it would be up to any beholder to treat people as non-mental and, for example, hurricanes as mental if he likes. But quite often in the AI literature the distinction is blurred in ways that would in the long run prove disastrous to the claim that AI is a cognitive enquiry. McCarthy, for example, writes, 'Machines as simple as thermostats can be said to have beliefs, and having beliefs seems to be a characteristic of most machines capable of problem solving performances' (McCarthy 1979). Anyone who thinks strong AI has a chance as a theory of the mind ought to ponder the implications of that remark. We are asked to accept it as a discovery of strong AI that the hunk of metal on the wall that we use to regulate the temperature has beliefs in exactly the same sense that we, our spouses, and our children have beliefs, and furthermore that 'most' of the other machines in the room — telephone, tape recorder, adding machine, electric light switch — also have beliefs in this literal sense. It is not the aim of this article to argue against McCarthy's point, so I will simply assert the following without argument. The study of the mind starts with such facts as that humans have belief, while thermostats, telephones, and adding machines don't. If you get a theory that denies this point you have produced a counter-example to the theory and the theory is false. One gets the impression that people in AI who write this sort of thing think they can get away with it because they don't really take it seriously, and they don't think anyone else will either. I propose for a moment at least, to take it seriously. Think hard for one minute about what would be necessary to establish that that hunk of metal on the wall over there had real beliefs, beliefs with direction of fit, propositional content, and conditions of satisfaction; beliefs that had the possibility of being strong beliefs or weak beliefs; nervous, anxious, or secure beliefs; dogmatic, rational, or superstitious beliefs; blind faiths or hesitant cognitions; any kind of beliefs. The thermostat is not a candidate. Neither is stomach, liver, adding machine, or telephone. However, since we are taking the idea seriously, notice that its truth would be fatal to strong AI's claim to be a science of the mind. For now the mind is everywhere. What we wanted to know is what distinguishes the mind from thermostats and livers. And if McCarthy were right, strong AI wouldn't have a hope of telling us that.

2. THE ROBOT REPLY (YALE)

'Suppose we wrote a different kind of program from Schank's program. Suppose we put a computer inside a robot, and this computer would not just take in formal symbols as input and give out formal symbols as output, but rather would actually operate the robot in such a way that the robot does something very much like perceiving, walking, moving about, hammering nails, eating, drinking—anything you like. The robot would, for example, have a television camera attached to it that enabled it to "see", it would have arms and legs that enabled it to "act", and all of this would be controlled by its computer "brain". Such a robot would, unlike Schank's computer, have genuine understanding and other mental states.'

The first thing to notice about the robot reply is that it tacitly concedes that cognition is not solely a matter of formal symbol-manipulation, since this reply adds a set of causal relations with the outside world (cf. Fodor 1980). But the answer to the robot reply is that the addition of such 'perceptual' and 'motor' capacities adds nothing by way of understanding, in particular, or intentionality, in general, to Schank's original program. To see this, notice that the same thought-experiment applies to the robot case. Suppose that instead of the computer inside the robot, you put me inside the room and, as in the original Chinese case, you give me more Chinese symbols with more instructions in English for matching Chinese symbols to Chinese symbols and feeding back Chinese symbols to the outside. Suppose, unknown to me, some of the Chinese symbols that come to me come from a television camera attached to the robot and other Chinese symbols that I am giving out serve to make the motors inside the robot move the robot's legs or arms. It is important to emphasize that all I am doing is manipulating formal symbols: I know none of these other facts. I am receiving 'information' from the robot's 'perceptual' apparatus, and I am giving out 'instructions' to its motor apparatus without knowing either of these facts. I am the robot's homunculus, but unlike the traditional homunculus, I don't know what's going on. I don't understand anything except the rules for symbol manipulation. Now in this case I want to say that the robot has no intentional states at all; it is simply moving about as a result of its electrical wiring and its program. And furthermore, by instantiating the program I have no intentional states of the relevant type. All I do is follow instructions about manipulating formal symbols.

3. THE BRAIN SIMULATOR REPLY (BERKELEY AND MIT)

'Suppose we design a program that doesn't represent information that we have about the world, such as the information in Schank's scripts, but simulates the actual sequence of neurone firings at the synapses of the brain of a native Chinese speaker when he understands stories in Chinese and gives answers to them. The machine takes in Chinese stories and questions about them as input, it simulates the formal structure of actual Chinese brains in processing these stories, and it gives out Chinese answers as outputs. We can even imagine that the machine operates, not with a single serial program, but with a whole set of programs operating in parallel, in the manner that actual human brains presumably operate when they process natural language. Now surely in such a case we would have to say that the machine understood the stories; and if we refuse to say that, wouldn't we also have to deny that native Chinese speakers understood the stories? At the level of the synapses, what would or could be different about the program of the computer and the program of the Chinese brain?'

Before countering this reply I want to digress to note that it is an odd reply for any partisan of artificial intelligence (or functionalism, etc.) to make: I thought the whole idea of strong AI is that we don't need to know how the brain works to know how the mind works. The basic hypothesis, or so I had supposed, was that there is a level of mental operations consisting of computational processes over formal elements that constitute the essence of the mental and can be realized in all sorts of different brain processes, in the same way that any computer program can be realized in different computer hardwares: on the assumptions of strong AI, the mind is to the brain as the program is to the hardware, and thus we can understand the mind without doing neurophysiology. If we had to know how the brain worked to do AI, we wouldn't bother with AI. However, even getting this close to the operation of the brain is still not sufficient to produce understanding. To see this, imagine that instead of a monolingual man in a room shuffling symbols we have the man operate an elaborate set of water pipes with valves connecting them. When the man receives the Chinese symbols, he looks up in the program, written in English, which valves he has to turn on and off. Each water connection corresponds to a synapse in the Chinese brain, and the whole system is rigged up so that after all the right firings, that is after turning on all the right faucets, the Chinese answers pop out at the output end of the series of pipes.

Now where is the understanding in this system? It takes Chinese as input, it simulates the formal structure of the synapses of the Chinese brain, and it gives Chinese as output. But the man certainly doesn't understand Chinese, and neither do the water pipes, and if we are tempted to adopt what I think is the absurd view that somehow the *conjunction* of man *and* water pipes understands, remember that in principle the man can internalize the formal structure of the water pipes and do all the 'neurone firings' in his imagination. The problem with the brain simulator is that it is simulating the wrong things about the brain. As long as it simulates only the formal structure of the sequence of neurone firings at the synapses, it won't have simulated what matters about the brain, namely its causal properties, its ability to produce intentional states. And that the formal properties are not sufficient for the causal properties is shown by the water pipe example: we can have all the formal properties carved off from the relevant neurobiological causal properties.

4. THE COMBINATION REPLY (BERKELEY AND STANFORD)

'While each of the previous three replies might not be completely convincing by itself as a refutation of the Chinese room counter-example, if you take all three together they are collectively much more convincing and even decisive. Imagine a robot with a brain-shaped computer lodged in its cranial cavity, imagine the computer programmed with all the synapses of a human brain, imagine the whole behaviour of the robot is undistinguishable from human behaviour, and now think of the whole thing as a unified system and not just as a computer with inputs and outputs. Surely in such a case we would have to ascribe intentionality to the system.'

I entirely agree that in such a case we would find it rational and indeed irresistible to accept the hypothesis that the robot had intentionality, as long as we knew nothing more about it. Indeed, besides appearance and behaviour, the other elements of the combination are really irrelevant. If we could build a robot whose behaviour was indistinguishable over a large range from human behaviour, we would attribute intentionality to it, pending some reason not to. We wouldn't need to know in advance that its computer brain was a formal analogue of the human brain.

But I really don't see that this is any help to the claims of strong AI; and here's why: According to strong AI, instantiating a formal program with the right input and output is a sufficient condition of, indeed is constitutive of, intentionality. As Newell (1979) puts it, the essence of the mental is the operation of a physical-symbol system. But the attributions of intentionality that we make to the robot in this example have nothing to do with formal programs. They are simply based on the assumption that if the robot looks and behaves sufficiently like us, then we would suppose, until proven otherwise, that it must have mental states like ours that cause and are expressed by its behaviour and it must have an inner mechanism capable of producing such mental states. If we knew independently how to account for its behaviour without such assumptions we would not attribute intentionality to it, especially if we knew it had a formal program. And this is precisely the point of my earlier reply to objection 2.

Suppose we knew that the robot's behaviour was entirely accounted for by the fact that a man inside it was receiving uninterpreted formal symbols from the robot's sensory receptors and sending out uninterpreted formal symbols to its motor mechanisms, and the man was doing this symbol manipulation in accordance with a bunch of rules. Furthermore, suppose the man knows none of these facts about the robot, all he knows is which operations to perform on which meaningless symbols. In such a case we would regard the robot as an ingenious mechanical dummy. The hypothesis that the dummy has a mind would now be unwarranted and unnecessary, for there is now no longer any reason to ascribe intentionality to the robot or to the system of which it is a part (except of course for the man's intentionality in manipulating the symbols). The formal symbol-manipulations go on, the input and output are correctly matched, but the only real locus of intentionality is the man, and he doesn't know any of the relevant intentional states; he doesn't, for example, *see* what comes into the robot's eyes, he doesn't *intend* to move the robot's arm, and he doesn't *understand* any of the remarks made to or by the robot. Nor, for the reasons stated earlier, does the system of which man and robot are a part.

To see this point, contrast this case with cases in which we find it completely natural to ascribe intentionality to members of certain other primate species such as apes and monkeys and to domestic animals such as dogs. The reasons we find it natural are, roughly, two: we can't make sense of the animal's behaviour without the ascription of intentionality, and we can see that the beasts are made of similar stuff to ourselves—that is an eye, that a nose, this is its skin, and so on. Given the coherence of the animal's behaviour and the assumption of the same causal stuff underlying it, we assume both that the animal must have mental states underlying its behaviour, and that the mental states must be produced by mechanisms made out of the stuff that is like our stuff. We would certainly make similar assumptions about the robot unless we had some reason not to, but as soon as we knew that the behaviour was the result of a formal program, and that the actual causal properties of the physical substance were irrelevant we would abandon the assumption of intentionality. (See Multiple authors 1978.)

There are two other responses to my example that come up frequently (and so are worth discussing) but really miss the point.

5. THE OTHER MINDS REPLY (YALE)

'How do you know that other people understand Chinese or anything else? Only by their behaviour. Now the computer can pass the behavioural tests as well as they can (in principle), so if you are going to attribute cognition to other people you must in principle also attribute it to computers.'

This objection really is only worth a short reply. The problem in this discussion is not about how I know that other people have cognitive states, but rather what it is that I am attributing to them when I attribute cognitive states to them. The thrust of the argument is that it couldn't be just computational processes and their output because the computational processes and their output can exist without the cognitive state. It is no answer to this argument to feign anesthesia. In 'cognitive sciences' one presupposes the reality and knowability of the mental in the same way that in physical sciences one has to presuppose the reality and knowability of physical objects.

6. THE MANY MANSIONS REPLY (BERKELEY)

'Your whole argument presupposes that AI is only about analogue and digital computers. But that just happens to be the present state of technology. Whatever these causal processes are that you say are essential for intentionality (assuming you are right), eventually we will be able to build devices that have these causal processes, and that will be artificial intelligence. So your arguments are in no way directed at the ability of artificial intelligence to produce and explain cognition.'

I really have no objection to this reply save to say that it in effect trivializes the project of strong AI by redefining it as whatever artificially produces and explains cognition. The interest of the original claim made on behalf of artificial intelligence is that it was a precise, well-defined thesis: mental processes are computational processes over formally defined elements. I have been concerned to challenge that thesis. If the claim is redefined so that it is no longer that thesis, my objections no longer apply because there is no longer a testable hypothesis for them to apply to.

Let us now return to the question I promised I would try to answer: granted that in my original example I understand the English and I do not understand the Chinese, and granted therefore that the machine doesn't understand either English or Chinese, still there must be something about me that makes it the case that I understand English and a corresponding something lacking in me makes it the case that I fail to understand Chinese. Now why couldn't we give those somethings, whatever they are, to a machine?

I see no reason in principle why we couldn't give a machine the capacity to understand English or Chinese, since in an important sense our bodies with our brains are precisely such machines. But I do see very strong arguments for saying that we could not give such a thing to a machine where the operation of the machine is defined solely in terms of computational processes over formally defined elements; that is, where the operation of the machine is defined as an instantiation of a computer program. It is not because I am the instantiation of a computer program that I am able to understand English and have other forms of intentionality (I am, I suppose, the instantiation of any number of computer programs), but as far as we know it is because I am a certain sort of organism with a certain biological (i.e. chemical and physical) structure, and this structure, under certain conditions, is causally capable of producing perception, action, understanding, learning, and other intentional phenomena. And part of the point of the present argument is that only something that had those causal powers could have that intentionality. Perhaps other physical and chemical processes could produce exactly these effects; perhaps, for example, Martians also have intentionality but their brains are made of different stuff. That is an empirical question, rather like the question whether photosynthesis can be done by something with a chemistry different from that of chlorophyll.

But the main point of the present argument is that no purely formal model will ever be sufficient by itself for intentionality because the formal properties are not by themselves constitutive of intentionality, and they have by themselves no causal power except the power, when instantiated, to produce the next stage of the formalism when the machine is running. And any other causal properties that particular realizations of the formal model have, are irrelevant to the formal model because we can always put the same formal model in a different realization where those causal properties are obviously absent. Even if, by some miracle, Chinese speakers exactly realize Schank's program, we can put the same program in English speakers, water pipes, or computers, none of which understand Chinese, the program notwithstanding.

What matters about brain operations is not the formal shadow cast by the sequence of synapses but rather the actual properties of the sequences. All the arguments for the strong version of artificial intelligence that I have seen insist on drawing an outline around the shadows cast by cognition and then claiming that the shadows are the real thing.

By way of concluding I want to try to state some of the general philosophical points implicit in the argument. For clarity I will try to do it in a question and answer fashion, and I begin with that old chestnut of a question:

'Could a machine think?'

The answer is, obviously, yes. We are precisely such machines.

'Yes, but could an artefact, a man-made machine, think?'

Assuming it is possible to produce artificially a machine with a nervous system, neurones with axons and dendrites, and all the rest of it, sufficiently like ours, again the answer to the question seems to be obviously, yes. If you can exactly duplicate the causes, you could duplicate the effects. And indeed it might be possible to produce consciousness, intentionality, and all the rest of it using some other sorts of chemical principles than those that human beings use. It is, as I said, an empirical question.

'OK, but could a digital computer think?'

If by 'digital computer' we mean anything at all that has a level of description where it can correctly be described as the instantiation of a computer program, then again the answer is, of course, yes, since we are the instantiations of any number of computer programs, and we can think.

'But could something think, understand, and so on *solely* in virtue of being a computer with the right sort of program? Could instantiating a program, the right program of course, by itself be a sufficient condition of understanding?'

This I think is the right question to ask, though it is usually confused with one or more of the earlier questions, and the answer to it is no.

'Why not?'

Because the formal symbol-manipulations by themselves don't have any intentionalty; they are quite meaningless; they aren't even *symbol* manipulations, since the symbols don't symbolize anything. In the linguistic jargon, they have only a syntax but no semantics. Such intentionality as computers appear to have is solely in the minds of those who program them and those who use them, those who send in the input and those who interpret the output.

The aim of the Chinese room example was to try to show this by showing that as soon as we put something into the system that really does have intentionality (a man), and we program him with the formal program, you can see that the formal program carries no additional intentionality. It adds nothing, for example, to a man's ability to understand Chinese.

Precisely that feature of AI that seemed so appealing—the distinction between the program and the realization—proves fatal to the claim that simulation could be duplication. The distinction between the program and its realization in the hardware seems to be parallel to the distinction between the level of mental operations and the level of brain operations. And if we could describe the level of mental operations as a formal program, then it seems we could describe what was essential about the mind without doing either introspective psychology or neurophysiology of the brain. But the equation, 'mind is to brain as program is to hardware' breaks down at several points, among them the following three:

First, the distinction between program and realization has the consequence that the same program could have all sorts of crazy realizations that had no form of intentionality. Wiezenbaum (1976: ch. 2), for example, shows in detail how to construct a computer using a roll of toilet paper and a pile of small stones. Similarly, the Chinese story-understanding program can be programmed into a sequence of water pipes, a set of wind machines, or a monolingual English speaker, none of which thereby acquires an understanding of Chinese. Stones, toilet paper, wind, and water pipes are the wrong kind of stuff to have intentionality in the first place—only something that has the same causal powers as brains can have intentionality—and though the English speaker has the right kind of stuff for intentionality you can easily see that he doesn't get any extra intentionality by memorizing the program, since memorizing it won't teach him Chinese.

Second, the program is purely formal, but the intentional states are not in that way formal. They are defined in terms of their content, not their form. The belief that it is raining, for example, is not defined as a certain formal shape, but as a certain mental content with conditions of satisfaction, a direction of fit (see Searle 1979a), and the like. Indeed the belief as such hasn't even got a formal shape in this syntactic sense, since one and the same belief can be given an indefinite number of different syntactic expressions in different linguistic systems.

Third, as I mentioned before, mental states and events are literally a product of the operation of the brain, but the program is not in that way a product of the computer.

'Well if programs are in no way constitutive of mental processes, why have so many people believed the converse? That at least needs some explanation.'

I don't really know the answer to that one. The idea that computer simulations could be the real thing ought to have seemed suspicious in the first place because the computer isn't confined to simulating mental operations, by any means. No one supposes that computer simulations of a five-alarm fire will burn the neighbourhood down or that a computer simulation of a rainstorm will leave us all drenched. Why on earth would anyone suppose that a computer simulation of understanding actually understood anything? It is sometimes said that it would be frightfully hard to get computers to feel pain or fall in love, but love and pain are neither harder nor easier than cognition or anything else. For simulation, all you need is the right input and output and program in the middle that transforms the former into the latter. That is all the computer has for anything it does. To confuse simulation with duplication is the same mistake, whether it is pain, love, cognition, fires, or rainstorms.

Still, there are several reasons why AI must have seemed—and to many people perhaps still does seem—in some way to reproduce and thereby explain mental phenomena, and I believe we will not succeed in removing these illusions until we have fully exposed the reasons that give rise to them.

First, and perhaps most important, is a confusion about the notion of 'information-processing': many people in cognitive science believe that the human brain, with its mind, does something called 'information-processing', and analogously the computer with its program does information-processing; but fires and rainstorms, on the other hand, don't do information-processing at all. Thus, though the computer can simulate the formal features of any process whatever, it stands in a special relation to the mind and brain because when the computer is properly programmed, ideally with the same program as the brain, the information-processing is identical in the two cases, and this information-processing is really the essence of the mental. But the trouble with this argument is that it rests on an ambiguity in the notion of 'information'. In the sense in which people 'process information' when they reflect, say, on problems in arithmetic or when they read and answer questions about stories, the programmed computer does not do 'information-processing'. Rather, what it does is manipulate formal symbols. The fact that the programmer and the interpreter of the computer output use the symbols to stand for objects in the world is totally beyond the scope of the computer. The computer, to repeat, has a syntax but no semantics. Thus, if you type into the computer '2 plus 2 equals?' it will type out '4'. But it has no idea that '4' means 4 or that it means anything at all. And the point is not that it lacks some second-order information about the interpretation of its first-order symbols, but rather that its first-order symbols don't have any interpretations as far as the computer is concerned. All the computer has is more symbols. The introduction of the notion of 'information-processing' therefore produces a dilemma: either we construe the notion of 'information-processing' in such a way that it implies intentionality as part of the process or we don't. If the former, then the programmed computer does not do information-processing, it only manipulates formal symbols. If the latter, then, though the computer does information-processing, it is only doing so in the sense in which adding machines, typewriters, stomachs, thermostats, rainstorms, and hurricanes do information-processing; namely, they have a level of description at which we can describe them as taking information in at one end, transforming it, and producing information as output. But in this case it is up to outside observers to interpret the input and output as information in the ordinary sense. And no similarity is established between the computer and the brain in terms of any similarity of information processing.

Second, in much of AI there is a residual behaviourism or operationalism. Since appropriately programmed computers can have input–output patterns similar to those of human beings, we are tempted to postulate mental states in the computer similar to human mental states. But once we see that it is both conceptually and empirically possible for a system to have human capacities in some realm without having any intentionality at all, we should be able to overcome this impulse. My desk adding machine has calculating capacities, but no intentionality, and in this paper I have tried to show that a system could have input and output capabilities that duplicated those of a native Chinese speaker and still not understand Chinese, regardless of how it was programmed. The Turing-test is typical of the tradition in being unashamedly behaviouristic and operationalistic, and I believe that if AI workers totally repudiated behaviourism and operationalism much of the confusion between simulation and duplication would be eliminated.

Third, this residual operationalism is joined to a residual form of dualism; indeed strong AI only makes sense given the dualistic assumption that, where the mind is concerned, the brain doesn't matter. In strong AI (and in functionalism, as well) what matters are programs, and programs are independent of their realization in machines; indeed, as far as AI is concerned, the same program could be realized by an electronic machine, a Cartesian mental substance, or a Hegelian world spirit. The single most surprising discovery that I have made in discussing these issues is that many AI workers are quite shocked by my idea that actual human mental phenomena might be dependent on actual physical-chemical properties of actual human brains. But if you think about it a minute you can see that I should not have been surprised; for unless you accept some form of dualism, the strong AI project hasn't got a chance. The project is to reproduce and explain the mental by designing programs, but unless the mind is not only conceptually but empirically independent of the brain you couldn't carry out the project, for the program is completely independent of any realization. Unless you believe that the mind is separable from the brain both conceptually and empirically—dualism in a strong form—you cannot hope to reproduce the mental by writing and running programs since programs must be independent of brains or any other particular forms of instantiation. If mental operations consist in computational operations on formal symbols, then it follows that they have no interesting connection with the brain; the only connection would be that the brain just happens to be one of the indefinitely many types of machines capable of instantiating the program. This form of dualism is not the traditional Cartesian variety that claims there are two sorts of *substances*, but it is Cartesian in the sense that it insists that what is specifically mental about the mind has no intrinsic connection with the actual properties of the brain. This underlying dualism is masked from us by the fact that AI literature contains frequent fulminations against 'dualism'; what the authors seem to be unaware of is that their position presupposes a strong version of dualism.

'Could a machine think?' My own view is that *only* a machine could think, and indeed only very special kinds of machines, namely brains and machines that had the same causal powers as brains. And that is the main reason strong AI has had little to tell us about thinking, since it has nothing to tell us about machines. By its own definition, it is about programs, and programs are not machines. Whatever else intentionality is, it is a biological phenomenon, and it is as likely to be as causally dependent on the specific biochemistry of its origins as lactation, photosynthesis, or any other biological phenomena. No one would suppose that we could produce milk and sugar by running a computer simulation of the formal sequences in lactation and photosynthesis, but where the mind is concerned many people are willing to believe in such a miracle because of a deep and abiding dualism: the mind they suppose is a matter of formal processes and is independent of quite specific material causes in the way that milk and sugar are not.

In defence of this dualism the hope is often expressed that the brain is a digital computer (early computers, by the way, were often called 'electronic brains'). But that is no help. Of course the brain is a digital computer. Since everything is a digital computer, brains are too. The point is that the brain's causal capacity to produce intentionality cannot consist in its instantiating a computer program, since for any program you like it is possible for something to instantiate that program and still not have any mental states. Whatever it is that the brain does to produce intentionality, it cannot consist in instantiating a program since no program, by itself, is sufficient for intentionality.[4]

[4] I am indebted to a rather large number of people for discussion of these matters and for their patient attempts to overcome my ignorance of artificial intelligence. I would especially like to thank Ned Block, Hubert Dreyfus, John Haugeland, Roger Schank, Robert Wilensky, and Terry Winograd.

REFERENCES

Fodor, J. A. (1980). 'Methodological Solipsism Considered as a Research Strategy in Cognitive Psychology.' *Behavioral and Brain Sciences* 3: 63–110.
McCarthy, J. (1979). 'Ascribing Mental Qualities to Machines.' In M. Ringle (ed.), *Philosophical Perspectives in Artificial Intelligence*, pp. 161–95. Atlantic Highlands, NJ: Humanities Press.
[Multiple authors] (1978). 'Cognition and Consciousness in Non-Human Species.' *Behavioral and Brain Sciences* 1(4): entire issue.
Newell, A. (1979). 'Physical Symbol Systems.' Lecture at the La Jolla Conference on Cognitive Science. Later published in *Cognitive Science* 4 (1980): 135–83.
—— and Simon, H. A. (1963). 'GPS—A Program that Simulates Human Thought.' In E. A. Feigenbaum and J. A. Feldman (eds.), *Computers and Thought*, pp. 279–96. New York: McGraw-Hill.
Pylyshyn, Z. W. (1980). 'Computation and Cognition: Issues in the Foundation of Cognitive Science.' *Behavioral and Brain Sciences* 3: 111–32.
Schank, R. C. and Abelson, R. P. (1977). *Scripts, Plans, Goals, and Understanding.* Hillsdale, NJ: Erlbaum.
Searle, J. R. (1979a). 'Intentionality and the Use of Language.' In A. Margolit (ed.), *Meaning and Use*. Dordrecht: Reidel.
—— (1979b). 'What is an Intentional State?' *Mind* 88: 74–92.

Weizenbaum, J. (1965). 'ELIZA—A Computer Program for the Study of Natural Language Communication Between Man and Machine.' *Commun. ACM* 9: 36–45.

—— (1976). *Computer Power and Human Reason*. San Francisco: W. H. Freeman.

Winograd, T. (1973). 'A Procedural Model of Language Understanding.' In R. C. Schank and K. M. Colby (eds.), *Computer Models of Thought and Language*, pp. 152–86. San Francisco: W. H. Freeman.

4

ESCAPING FROM THE CHINESE ROOM

MARGARET A. BODEN

John Searle, in his paper on 'Minds, Brains, and Programs' (1980), argues that computational theories in psychology are essentially worthless. He makes two main claims: that computational theories, being purely formal in nature, cannot possibly help us to understand mental processes; and that computer hardware—unlike neuroprotein—obviously lacks the right causal powers to generate mental processes. I shall argue that both these claims are mistaken.

His first claim takes for granted the widely-held (formalist) assumption that the 'computations' studied in computer science are purely syntactic, that they can be defined (in terms equally suited to symbolic logic) as *the formal manipulation of abstract symbols, by the application of formal rules*. It follows, he says, that formalist accounts—appropriate in explaining the meaningless 'information'-processing or 'symbol'-manipulations in computers—are unable to explain how human minds employ *information* or *symbols* properly so-called. Meaning, or intentionality, cannot be explained in computational terms.

Searle's point here is not that no machine can think. Humans can think, and humans—he allows—are machines; he even adopts the materialist credo that only machines can think. Nor is he saying that humans and programs are utterly incommensurable. He grants that, at some highly abstract level of description, people (like everything else) are instantiations of digital computers. His point, rather, is that nothing can think, mean, or understand *solely* in virtue of its instantiating a computer program.

To persuade us of this, Searle employs an ingenious thought-experiment. He imagines himself locked in a room, in which there are various slips of paper with doodles on them; a window through which people can pass further doodle-papers to him, and through which he can

Margaret A. Boden, 'Escaping from the Chinese Room' extracted from Chapter 8 of *Computer Models of Mind* (1988). Reprinted by permission of Cambridge University Press.

pass papers out; and a book of rules (in English) telling him how to pair the doodles, which are always identified by their shape or form. Searle spends his time, while inside the room, manipulating the doodles according to the rules.

One rule, for example, instructs him that when *squiggle-squiggle* is passed in to him, he should give out *squoggle-squoggle*. The rule-book also provides for more complex sequences of doodle-pairing, where only the first and last steps mention the transfer of paper into or out of the room. Before finding any rule directly instructing him to give out a slip of paper, he may have to locate a *blongle* doodle and compare it with a *blungle* doodle — in which case, it is the result of this comparison which determines the nature of the doodle he passes out. Sometimes many such doodle–doodle comparisons and consequent doodle-selections have to be made by him inside the room before he finds a rule allowing him to pass anything out.

So far as Searle-in-the-room is concerned, the *squiggles* and *squoggles* are mere meaningless doodles. Unknown to him, however, they are Chinese characters. The people outside the room, being Chinese, interpret them as such. Moreover, the patterns passed in and out at the window are understood by them as *questions* and *answers* respectively: the rules happen to be such that most of the questions are paired, either directly or indirectly, with what they recognize as a sensible answer. But Searle himself (inside the room) knows nothing of this.

The point, says Searle, is that Searle-in-the-room is clearly instantiating a computer program. That is, he is performing purely formal manipulations of uninterpreted patterns: he is all syntax and no semantics.

The doodle-pairing rules are equivalent to the IF–THEN rules, or 'productions', commonly used (for example) in expert systems. Some of the internal doodle-comparisons could be equivalent to what AI workers in natural-language processing call a script — for instance, the restaurant script described by R. C. Schank and R. P. Abelson (1977). In that case, Searle-in-the-room's paper-passing performance would be essentially comparable to the performance of a 'question-answering' Schankian text-analysis program. But 'question-answering' is not question-answering. Searle-in-the-room is not really *answering*: how could he, since he cannot understand the questions? Practice does not help (except perhaps in making the doodle-pairing swifter): if Searle-in-the-room ever escapes, he will be just as ignorant of Chinese as he was when he was first locked in.

Certainly, the Chinese people outside might find it useful to keep Searle-in-the-room fed and watered, much as in real life we are willing to spend large sums of money on computerized 'advice' systems. But the fact that people who already possess understanding may use an intrinsically meaningless formalist computational system to provide what they interpret (*sic*) as questions, answers, designations, interpretations, or symbols is irrelevant. They can do this only if they can externally specify a mapping between the formalism and matters of interest to them. In principle, one and the same formalism might be mappable onto several different domains, so could be used (by people) in answering questions about any of those domains. In itself, however, it would be meaningless—as are the Chinese symbols from the point of view of Searle-in-the-room.

It follows, Searle argues, that no system can understand anything solely in virtue of its instantiating a computer program. For if it could, then Searle-in-the-room would understand Chinese. Hence, theoretical psychology cannot properly be grounded in computational concepts.

Searle's second claim concerns what a proper explanation of understanding would be like. According to him, it would acknowledge that meaningful symbols must be embodied in something having 'the right causal powers' for generating understanding, or intentionality. Obviously, he says, brains do have such causal powers whereas computers do not. More precisely (since the brain's organization could be paralleled in a computer), neuroprotein does whereas metal and silicon do not: the biochemical properties of the brain matter are crucial.

A. Newell's (1980) widely cited definition of 'physical-symbol systems' is rejected by Searle, because it demands merely that symbols be embodied in some material that can implement formalist computations—which computers, admittedly, can do. In Searle's view, no electronic computer can really manipulate symbols, nor really designate or interpret anything at all—*irrespective* of any causal dependencies linking its internal physical patterns to its behaviour. (This strongly realist view of intentionality contrasts with the instrumentalism of D. C. Dennett (1971). For Dennett, an intentional system is one whose behaviour we can explain, predict, and control only by ascribing beliefs, goals, and rationality to it. On this criterion, some *existing* computer programs are intentional systems, and the hypothetical humanoids beloved of science-fiction would be intentional systems *a fortiori*.)

Intentionality, Searle declares, is a biological phenomenon. As such, it is just as dependent on the underlying biochemistry as are photosynthesis and lactation. He grants that neuroprotein may not be the only substance in the universe capable of supporting mental life, much as substances other than chlorophyll may be able (on Mars, perhaps) to catalyse the synthesis of carbohydrates. But he rejects metal or silicon as potential alternatives, even on Mars. He asks whether a computer made out of old beer-cans could possibly *understand*—a rhetorical question to which the expected answer is a resounding 'No!' In short, Searle takes it to be intuitively obvious that the inorganic substances with which (today's) computers are manufactured are essentially incapable of supporting mental functions.

In assessing Searle's two-pronged critique of computational psychology, let us first consider his view that intentionality must be biologically grounded. One might be tempted to call this a positive claim, in contrast with his (negative) claim that purely formalist theories cannot explain mentality. However, this would be to grant it more than it deserves, for its explanatory power is illusory. The biological analogies mentioned by Searle are misleading, and the intuitions to which he appeals are unreliable.

The brain's production of intentionality, we are told, is comparable to photosynthesis—but is it, really? We can define the *products* of photosynthesis, clearly distinguishing various sugars and starches within the general class of carbohydrates, and showing how these differ from other biochemical products such as proteins. Moreover, we not only *know that* chlorophyll supports photosynthesis, we also *understand how* it does so (and *why* various other chemicals cannot). We know that it is a catalyst rather than a raw material; and we can specify the point at which, and the subatomic process by which, its catalytic function is exercised. With respect to brains and understanding, the case is very different.

Our theory of what intentionality is (never mind how it is generated) does not bear comparison with our knowledge of carbohydrates: just what intentionality *is* is still philosophically controversial. We cannot even be entirely confident that we can recognize it when we see it. It is generally agreed that the propositional attitudes are intentional, and that feelings and sensations are not; but there is no clear consensus about the intentionality of emotions.

Various attempts have been made to characterize intentionality and to distinguish its subspecies as distinct intentional states (beliefs, desires, hopes, intentions, and the like). Searle himself has made a number of relevant contributions, from his early work on speech-acts (1969) to his more recent account (1983) of intentionality in general. A commonly used criterion (adopted by Brentano in the nineteenth century and also by Searle) is a *psychological* one. In Brentano's words, intentional states direct the mind on an object; in Searle's, they have intrinsic representational capacity, or 'aboutness'; in either case they relate the mind to the world, and to possible worlds. But some writers define intentionality in *logical* terms (Chisholm 1967). It is not even clear whether the logical and psychological definitions are precisely co-extensive (Boden 1970). In brief, no theory of intentionality is accepted as unproblematic, as the chemistry of carbohydrates is.

As for the brain's biochemical 'synthesis' of intentionality, this is even more mysterious. We have very good reason to believe *that* neuroprotein supports intentionality, but we have hardly any idea *how—qua* neuroprotein—it is able to do so.

In so far as we understand these matters at all, we focus on the neurochemical basis of certain *informational functions*—such as message-passing, facilitation, and inhibition—embodied in neurones and synapses. For example: how the sodium-pump at the cell-membrane enables an action potential to propagate along the axon; how electrochemical changes cause a neurone to enter into and recover from its refractory period; or how neuronal thresholds can be altered by neurotransmitters, such as acetylcholine.

With respect to a visual cell, for instance, a crucial psychological question may be *whether it can function so as to detect intensity-gradients*. If the neurophysiologist can tell us which molecules enable it to do so, so much the better. But from the psychological point of view, it is not the biochemistry as such which matters but the information-bearing functions grounded in it. (Searle apparently admits this when he says, 'The type of realizations that intentional states have in the brain may be describable at a much higher functional level than that of the specific biochemistry of the neurons involved' (1983: 272).)

As work in 'computer vision' has shown, metal and silicon are undoubtedly able to support some of the functions necessary for the 2D-to-3D mapping involved in vision. Moreover, they can embody specific mathematical functions for recognizing intensity-gradients (namely 'DOG-detectors', which compute the difference of Gaussians) which seem to be involved in many biological visual systems. Admittedly, it may be that metal and silicon cannot support all the functions involved in normal vision, or in understanding generally. Perhaps only neuroprotein can do so, so that only creatures with a 'terrestrial' biology can enjoy intentionality. But we have no specific reason, at present, to think so. Most important in this context, any such reasons we might have in the future must be grounded in empirical discovery: intuitions will not help.

If one asks which mind-matter dependencies are intuitively plausible, the answer must be that *none* is. Nobody who was puzzled about intentionality (as opposed to action-potentials) ever exclaimed 'Sodium—of course!' Sodium-pumps are no less 'obviously' absurd than silicon chips, electrical polarities no less 'obviously' irrelevant than old beer-cans, acetylcholine hardly less surprising than beer. The fact that the first member of each of these three pairs is *scientifically* compelling does not make any of them *intuitively* intelligible: our initial surprise persists.

Our intuitions might change with the advance of science. Possibly we shall eventually see neuroprotein (and perhaps silicon too) as obviously capable of embodying mind, much as we now see biochemical substances in general (including chlorophyll) as obviously capable of producing other such substances—an intuition that was not obvious, even to chemists, prior to the synthesis of urea. At present, however, our intuitions have nothing useful to say about the material basis of intentionality. Searle's 'positive' claim, his putative alternative explanation of intentionality, is at best a promissory note, at worst mere mystery-mongering.

Searle's negative claim—that formal-computational theories cannot explain understanding—is less quickly rebutted. My rebuttal will involve two parts: the first directly addressing his example of the Chinese room, the second dealing with his background assumption (on which his example depends) that computer programs are pure syntax.

The Chinese-room example has engendered much debate, both within and outside the community of cognitive science. Some criticisms were anticipated by Searle himself in his original paper, others appeared as the accompanying peer-commentary (together with his Reply), and more have been published since. Here, I shall concentrate on only two points: what Searle calls the Robot reply, and what I shall call the English reply.

The Robot reply accepts that the only understanding of Chinese which exists in Searle's example is that enjoyed by the Chinese people outside the room. Searle-in-the-room's inability to connect Chinese characters with events in the outside world shows that he does not understand Chinese. Likewise, a Schankian teletyping computer that cannot recognize a restaurant, hand money to a waiter, or chew a morsel of food understands nothing of restaurants—even if it can usefully 'answer' our questions about them. But a robot, provided not only with a restaurant-script but also with camera-fed visual programs and limbs capable of walking and picking things up, would be another matter. If the input–output behaviour of such a robot were identical with that of human beings, then it would demonstrably understand both restaurants and the natural language—Chinese, perhaps—used by people to communicate with it.

Searle's first response to the Robot reply is to claim a victory already, since the reply concedes that cognition is not solely a matter of formal symbol-manipulation but requires in addition a set of causal relations with the outside world. Second, Searle insists that to add perceptuomotor capacities to a computational system is not to add intentionality, or under-standing.

He argues this point by imagining a robot which, instead of being provided with a computer program to make it work, has a miniaturized Searle inside it — in its skull, perhaps. Searle-in-the-robot, with the aid of a (new) rule-book, shuffles paper and passes *squiggles* and *squoggles* in and out, much as Searle-in-the-room did before him. But now some or all of the incoming Chinese characters are not handed in by Chinese people, but are triggered by causal processes in the cameras and audio-equipment in the robot's eyes and ears. And the outgoing Chinese characters are not received by Chinese hands, but by motors and levers attached to the robot's limbs — which are caused to move as a result. In short, this robot is apparently able not only to answer questions in Chinese, but also to see and do things accordingly: it can recognize raw beansprouts and, if the recipe requires it, toss them into a wok as well as the rest of us.

(The work on computer vision mentioned above suggests that the vocabulary of Chinese would require considerable extension for this example to be carried through. And the large body of AI research on language-processing suggests that the same could be said of the English required to express the rules in Searle's initial 'question-answering' example. In either case, what Searle-in-the-room needs is not so much Chinese, or even English, as a programming-language. We shall return to this point presently.)

Like his roombound predecessor, however, Searle-in-the-robot knows nothing of the wider context. He is just as ignorant of Chinese as he ever was, and has no more purchase on the outside world than he did in the original example. To him, beansprouts and woks are invisible and intang-ible: all Searle-in-the-robot can see and touch, besides the rule-book and the doodles, are his own body and the inside walls of the robot's skull. Consequently, Searle argues, the robot cannot be credited with under-standing of any of these worldly matters. In truth, it is not *seeing* or *doing* anything at all: it is 'simply moving about as a result of its electrical wiring and its program', which latter is instantiated by the man inside it, who 'has no intentional states of the relevant type' (1980: 420).

Searle's argument here is unacceptable as a rebuttal of the Robot reply, because it draws a false analogy between the imagined example and what is claimed by computational psychology.

Searle-in-the-robot is supposed by Searle to be performing the functions performed (according to computational theories) by the human brain. But, whereas most computationalists do not ascribe intentionality to the brain (and those who do, as we shall see presently, do so only in a very limited way), Searle characterizes Searle-in-the-robot as enjoying full-blooded intentionality, just as he does himself. Computational psychology does not credit the brain with *seeing beansprouts* or *understanding English*: intentional states such as these are properties of people, not of brains. In general, although representations and mental processes are assumed (by computationalists and Searle alike) to be embodied in the brain, the sensorimotor capacities and propositional attitudes which they make possible are ascribed to the person as a whole. So Searle's description of the system inside the robot's skull as one which can understand English does not truly parallel what computationalists say about the brain.

Indeed, the specific procedures hypothesized by computational psychologists, and embodied by them in computer models of the mind, are relatively stupid—and they become more and more stupid as one moves to increasingly basic theoretical levels. Consider theories of natural-language parsing, for example. A parsing procedure that searches for a determiner does not understand English, and nor does a procedure for locating the reference of a personal pronoun: only the person whose brain performs these interpretive processes, and many others associated with them, can do that. The capacity to understand English involves a host of interacting information processes, each of which performs only a very limited function but which together provide the capacity to take English sentences as input and give appropriate English sentences as output. Similar remarks apply to the individual components of computational theories of vision, problem-solving, or learning. Precisely because psychologists wish to *explain* human language, vision, reasoning, and learning, they posit underlying processes which lack the capacities.

In short, Searle's description of the robot's pseudo-brain (that is, of Searle-in-the-robot) as understanding English involves a category-mistake comparable to treating the brain as the bearer—as opposed to the causal basis—of intelligence.

Someone might object here that I have contradicted myself, that I am claiming that one cannot ascribe intentionality to brains and yet am implicitly doing just that. For I spoke of the brain's effecting 'stupid' component-procedures—but stupidity is virtually a *species* of intelligence. To be stupid is to be intelligent, but not very (a person or a fish can be stupid, but a stone or a river cannot).

My defence would be twofold. First, the most basic theoretical level of all would be at the neuroscientific equivalent of the machine-code, a level 'engineered' by evolution. The facts that a certain light-sensitive cell *can* respond to intensity-gradients by acting as a DOG-detector and that one neurone *can* inhibit the firing of another, are explicable by the biochemistry of the brain. The notion of stupidity, even in scare-quotes, is wholly inappropriate in discussing such facts. However, these very basic information-processing functions (DOG-detecting and synaptic inhibition) *could* properly be described as 'very, very, very . . . stupid'. This of course implies that intentional language, if only of a highly grudging and uncomplimentary type, is applicable to brain processes after all—which prompts the second point in my defence. I did not say that intentionality cannot be ascribed to brains, but that full-blooded intentionality cannot. Nor did I say that brains cannot understand anything at all, in howsoever limited a fashion, but that they cannot (for example) understand English. I even hinted, several paragraphs ago, that a few computationalists do ascribe some degree of intentionality to the brain (or to the computational processes going on in the brain). These two points will be less obscure after we have considered the English reply and its bearing on Searle's background assumption that formal-syntactic computational theories are purely syntactic.

The crux of the English reply is that the instantiation of a computer program, whether by man or by manufactured machine, does involve understanding—at least of the rule-book. Searle's initial example depends critically on Searle-in-the-room's being able to understand the language in which the rules are written, namely English; similarly, without Searle-in-the-robot's familiarity with English, the robot's beansprouts would never get thrown into the wok. Moreover, as remarked above, the vocabulary of English (and, for Searle-in-the-robot, of Chinese too) would have to be significantly modified to make the example work.

An unknown language (whether Chinese or Linear B) can be dealt with only as an aesthetic object or a set of systematically related forms. Artificial languages can be designed and studied, by the logician or the pure mathematician, with only their structural properties in mind (although D. R. Hofstadter's (1979) example of the quasi-arithmetical pq-system shows that a psychologically compelling, and predictable, interpretation of a formal calculus may arise spontaneously). But one normally responds in a very different way to the symbols of one's native tongue; indeed, it is very difficult to 'bracket' (ignore) the meanings of familiar words. The view held by computational psychologists, that natural languages can be characterized in procedural terms, is relevant here: words, clauses, and sentences can be seen as mini-programs. The symbols in a natural language one understands initiate mental activity of various kinds. To learn a language is to set up the relevant causal connections, not only between words and the world ('cat' and the thing on the mat) but between words and the many non-introspectible procedures involved in interpreting them.

Moreover, we do not need to be told *ex hypothesi* (by Searle) that Searle-in-the-room understands English: his behaviour while in the room shows clearly that he does. Or, rather, it shows that he understands a *highly limited subset* of English.

Searle-in-the-room could be suffering from total amnesia with respect to 99 per cent of Searle's English vocabulary, and it would make no difference. The only grasp of English he needs is whatever is necessary to interpret (*sic*) the rule-book—which specifies how to accept, select, compare, and give out different patterns. Unlike Searle, Searle-in-the-room does not require words like 'catalyse', 'beer-can', chlorophyll', and 'restaurant'. But he may need 'find', 'compare', 'two', 'triangular', and 'window' (although his understanding of these words could be much less full than Searle's). He must understand conditional sentences, if any rule states that if he sees a *squoggle* he should give out a *squiggle*. Very likely, he must understand some way of expressing negation, temporal ordering, and (especially if he is to learn to do his job faster) generalization. If the rules he uses include some which parse the Chinese sentences, then he will need words for grammatical categories too. (He will not need explicit rules for parsing English sentences, such as the parsing procedures employed in AI programs for language-processing, because he already understands English.)

In short, Searle-in-the-room needs to understand only that subset of Searle's English which is equivalent to the programming-language understood by a computer generating the same 'question-answering' input–output behaviour at the window. Similarly, Searle-in-the-robot must be able to understand whatever subset of English is equivalent to the programming-language understood by a fully computerized visuomotor robot.

The two preceding sentences may seem to beg the very question at issue. Indeed, to speak thus of the programming-language understood by a computer is seemingly self-contradictory. For Searle's basic premiss— which he assumes is accepted by all participants in the debate—is that a computer program is purely formal in nature: the computation it specifies is purely syntactic and has no intrinsic meaning or semantic content to be understood.

If we accept this premiss, the English reply sketched above can be dismissed forthwith for seeking to draw a parallel where no parallel can properly be drawn. But if we do not, if—*pace* Searle (and others (Fodor 1980; Stich 1983))—computer programs are not concerned only with syntax, then the English reply may be relevant after all. We must now turn to address this basic question.

Certainly, one can for certain purposes think of a computer program as an uninterpreted logical calculus. For example, one might be able to prove, by purely formal means, that a particular well-formed formula is derivable from the program's data-structures and inferential rules. Moreover, it is true that a so-called interpreter program that could take as input the list-structure '(FATHER (MAGGIE))' and return '(LEONARD)' would do so on formal criteria alone, having no way of interpreting these patterns as possibly denoting real people. Likewise, as Searle points out, programs provided with restaurant-scripts are not thereby provided with knowledge of restaurants. The existence of a mapping between a formalism and a certain domain does not in itself provide the manipulator of the formalism with any understanding of that domain.

But what must not be forgotten is that a computer program is *a program for a computer*: when a program is run on suitable hardware, the machine *does* something as a result (hence the use in computer science of the words 'instruction' and 'obey'). At the level of the machine-code the effect of the program on the computer is direct, because the machine is engineered so that a given instruction elicits a unique operation (instructions in high-level languages must be converted into machine-code instructions before they can be obeyed). A programmed instruction, then, is not a mere formal pattern—nor even a declarative statement (although it may for some purposes be thought of under either of those descriptions). It is a procedure specification that, given a suitable hardware context, can cause the procedure in question to be executed.

One might put this by saying that a programming-language is a medium not only for expressing *representations* (structures that can be written on a page or provided to a computer, some of which structures may be iso-morphic with things that interest people) but also for bringing about the *representational activity* of certain machines.

One might even say that a representation *is* an activity rather than a structure. Many philosophers and psychologists have supposed that mental representations are intrinsically active. Among those who have recently argued for this view is Hofstadter (1985: 648), who specifically criticizes Newell's account of *symbols* as manipulable formal tokens. In his words, 'The brain itself does not "manipulate symbols"; the brain is the medium in which the symbols are floating and in which they trigger each other.' Hofstadter expresses more sympathy for 'connectionist' than for 'formalist' psychological theories. Connectionist approaches involve parallel-processing systems broadly reminiscent of the brain, and are well suited to model cerebral representations, symbols, or concepts, as *dynamic*. But it is not only connectionists who can view concepts as intrinsically active, and not only *cerebral* representations which can be thought of in this way: this claim has been generalized to cover traditional computer programs, specifically designed for von Neumann machines. The computer scientist B. C. Smith (1982) argues that programmed representations, too, are inherently active — and that an adequate theory of the semantics of programming-languages would recognize the fact.

At present, Smith claims, computer scientists have a radically inadequate understanding of such matters. He reminds us that, as remarked above, there is no general agreement — either within or outside computer science — about what *intentionality* is, and deep unclarities about *representation* as well. Nor can unclarities be avoided by speaking more technically, in terms of *computation* and *formal symbol-manipulation*. For the computer scientist's understanding of what these phenomena really are is also largely intuitive. Smith's discussion of programming-languages identifies some fundamental confusions within computer science. Especially relevant here is his claim that computer scientists commonly make too complete a theoretical separation between a program's control-functions and its nature as a formal-syntactic system.

The theoretical divide criticized by Smith is evident in the widespread 'dual-calculus' approach to programming. The dual-calculus approach posits a sharp theoretical distinction between a declarative (or denotational) representational structure and the procedural language that interprets it when the program is run. Indeed, the knowledge-representation and the interpreter are sometimes written in two quite distinct formalisms (such as predicate calculus and LISP, respectively). Often, however, they are both expressed in the same formalism; for example, LISP (an acronym for LISt-Processing language) allows facts and procedures to be expressed in formally similar ways, and so does PROLOG (PROgramming-in-LOGic). In such cases, the dual-calculus approach dictates that the (single) programming-language concerned be theoretically described in two quite different ways.

To illustrate the distinction at issue here, suppose that we wanted a representation of family relationships which could be used to provide answers to questions about such matters. We might decide to employ a list-structure to represent such facts as that Leonard is the father of Maggie. Or we might prefer a frame-based representation, in which the relevant name-slots in the FATHER-frame could be simultaneously filled by 'LEONARD' and 'MAGGIE'. Again, we might choose a formula of the predicate calculus, saying that there exist two people (namely, Leonard and

Maggie), and Leonard is the father of Maggie. Last, we might employ the English sentence 'Leonard is the father of Maggie.'

Each of these four representations could be written/drawn on paper (as are the rules in the rule-book used by Searle-in-the-room), for us to interpret *if* we have learnt how to handle the relevant notation. Alternatively, they could be embodied in a computer database. But to make them usable by the computer, there has to be an interpreter-program which (for instance) can find the item 'LEONARD' when we 'ask' it who is the father of Maggie. No one with any sense would embody list-structures in a computer without providing it also with a *list-processing* facility, nor give it frames without a *slot-filling* mechanism, logical formulae without *rules of inference*, or English sentences without *parsing procedures*. (Analogously, people who knew that Searle speaks no Portuguese would not give Searle-in-the-room a Portuguese rule-book unless they were prepared to teach him the language first.)

Smith does not deny that there is an important distinction between the *denotational import* of an expression (broadly: what actual or possible worlds can be mapped onto it) and its *procedural consequence* (broadly: what it does, or makes happen). The fact that the expression '(FATHER (MAGGIE))' is isomorphic with a certain parental relationship between two actual people (and so might be mapped onto that relationship by us) is one thing. The fact that the expression '(FATHER (MAGGIE))' can cause a certain computer to locate 'LEONARD' is quite another thing. Were it not so, the dual-calculus approach would not have developed. But he argues that, rather than persisting with the dual-calculus approach, it would be more elegant and less confusing to adopt a 'unified' theory of programming-languages, designed to cover both denotative and procedural aspects.

He shows that many basic terms on either side of the dual-calculus divide have deep theoretical commonalities as well as significant differences. The notion of *variable*, for instance, is understood in somewhat similar fashion by the logician and the computer scientist: both allow that a variable can have different *values* assigned to it at different times. That being so, it is redundant to have two distinct theories of what a variable is. To some extent, however, logicians and computer scientists understand different things by this term: the value of a variable in the LISP programming-language (for example) is another LISP-expression, whereas the value of a variable in logic is usually some object external to the formalism itself. These differences should be clarified — not least to avoid confusion when a system attempts to reason *about* variables by *using* variables. In short, we need a single definition of 'variable', allowing both for its declarative use (in logic) and for its procedural use (in programming). Having shown that similar remarks apply to other basic computational terms, Smith outlines a unitary account of the semantics of LISP and describes a new calculus (MANTIQ) designed with the unified approach in mind.

As the example of using variables to reason about variables suggests, a unified theory of computation could illuminate how *reflective* knowledge is possible. For, given such a theory, a system's representations of data and of processes—including processes internal to the system itself—would be essentially comparable. This theoretical advantage has psychological relevance (and was a major motivation behind Smith's work).

For our present purposes, however, the crucial point is that a fundamental theory of *programs*, and of *computation*, should acknowledge that an essential function of a computer program is to make things happen. Whereas symbolic logic can be viewed as mere playing around with uninterpreted formal calculi (such as the predicate calculus), and computational logic can be seen as the study of abstract timeless relations in mathematically specified 'machines' (such as Turing machines), computer science cannot properly be described in either of these ways.

It follows from Smith's argument that the familiar characterization of computer programs as all syntax and no semantics is mistaken. The inherent procedural consequences of any computer program give it a toehold in semantics, where the semantics in question is not denotational, but causal. The analogy is with Searle-in-the-room's understanding of English, not his understanding of Chinese.

This is implied also by A. Sloman's (1986*a*; 1986*b*) discussion of the sense in which programmed instructions and computer symbols must be thought of as having some semantics, howsoever restricted. In a causal semantics, the meaning of a symbol (whether simple or complex) is to be sought by reference to its causal links with other phenomena. The central questions are 'What causes the symbol to be built and/or activated?' and 'What happens as a result of it?' The answers will sometimes mention external objects and events visible to an observer, and sometimes they will not.

If the system is a human, animal, or robot, it may have causal powers which enable it to refer to restaurants and beansprouts (the philosophical complexities of reference to external, including unobservable, objects may be ignored here, but are helpfully discussed by Sloman). But whatever the information-processing system concerned, the answers will sometimes describe purely *internal* computational processes—whereby other symbols are built, other instructions activated. Examples include the interpretative processes inside Searle-in-the-room's mind (comparable perhaps to the parsing and semantic procedures defined for automatic natural-language processing) that are elicited by English words, and the computational processes within a Schankian text-analysis program. Although such a program cannot use the symbol 'restaurant' to mean *restaurant* (because it has no causal links with restaurants, food and so forth), its internal symbols and procedures do embody some minimal understanding of certain other matters—of what it is to compare two formal structures, for example.

One may feel that the 'understanding' involved in such a case is *so* minimal that this word should not be used at all. So be it. As Sloman makes clear, the important question is not *'When does a machine understand something?'* (a question which misleadingly implies that there is some clear cut-off point at which understanding ceases) but *'What things does a machine (whether biological or not) need to be able to do in order to be able to understand?'* This question is relevant not only to the *possibility* of a computational psychology, but to its *content* also.

In sum, my discussion has shown Searle's attack on computational psychology to be ill founded. To view Searle-in-the-room as an instantiation of a computer program is not to say that he lacks all understanding. Since the theories of a formalist-computational psychology should be likened to computer programs rather than to formal logic, computational psychology is not in principle incapable of explaining how meaning attaches to mental processes.

REFERENCES

Boden, M. A (1970). 'Intentionality and Physical Systems.' *Philosophy of Science* 37: 200–14.

Chisholm, R. M. (1967). 'Intentionality.' In P. Edwards (ed.), *The Encyclopedia of Philosophy*. Vol. IV, pp. 201–4. New York: Macmillan.

Dennett, D. C. (1971). 'Intentional Systems.' *J. Philosophy* 68: 87–106. Repr. in D. C. Dennett, *Brainstorms: Philosophical Essays on Mind and Psychology*, pp. 3–22. Cambridge, Mass.: MIT Press, 1978.

Fodor, J. A. (1980). 'Methodological Solipsism Considered as a Research Strategy in Cognitive Psychology.' *Behavioral and Brain Sciences* 3: 63–110. Repr. in J. A. Fodor, *Representations: Philosophical Essays on the Foundations of Cognitive Science*, pp. 225–56. Brighton: Harvester Press, 1981.

Hofstadter, D. R. (1979). *Godel, Escher, Bach: An Eternal Golden Braid.* New York: Basic Books.

—— (1985). 'Waking Up from the Boolean Dream; Or, Subcognition as Computation.' In D. R. Hofstadter, *Metamagical Themas: Questing for the Essence of Mind and Pattern*, pp. 631–65. New York: Viking.

Newell, A. (1980). 'Physical Symbol Systems.' *Cognitive Science* 4: 135–83.

Schank, R. C., and Abelson, R. P. (1977). *Scripts, Plans, Goals, and Understanding*. Hillsdale, NJ: Erlbaum.

Searle, J. R. (1969). *Speech Acts: An Essay in the Philosophy of Language*. Cambridge: Cambridge University Press.

—— (1980). 'Minds, Brains, and Programs.' *Behavioral and Brain Sciences* 3: 417–24.

—— (1983). *Intentionality: An Essay in the Philosophy of Mind*. Cambridge: Cambridge University Press.

Sloman, A. (1986*a*). 'Reference Without Causal Links.' In B. du Boulay and L. J. Steels (eds.), *Seventh European Conference on Artificial Intelligence*, pp. 369–81. Amsterdam: North-Holland.

—— (1986*b*). 'What Sorts of Machines Can Understand the Symbols They Use?' *Proc. Aristotelian Soc.* Supp. 60: 61–80.

Smith, B. C. (1982). *Reflection and Semantics in a Procedural Language*. Cambridge, Mass.: MIT Ph.D. dissertation and Technical Report LCS/TR–272.

Stich, S. C. (1983). *From Folk Psychology to Cognitive Science: The Case Against Belief*. Cambridge, Mass.: MIT Press/Bradford Books.

5

COMPUTER SCIENCE AS EMPIRICAL ENQUIRY: SYMBOLS AND SEARCH

ALLEN NEWELL and HERBERT A. SIMON

Computer science is the study of the phenomena surrounding computers. The founders of this society understood this very well when they called themselves the Association for Computing Machinery. The machine—not just the hardware, but the programmed living machine—is the organism we study.

This is the tenth Turing Lecture. The nine persons who preceded us on this platform have presented nine different views of computer science. For our organism, the machine, can be studied at many levels and from many sides. We are deeply honoured to appear here today and to present yet another view, the one that has permeated the scientific work for which we have been cited. We wish to speak of computer science as empirical enquiry.

Our view is only one of many; the previous lectures make that clear. However, even taken together the lectures fail to cover the whole scope of our science. Many fundamental aspects of it have not been represented in these ten awards. And if the time ever arrives, surely not soon, when the compass has been boxed, when computer science has been discussed from every side, it will be time to start the cycle again. For the hare as lecturer will have to make an annual sprint to overtake the cumulation of small, incremental gains that the tortoise of scientific and technical development has achieved in his steady march. Each year will create a new gap and call for a new sprint, for in science there is no final word.

Computer science is an empirical discipline. We would have called it an experimental science, but like astronomy, economics, and geology, some of its unique forms of observation and experience do not fit a narrow stereotype of the experimental method. Nonetheless, they are

A. Newell and H. A. Simon, 'Computer Science as Empirical Inquiry: Symbols and Search', the Tenth Turing Lecture, first published in *Communications of the Association for Computing Machinery* 19 (Mar. 1976). © 1976 Association for Computing Machinery, Inc. Reprinted by permission.

experiments. Each new machine that is built is an experiment. Actually constructing the machine poses a question to nature; and we listen for the answer by observing the machine in operation and analysing it by all analytical and measurement means available. Each new program that is built is an experiment. It poses a question to nature, and its behaviour offers clues to an answer. Neither machines nor programs are black boxes; they are artefacts that have been designed, both hardware and software, and we can open them up and look inside. We can relate their structure to their behaviour and draw many lessons from a single experiment. We don't have to build 100 copies of, say, a theorem prover, to demonstrate statistically that it has not overcome the combinatorial explosion of search in the way hoped for. Inspection of the program in the light of a few runs reveals the flaw and lets us proceed to the next attempt.

We build computers and programs for many reasons. We build them to serve society and as tools for carrying out the economic tasks of society. But as basic scientists we build machines and programs as a way of discovering new phenomena and analysing phenomena we already know about. Society often becomes confused about this, believing that computers and programs are to be constructed only for the economic use that can be made of them (or as intermediate items in a developmental sequence leading to such use). It needs to understand that the phenomena surrounding computers are deep and obscure, requiring much experimentation to assess their nature. It needs to understand that, as in any science, the gains that accrue from such experimentation and understanding pay off in the permanent acquisition of new techniques; and that it is these techniques that will create the instruments to help society in achieving its goals.

Our purpose here, however, is not to plead for understanding from an outside world. It is to examine one aspect of our science, the development of new basic understanding by empirical enquiry. This is best done by illustrations. We will be pardoned if, presuming upon the occasion, we choose our examples from the area of our own research. As will become apparent, these examples involve the whole development of artificial intelligence, especially in its early years. They rest on much more than our own personal contributions. And even where we have made direct contributions, this has been done in co-operation with others. Our collaborators have included especially Cliff Shaw, with whom we formed a team of three through the exciting period of the late fifties. But we have also worked with a great many colleagues and students at Carnegie-Mellon University.

Time permits taking up just two examples. The first is the development of the notion of a symbolic system. The second is the development of the notion of heuristic search. Both conceptions have deep significance for understanding how information is processed and how intelligence is achieved. However, they do not come close to exhausting the full scope of artificial intelligence, though they seem to us to be useful for exhibiting the nature of fundamental knowledge in this part of computer science.

1. SYMBOLS AND PHYSICAL-SYMBOL SYSTEMS

One of the fundamental contributions to knowledge of computer science has been to explain, at a rather basic level, what symbols are. This explanation is a scientific proposition about Nature. It is empirically derived, with a long and gradual development.

Symbols lie at the root of intelligent action, which is, of course, the primary topic of artificial intelligence. For that matter, it is a primary question for all of computer science. For all information is processed by computers in the service of ends, and we measure the intelligence of a system by its ability to achieve stated ends in the face of variations, difficulties, and complexities posed by the task environment. This general investment of computer science in attaining intelligence is obscured when the tasks being accomplished are limited in scope, for then the full variations in the environment can be accurately foreseen. It becomes more obvious as we extend computers to more global, complex, and knowledge-intensive tasks—as we attempt to make them our agents, capable of handling on their own the full contingencies of the natural world.

Our understanding of the system's requirements for intelligent action emerges slowly. It is composite, for no single elementary thing accounts for intelligence in all its manifestations. There is no 'intelligence principle', just as there is no 'vital principle' that conveys by its very nature the essence of life. But the lack of a simple *deus ex machina* does not imply that there are no structural requirements for intelligence. One such requirement is the ability to store and manipulate symbols. To put the scientific question, we may paraphrase the title of a famous paper by Warren McCulloch (1961): What is a symbol, that intelligence may use it, and intelligence, that it may use a symbol?

Laws of Qualitative Structure

All sciences characterize the essential nature of the systems they study. These characterizations are invariably qualitative in nature, for they set the terms within which more detailed knowledge can be developed. Their essence can often be captured in very short, very general statements. One might judge these general laws, because of their limited specificity, as making relatively little contribution to the sum of a science, were it not for the historical evidence that shows them to be results of the greatest importance.

The Cell Doctrine in Biology.   A good example of a law of qualitative structure is the cell doctrine in biology, which states that the basic building block of all living organisms is the cell. Cells come in a large variety of forms, though they all have a nucleus surrounded by protoplasm, the whole encased by a membrane. But this internal structure was not, historically, part of the specification of the cell doctrine; it was subsequent specificity developed by intensive investigation. The cell doctrine can be conveyed almost entirely by the statement we gave above, along with some vague notions about what size a cell can be. The impact of this law on biology, however, has been tremendous, and the lost motion in the field prior to its gradual acceptance was considerable.

Plate Tectonics in Geology.   Geology provides an interesting example of a qualitative structure law, interesting because it has gained acceptance in the last decade and so its rise in status is still fresh in our memory. The theory of plate tectonics asserts that the surface of the globe is a collection of huge plates—a few dozen in all—which move (at geological speeds) against, over, and under each other into the centre of the earth, where they lose their identity. The movements of the plates account for the shapes and relative locations of the continents and oceans, for the areas of volcanic and earthquake activity, for the deep sea ridges, and so on. With a few additional particulars as to speed and size, the essential theory has been specified. It was of course not accepted until it succeeded in explaining a number of details, all of which hung together (e.g. accounting for flora, fauna, and stratification agreements between West Africa and Northeast South America). The plate tectonics theory is highly qualitative. Now that it is accepted, the whole earth seems to offer evidence for it everywhere, for we see the world in its terms.

The Germ Theory of Disease.   It is little more than a century since Pasteur enunciated the germ theory of disease, a law of qualitative structure that produced a revolution in medicine. The theory proposes that most diseases are caused by the presence and multiplication in the body of tiny single-celled living organisms, and that contagion consists in the transmission of these organisms from one host to another. A large part of the elaboration of the theory consisted in identifying the organisms associated with specific diseases, describing them, and tracing their life histories. The fact that the law has many exceptions—that many diseases are not produced by germs—does not detract from its importance. The law tells us to look for a particular kind of cause; it does not insist that we will always find it.

The Doctrine of Atomism.   The doctrine of atomism offers an interesting contrast to the three laws of qualitative structure we have just described. As it emerged from the work of Dalton and his demonstrations that the chemicals combined in fixed proportions, the law provided a typical example of qualitative structure: the elements are composed of small, uniform particles, differing from one element to another. But because the underlying species of atoms are so simple and limited in their variety, quantitative theories were soon formulated which assimilated all the general structure in the original qualitative hypothesis. With cells, tectonic plates, and germs, the variety of structure is so great that the underlying qualitative principle remains distinct, and its contribution to the total theory clearly discernible.

Conclusion.   Laws of qualitative structure are seen everywhere in science. Some of our greatest scientific discoveries are to be found among them. As the examples illustrate, they often set the terms on which a whole science operates.

Physical-Symbol Systems

Let us return to the topic of symbols, and define a *physical-symbol system.* The adjective 'physical' denotes two important features:

1. Such systems clearly obey the laws of physics—they are realizable by engineered systems made of engineered components;
2. although our use of the term 'symbol' prefigures our intended interpretation, it is not restricted to human symbol systems.

A physical-symbol system consists of a set of entities, called symbols, which are physical patterns that can occur as components of another type of entity called an expression (or symbol structure). Thus a symbol structure is composed of a number of instances (or tokens) of symbols related in some physical way (such as one token being next to another). At any instant of time the system will contain a collection of these symbol structures. Besides these structures, the system also contains a collection of processes that operate on expressions to produce other expressions: processes of creation, modification, reproduction, and destruction. A physical-symbol system is a machine that produces through time an evolving collection of symbol structures. Such a system exists in a world of objects wider than just these symbolic expressions themselves.

Two notions are central to this structure of expressions, symbols, and objects: designation and interpretation.

Designation. An expression designates an object if, given the expression, the system can either affect the object itself or behave in ways depending on the object.

In either case, access to the object via the expression has been obtained, which is the essence of designation.

Interpretation. The system can interpret an expression if the expression designates a process and if, given the expression, the system can carry out the process.

Interpretation implies a special form of dependent action: given an expression, the system can perform the indicated process, which is to say, it can evoke and execute its own processes from expressions that designate them.

A system capable of designation and interpretation, in the sense just indicated, must also meet a number of additional requirements, of completeness and closure. We will have space only to mention these briefly; all of them are important and have far-reaching consequences.

1. A symbol may be used to designate any expression whatsoever. That is, given a symbol, it is not prescribed a priori what expressions it can designate. This arbitrariness pertains only to symbols: the symbol tokens and their mutual relations determine what object is designated by a complex expression.

2. There exist expressions that designate every process of which the machine is capable.

3. There exist processes for creating any expression and for modifying any expression in arbitrary ways.

4. Expressions are stable; once created, they will continue to exist until explicitly modified or deleted.

5. The number of expressions that the system can hold is essentially unbounded.

The type of system we have just defined is not unfamiliar to computer scientists. It bears a strong family resemblance to all general-purpose computers. If a symbol-manipulation language, such as LISP, is taken as defining a machine, then the kinship becomes truly brotherly. Our intent in laying out such a system is not to propose something new. Just the opposite: it is to show what is now known and hypothesized about systems that satisfy such a characterization.

We can now state a general scientific hypothesis—a law of qualitative structure for symbol systems:

The Physical-Symbol System Hypothesis. A physical-symbol system has the necessary and sufficient means for general intelligent action.

By 'necessary' we mean that any system that exhibits general intelligence will prove upon analysis to be a physical-symbol system. By 'sufficient' we mean that any physical-symbol system of sufficient size can be organized further to exhibit general intelligence. By 'general intelligent action' we wish to indicate the same scope of intelligence as we see in human action: that in any real situation behaviour appropriate to the ends of the system and adaptive to the demands of the environment can occur, within some limits of speed and complexity.

The Physical-Symbol System Hypothesis clearly is a law of qualitative structure. It specifies a general class of systems within which one will find those capable of intelligent action.

This is an empirical hypothesis. We have defined a class of systems; we wish to ask whether that class accounts for a set of phenomena we find in the real world. Intelligent action is everywhere around us in the biological world, mostly in human behaviour. It is a form of behaviour we can recognize by its effects whether it is performed by humans or not. The hypothesis could indeed be false. Intelligent behaviour is not so easy to produce that any system will exhibit it willy-nilly. Indeed, there are people whose analyses lead them to conclude either on philosophical or on scientific grounds that the hypothesis *is* false. Scientifically, one can attack or defend it only by bringing forth empirical evidence about the natural world.

We now need to trace the development of this hypothesis and look at the evidence for it.

Development of the Symbol-System Hypothesis

A physical-symbol system is an instance of a universal machine. Thus the symbol-system hypothesis implies that intelligence will be realized by a universal computer. However, the hypothesis goes far beyond the argument, often made on general grounds of physical determinism, that any computation that is realizable can be realized by a universal machine, provided that it is specified. For it asserts specifically that the intelligent machine is a symbol system, thus making a specific architectural assertion about the nature of intelligent systems. It is important to understand how this additional specificity arose.

Formal Logic. The roots of the hypothesis go back to the program of Frege and of Whitehead and Russell for formalizing logic: capturing the basic conceptual notions of mathematics in logic and putting the notions of proof and deduction on a secure footing. This effort culminated in mathematical logic—our familiar propositional, first-order, and higher-order logics. It developed a characteristic view, often referred to as the 'symbol game'. Logic, and by incorporation all of mathematics, was a game played with meaningless tokens according to certain purely syntactic rules. All meaning had been purged. One had a mechanical, though permissive (we would now say non-deterministic), system about which various things could be proved. Thus progress was first made by walking away from all that seemed relevant to meaning and human symbols. We could call this the stage of formal symbol-manipulation.

This general attitude is well reflected in the development of information theory. It was pointed out time and again that Shannon had defined a system that was useful only for communication and selection, and which had nothing to do with meaning. Regrets were expressed that such a general name as 'information theory' had been given to the field, and attempts were made to rechristen it as 'the theory of selective information'—to no avail, of course.

Turing Machines and the Digital Computer. The development of the first digital computers and of automata theory, starting with Turing's own work in the thirties, can be treated together. They agree in their view of what is essential. Let us use Turing's own model, for it shows the features well.

A Turing machine consists of two memories: an unbounded tape and a finite-state control. The tape holds data, i.e. the famous zeros and ones. The machine has a very small set of proper operations—read, write, and scan operations—on the tape. The read operation is not a data operation, but provides conditional branching to a control state as a function of the data under the read head. As we all know, this model contains the essentials of all computers, in terms of what they can do, though other computers with different memories and operations might carry out the same computations with different requirements of space and time. In particular, the model of a Turing machine contains within it the notions both of what cannot be computed and of universal machines—computers that can do anything that can be done by any machine.

We should marvel that two of our deepest insights into information processing were achieved in the thirties, before modern computers came into being. It is a tribute to the genius of Alan Turing. It is also a tribute to the development of mathematical logic at the time, and testimony to the depth of computer science's obligation to it. Concurrently with Turing's work appeared the work of the logicians Emil Post and (independently) Alonzo Church. Starting from independent notions of logistic systems (Post productions and recursive functions, respectively), they arrived at analogous results on undecidability and universality—results that were soon shown to imply that all three systems were equivalent. Indeed, the convergence of all these attempts to define the most general class of information-processing systems provides some of the force of our conviction that we have captured the essentials of information-processing in these models.

In none of these systems is there, on the surface, a concept of the symbol as something that *designates*. The data are regarded as just strings of zeros and ones—indeed that data be inert is essential to the reduction of computation to physical process. The finite-state control system was always viewed as a small controller, and logical games were played to see how small a state system could be used without destroying the universality of the machine. No games, as far as we can tell, were ever played to add new states dynamically to the finite control—to think of the control memory as holding the bulk of the system's knowledge. What was accomplished at this stage was half the principle of interpretation— showing that a machine could be run from a description. Thus this is the stage of automatic formal symbol-manipulation.

The Stored Program Concept.   With the development of the second generation of electronic machines in the mid-forties (after the Eniac) came the stored program concept. This was rightfully hailed as a milestone, both conceptually and practically. Programs now can be data, and can be operated on as data. This capability is, of course, already implicit in the model of Turing: the descriptions are on the very same tape as the data. Yet the idea was realized only when machines acquired enough memory to make it practicable to locate actual programs in some internal place. After all, the Eniac had only twenty registers.

The stored program concept embodies the second half of the interpretation principle, the part that says that the system's own data can be interpreted. But it does not yet contain the notion of designation—of the physical relation that underlies meaning.

List-Processing.   The next step, taken in 1956, was list-processing. The contents of the data-structures were now symbols, in the sense of our physical-symbol system: patterns that designated, that had referents. Lists held addresses which permitted access to other lists—thus the notion of list-structures. That this was a new view was demonstrated to us many times in the early days of list-processing when colleagues would ask where the data were—that is, which list finally held the collection of bits that were the content of the system. They found it strange that there were no such bits, there were only symbols that designated yet other symbol structures.

List-processing is simultaneously three things in the development of computer science.

1. It is the creation of a genuine dynamic memory structure in a machine that had heretofore been perceived as having fixed structure. It added to our ensemble of operations those that built and modified structure in addition to those that replaced and changed content.

2. It was an early demonstration of the basic abstraction that a computer consists of a set of data types and a set of operations proper to these data types, so that a computational system should employ whatever data types are appropriate to the application, independent of the underlying machine.

3. List-processing produced a model of designation, thus defining symbol-manipulation in the sense in which we use this concept in computer science today.

As often occurs, the practice of the time already anticipated all the elements of list-processing: addresses are obviously used to gain access, the drum machines used linked programs (so-called one-plus-one addressing), and so on. But the conception of list-processing as an abstraction created a new world in which designation and dynamic symbolic structure were defining characteristics. The embedding of the early list-processing systems in languages (the IPLs, LISP) is often decried as having been a barrier to the diffusion of list-processing techniques throughout programming practice; but it was the vehicle that held the abstraction together.

LISP. One more step is worth noting: McCarthy's creation of LISP in 1959–60 (McCarthy 1960). It completed the act of abstraction, lifting list-structures out of their embedding in concrete machines, creating a new formal system with S-expressions, which could be shown to be equivalent to the other universal schemes of computation.

Conclusion. That the concept of the designating symbol and symbol-manipulation does not emerge until the mid-fifties does not mean that the earlier steps were either inessential or less important. The total concept is the join of computability, physical realizability (and by multiple technologies), universality, the symbolic representation of processes (i.e. interpretability), and, finally, symbolic structure and designation. Each of the steps provided an essential part of the whole.

The first step in this chain, authored by Turing, is theoretically motivated, but the others all have deep empirical roots. We have been led by the evolution of the computer itself. The stored program principle arose out of the experience with Eniac. List-processing arose out of the attempt to construct intelligent programs. It took its cue from the emergence of random access memories, which provided a clear physical realization of a designating symbol in the address. LISP arose out of the evolving experience with list-processing.

The Evidence

We come now to the evidence for the hypothesis that physical-symbol systems are capable of intelligent action, and that general intelligent action calls for a physical-symbol system. The hypothesis is an empirical generalization and not a theorem. We know of no way of demonstrating the connection between symbol systems and intelligence on purely logical grounds. Lacking such a demonstration, we must look at the facts. Our central aim, however, is not to review the evidence in detail, but to use the example before us to illustrate the proposition that computer science is a field of empirical enquiry. Hence, we will only indicate what kinds of evidence there are, and the general nature of the testing process.

The notion of physical-symbol system had taken essentially its present form by the middle of the 1950s, and one can date from that time the growth of artificial intelligence as a coherent subfield of computer science. The twenty years of work since then has seen a continuous accumulation of empirical evidence of two main varieties. The first addresses itself to the *sufficiency* of physical-symbol systems for producing intelligence, attempting to construct and test specific systems that have such a capability. The second kind of evidence addresses itself to the *necessity* of having a physical-symbol system wherever intelligence is exhibited. It starts with Man, the intelligent system best known to us, and attempts to discover whether his cognitive activity can be explained as the working of a physical-symbol system. There are other forms of evidence, which we will comment upon briefly later, but these two are the important ones. We will consider them in turn. The first is generally called artificial intelligence, the second, research in cognitive psychology.

Constructing Intelligent Systems. The basic paradigm for the initial testing of the germ theory of disease was: identify a disease, then look for the germ. An analogous paradigm has inspired much of the research in artificial intelligence: identify a task-domain calling for intelligence, then construct a program for a digital computer that can handle tasks in that domain. The easy and well structured tasks were looked at first: puzzles and games, operations-research problems of scheduling and allocating resources, simple induction tasks. Scores, if not hundreds, of programs of these kinds have by now been constructed, each capable of some measure of intelligent action in the appropriate domain.

Of course intelligence is not an all-or-none matter, and there has been steady progress towards higher levels of performance in specific domains, as well as towards widening the range of those domains. Early chess programs, for example, were deemed successful if they could play the game legally and with some indication of purpose; a little later, they reached the level of human beginners; within ten or fifteen years, they began to compete with serious amateurs. Progress has been slow (and the total programming effort invested small) but continuous, and the paradigm of construct-and-test proceeds in a regular cycle—the whole research activity mimicking at a macroscopic level the basic generate-and-test cycle of many of the AI programs.

There is a steadily widening area within which intelligent action is attainable. From the original tasks, research has extended to building systems that handle and understand natural language in a variety of ways, systems for interpreting visual scenes, systems for hand–eye co-ordination, systems that design, systems that write computer programs, systems for speech understanding—the list is, if not endless, at least very long. If there are limits beyond which the hypothesis will not carry us, they have not yet become apparent. Up to the present, the rate of progress has been governed mainly by the rather modest quantity of scientific resources that have been applied and the inevitable requirement of a substantial system-building effort for each new major undertaking.

Much more has been going on, of course, than simply a piling up of examples of intelligent systems adapted to specific task-domains. It would be surprising and unappealing if it turned out that the AI programs performing these diverse tasks had nothing in common beyond their being instances of physical-symbol systems. Hence, there has been great interest in searching for mechanisms possessed of generality, and for common components among programs performing a variety of tasks. This search carries the theory beyond the initial symbol-system hypothesis to a more complete characterization of the particular kinds of symbol systems that are effective in artificial intelligence. In the second section of this paper, we will discuss one example of a hypothesis at this second level of specificity: the heuristic-search hypothesis.

The search for generality spawned a series of programs designed to separate out general problem-solving mechanisms from the requirements of particular task-domains. The General Problem Solver (GPS) was perhaps the first of these; while among its descendants are such contemporary systems as PLANNER and CONNIVER. The search for common components has led to generalized schemes of representation for goals and plans, methods for constructing discrimination nets, procedures for the control of tree search, pattern-matching mechanisms, and language-parsing systems. Experiments are at present under way to find convenient devices for representing sequences of time and tense, movement, causality, and the like. More and more, it becomes possible to assemble large intelligent systems in a modular way from such basic components.

We can gain some perspective on what is going on by turning, again, to the analogy of the germ theory. If the first burst of research stimulated by that theory consisted largely in finding the germ to go with each disease, subsequent effort turned to learning what a germ was—to building on the basic qualitative law a new level of structure. In artificial intelligence, an initial burst of activity aimed at building intelligent programs for a wide variety of almost randomly selected tasks is giving way to more sharply targeted research aimed at understanding the common mechanisms of such systems.

The Modelling of Human Symbolic Behaviour. The symbol-system hypothesis implies that the symbolic behaviour of man arises because he has the characteristics of a physical-symbol system. Hence, the results of efforts to model human behaviour with symbol systems become an important part of the evidence for the hypothesis, and research in artificial intelligence goes on in close collaboration with research in information-processing psychology, as it is usually called.

The search for explanations of man's intelligent behaviour in terms of symbol systems has had a large measure of success over the past twenty years; to the point where information-processing theory is the leading contemporary point of view in cognitive psychology. Especially in the areas of problem-solving, concept attainment, and long-term memory, symbol-manipulation models now dominate the scene.

Research in information-processing psychology involves two main kinds of empirical activity. The first is the conduct of observations and experiments on human behaviour in tasks requiring intelligence. The second, very similar to the parallel activity in artificial intelligence, is the programming of symbol systems to model the observed human behaviour. The psychological observations and experiments lead to the formulation of hypotheses about the symbolic processes the subjects are using, and these are an important source of the ideas that go into the construction of the programs. Thus many of the ideas for the basic mechanisms of GPS were derived from careful analysis of the protocols that human subjects produced while thinking aloud during the performance of a problem-solving task.

The empirical character of computer science is nowhere more evident than in this alliance with psychology. Not only are psychological experiments required to test the veridicality of the simulation models as explanations of the human behaviour, but out of the experiments come new ideas for the design and construction of physical-symbol systems.

Other Evidence. The principal body of evidence for the symbol-system hypothesis that we have now considered is negative evidence: the absence of specific competing hypotheses as to how intelligent activity might be accomplished—whether by man or by machine. Most attempts to build such hypotheses have taken place within the field of psychology. Here we have had a continuum of theories from the points of view usually labelled 'behaviourism' to those usually labelled 'Gestalt theory'. Neither of these points of view stands as a real competitor to the symbol-system hypothesis, and for two reasons. First, neither behaviourism nor Gestalt theory has demonstrated, or even shown how to demonstrate, that the explanatory mechanisms it postulates are sufficient to account for intelligent behaviour in complex tasks. Second, neither theory has been formulated with anything like the specificity of artificial programs. As a matter of fact, the alternative theories are so vague that it is not terribly difficult to give them information-processing interpretations, and thereby assimilate them to the symbol-system hypothesis.

Conclusion

We have tried to use the example of the Physical-Symbol System Hypothesis to illustrate concretely that computer science is a scientific enterprise in the usual meaning of that term: it develops scientific hypotheses which it then seeks to verify by empirical enquiry. We had a second reason, however, for choosing this particular example to illustrate our point. The Physical-Symbol System Hypothesis is itself a substantial scientific hypothesis of the kind that we earlier dubbed 'laws of qualitative structure'. It represents an important discovery of computer science, which if borne out by the empiricial evidence, as in fact appears to be occurring, will have major continuing impact on the field.

We turn now to a second example, the role of search in intelligence. This topic, and the particular hypothesis about it that we shall examine, have also played a central role in computer science, in general, and artificial intelligence, in particular.

2. HEURISTIC SEARCH

Knowing that physical-symbol systems provide the matrix for intelligent action does not tell us how they accomplish this. Our second example of a law of qualitative structure in computer science addresses this latter question, asserting that symbol systems solve problems by using the processes of heuristic search. This generalization, like the previous one, rests on empirical evidence, and has not been derived formally from other premises. We shall see in a moment, however, that it does have some logical connection with the symbol-system hypothesis, and perhaps we can expect to formalize the connection at some time in the future. Until that time arrives, our story must again be one of empirical enquiry. We will describe what is known about heuristic search and review the empirical findings that show how it enables action to be intelligent. We begin by stating this law of qualitative structure, the Heuristic Search Hypothesis.

Heuristic Search Hypothesis. The solutions to problems are represented as symbol structures. A physical-symbol system exercises its intelligence in problem-solving by search—that is, by generating and progressively modifying symbol structures until it produces a solution structure.

Physical-symbol systems must use heuristic search to solve problems because such systems have limited processing resources; in a finite number of steps, and over a finite interval of time, they can execute only a finite number of processes. Of course that is not a very strong limitation, for all universal Turing machines suffer from it. We intend the limitation, however, in a stronger sense: we mean *practically* limited. We can conceive of systems that are not limited in a practical way but are capable, for example, of searching in parallel the nodes of an exponentially expanding tree at a constant rate for each unit advance in depth. We will not be concerned here with such systems, but with systems whose computing resources are scarce relative to the complexity of the situations with which they are confronted. The restriction will not exclude any real symbol systems, in computer or man, in the context of real tasks. The fact of limited resources allows us, for most purposes, to view a symbol system as though it were a serial, one-process-at-a-time device. If it can accomplish only a small amount of processing in any short time interval, then we might as well regard it as doing things one at a time. Thus 'limited resource symbol system' and 'serial symbol system' are practically synonymous. The problem of allocating a scarce resource from moment to moment can usually be treated, if the moment is short enough, as a problem of scheduling a serial machine.

Problem-Solving

Since ability to solve problems is generally taken as a prime indicator that a system has intelligence, it is natural that much of the history of artificial intelligence is taken up with attempts to build and understand problem-solving systems. Problem-solving has been discussed by philosophers and psychologists for two millennia, in discourses dense with a feeling of mystery. If you think there is nothing problematic or mysterious about a symbol system solving problems, you are a child of today, whose views have been formed since mid-century. Plato (and, by his account, Socrates) found difficulty understanding even how problems could be *entertained*, much less how they could be solved. Let me remind you of how he posed the conundrum in the *Meno*:

MENO: And how will you inquire, Socrates, into that which you know not? What will you put forth as the subject of inquiry? And if you find what you want, how will you ever know that this is what you did not know?

To deal with this puzzle, Plato invented his famous theory of recollection: when you think you are discovering or learning something, you are really just recalling what you already knew in a previous existence. If you find this explanation preposterous, there is a much simpler one available today, based upon our understanding of symbol systems. An approximate statement of it is:

To state a problem is to designate (1) a *test* for a class of symbol structures (solutions of the problem), and (2) a *generator* of symbol structures (potential solutions). To solve a problem is to generate a structure, using (2), that satisfies the test of (1).

We have a problem if we know what we want to do (the test), and if we don't know immediately how to do it (our generator does not immediately produce a symbol structure satisfying the test). A symbol system can state and solve problems (sometimes) because it can generate and test.

If that is all there is to problem-solving, why not simply generate at once an expression that satisfies the test? This is, in fact, what we do when we wish and dream. 'If wishes were horses, beggars might ride.' But outside the world of dreams, it isn't possible. To know how we would test something, once constructed, does not mean that we know how to construct it — that we have any generator for doing so.

For example, it is well known what it means to 'solve' the problem of playing winning chess. A simple test exists for noticing winning positions, the test for checkmate of the enemy king. In the world of dreams one simply generates a strategy that leads to checkmate for all counter strategies of the opponent. Alas, no generator that will do this is known to existing symbol systems (man or machine). Instead, good moves in chess are sought by generating various alternatives, and painstakingly evaluating them with the use of approximate, and often erroneous, measures that are supposed to indicate the likelihood that a particular line of play is on the route to a winning position. Move generators there are; winning-move generators there are not.

Before there can be a move generator for a problem, there must be a problem space: a space of symbol structures in which problem situations, including the initial and goal situations, can be represented. Move generators are processes for modifying one situation in the problem space into another. The basic characteristics of physical-symbol systems guarantee that they can represent problem spaces and that they possess move generators. How, in any concrete situation they synthesize a problem space and move generators appropriate to that situation is a question that is still very much on the frontier of artificial intelligence research.

The task that a symbol system is faced with, then, when it is presented with a problem and a problem space, is to use its limited processing resources to generate possible solutions, one after another, until it finds one that satisfies the problem-defining test. If the system had some control over the order in which potential solutions were generated, then it would be desirable to arrange this order of generation so that actual solutions would have a high likelihood of appearing early. A symbol system would exhibit intelligence to the extent that it succeeded in doing this. Intelligence for a system with limited processing resources consists in making wise choices of what to do next.

Search in Problem-Solving

During the first decade or so of artificial intelligence research, the study of problem-solving was almost synonymous with the study of search processes. From our characterization of problems and problem-solving, it is easy to see why this was so. In fact, it might be asked whether it could be otherwise. But before we try to answer that question, we must explore further the nature of search processes as it revealed itself during that decade of activity.

Extracting Information from the Problem Space. Consider a set of symbol structures, some small subset of which are solutions to a given problem. Suppose, further, that the solutions are distributed randomly through the entire set. By this we mean that no information exists that would enable any search generator to perform better than a random search. Then no symbol system could exhibit more intelligence (or less intelligence) than any other in solving the problem, although one might experience better luck than another.

A condition, then, for the appearance of intelligence is that the distribution of solutions be not entirely random, that the space of symbol structures exhibit at least some degree of order and pattern. A second condition is that pattern in the space of symbol structures be more or less detectible. A third condition is that the generator of potential solutions be able to behave differentially, depending on what pattern it detected. There must be information in the problem space, and the symbol system must be capable of extracting and using it. Let us look first at a very simple example, where the intelligence is easy to come by.

Consider the problem of solving a simple algebraic equation:

$$AX + B = CX + D$$

The test defines a solution as any expression of the form, $X = E$, such that $AE + B = CE + D$. Now one could use as generator any process that would produce numbers which could then be tested by substituting in the latter equation. We would not call this an intelligent generator.

Alternatively, one could use generators that would make use of the fact that the original equation can be modified—by adding or subtracting equal quantities from both sides, or multiplying or dividing both sides by the same quantity—without changing its solutions. But, of course, we can obtain even more information to guide the generator by comparing the original expression with the form of the solution, and making precisely those changes in the equation that leave its solution unchanged, while at the same time bringing it into the desired form. Such a generator could notice that there was an unwanted CX on the right-hand side of the original equation, subtract it from both sides, and collect terms again. It could then notice that there was an unwanted B on the left-hand side and subtract that. Finally, it could get rid of the unwanted coefficient $(A - C)$ on the left-hand side by dividing.

Thus by this procedure, which now exhibits considerable intelligence, the generator produces successive symbol structures, each obtained by modifying the previous one; and the modifications are aimed at reducing the differences between the form of the input structure and the form of the test expression, while maintaining the other conditions for a solution.

This simple example already illustrates many of the main mechanisms that are used by symbol systems for intelligent problem-solving. First, each successive expression is not generated independently, but is produced by modifying one produced previously. Second, the modifications are not haphazard, but depend upon two kinds of information. They depend on information that is constant over this whole class of algebra problems, and that is built into the structure of the generator itself: all modifications of expressions must leave the equation's solution unchanged. They also depend on information that changes at each step: detection of the differences in form that remain between the current expression and the desired expression. In effect, the generator incorporates some of the tests the solution must satisfy, so that expressions that don't meet these tests will never be generated. Using the first kind of information guarantees that only a tiny subset of all possible expressions is actually generated, but without losing the solution expression from this subset. Using the second kind of information arrives at the desired solution by a succession of approximations, employing a simple form of means-ends analysis to give direction to the search.

There is no mystery where the information that guided the search came from. We need not follow Plato in endowing the symbol system with a previous existence in which it already knew the solution. A moderately sophisticated generator-test system did the trick without invoking reincarnation.

Search Trees. The simple algebra problem may seem an unusual, even pathological, example of search. It is certainly not trial-and-error search, for though there were a few trials, there was no error. We are more accustomed to thinking of problem-solving search as generating lushly branching trees of partial solution possibilities which may grow to thousands, or even millions, of branches, before they yield a solution. Thus, if from each expression it produces, the generator creates B new branches, then the tree will grow as B^D, where D is its depth. The tree grown for the algebra problem had the peculiarity that its branchiness, B, equalled unity.

Programs that play chess typically grow broad search trees, amounting in some cases to a million branches or more. Although this example will serve to illustrate our points about tree search, we should note that the purpose of search in chess is not to generate proposed solutions, but to evaluate (test) them. One line of research into game-playing programs has been centrally concerned with improving the representation of the chess board, and the processes for making moves on it, so as to speed up search and make it possible to search larger trees. The rationale for this direction, of course, is that the deeper the dynamic search, the more accurate should be the evaluations at the end of it. On the other hand, there is good empirical evidence that the strongest human players, grand masters, seldom explore trees of more than one hundred branches. This economy is achieved not so much by searching less deeply than do chess-playing programs, but by branching very sparsely and selectively at each node. This is only possible, without causing a deterioration of the evaluations, by having more of the selectivity built into the generator itself, so that it is able to select for generation only those branches which are very likely to yield important relevant information about the position.

The somewhat paradoxical-sounding conclusion to which this discussion leads is that search—successive generation of potential solution structures—is a fundamental aspect of a symbol system's exercise of intelligence in problem-solving but that amount of search is not a measure of the amount of intelligence being exhibited. What makes a problem a problem is not that a large amount of search is required for its solution, but that a large amount *would* be required if a requisite level of intelligence were not applied. When the symbolic system that is endeavouring to solve a problem knows enough about what to do, it simply proceeds directly towards its goal; but whenever its knowledge becomes inadequate, when it enters *terra incognita,* it is faced with the threat of going through large amounts of search before it finds its way again.

The potential for the exponential explosion of the search tree that is present in every scheme for generating problem solutions warns us against depending on the brute force of computers—even the biggest and fastest computers—as a compensation for the ignorance and unselectivity of their generators. The hope is still periodically ignited in some human breasts that a computer can be found that is fast enough, and that can be programmed cleverly enough, to play good chess by brute-force search. There is nothing known in theory about the game of chess that rules out this possibility. But empirical studies on the management of search in sizable trees with only modest results make this a much less promising direction than it was when chess was first chosen as an appropriate task for artificial intelligence. We must regard this as one of the important empirical findings of research with chess programs.

The Forms of Intelligence. The task of intelligence, then, is to avert the ever-present threat of the exponential explosion of search. How can this be accomplished? The first route, already illustrated by the algebra example and by chess programs that only generate 'plausible' moves for further analysis, is to build selectivity into the generator: to generate only structures that show promise of being solutions or of being along the path towards solutions. The usual consequence of doing this is to decrease the rate of branching, not to prevent it entirely. Ultimate exponential explosion is not avoided—save in exceptionally highly structured situations like the algebra example—but only postponed. Hence, an intelligent system generally needs to supplement the selectivity of its solution generator with other information-using techniques to guide search.

Twenty years of experience with managing tree search in a variety of task environments has produced a small kit of general techniques which is part of the equipment of every researcher in artificial intelligence today. Since these techniques have been described in general works like that of Nilsson (1971), they can be summarized very briefly here.

In serial heuristic search, the basic question always is: What shall be done next? In tree search, that question, in turn, has two components: (1) from what node in the tree shall we search next, and (2) what direction shall we take from that node? Information helpful in answering the first question may be interpreted as measuring the relative distance of different nodes from the goal. Best-first search calls for searching next from the node that appears closest to the goal. Information helpful in answering the second question—in what direction to search—is often obtained, as in the algebra example, by detecting specific differences between the current nodal structure and the goal structure described by the test of a solution, and selecting actions that are relevant to reducing these particular kinds of differences. This is the technique known as means-ends analysis, which plays a central role in the structure of the General Problem Solver.

The importance of empirical studies as a source of general ideas in AI research can be demonstrated clearly by tracing the history, through large numbers of problem-solving programs, of these two central ideas: best-first search and means-ends analysis. Rudiments of best-first search were already present, though unnamed, in the Logic Theorist in 1955. The General Problem Solver, embodying means-ends analysis, appeared about 1957—but combined it with modified depth-first search rather than best-first search. Chess programs were generally wedded, for reasons of economy of memory, to depth-first search, supplemented after about 1958 by the powerful alpha-beta pruning procedure. Each of these techniques appears to have been reinvented a number of times, and it is hard to find general, task-independent, theoretical discussions of problem-solving in terms of these concepts until the middle or late 1960s. The amount of formal buttressing they have received from mathematical theory is still minuscule: some theorems about the reduction in search that can be secured from using the alpha-beta heuristic, a couple of theorems (reviewed by Nilsson 1971) about shortest-path search, and some very recent theorems on best-first search with a probabilistic evaluation function.

'Weak' and 'Strong' Methods. The techniques we have been discussing are dedicated to the control of exponential expansion rather than its prevention. For this reason, they have been properly called 'weak methods'—methods to be used when the symbol system's knowledge or the amount of structure actually contained in the problem space are inadequate to permit search to be avoided entirely. It is instructive to contrast a highly structured situation, which can be formulated, say, as a linear-programming problem, with the less structured situations of combinatorial problems like the travelling salesman problem or scheduling problems. ('Less structured' here refers to the insufficiency or non-existence of relevant theory about the structure of the problem space.)

In solving linear-programming problems, a substantial amount of computation may be required, but the search does not branch. Every step is a step along the way to a solution. In solving combinatorial problems or in proving theorems, tree search can seldom be avoided, and success depends on heuristic search methods of the sort we have been describing.

Not all streams of AI problem-solving research have followed the path we have been outlining. An example of a somewhat different point is provided by the work on theorem-proving systems. Here, ideas imported from mathematics and logic have had a strong influence on the direction of enquiry. For example, the use of heuristics was resisted when properties of completeness could not be proved (a bit ironic, since most interesting mathematical systems are known to be undecidable). Since completeness can seldom be proved for best-first search heuristics, or for many kinds of selective generators, the effect of this requirement was rather inhibiting. When theorem-proving programs were continually incapacitated by the combinatorial explosion of their search trees, thought began to be given to selective heuristics, which in many cases proved to be analogues of heuristics used in general problem-solving programs. The set-of-support heuristic, for example, is a form of working backward, adapted to the resolution theorem-proving environment.

A Summary of the Experience. We have now described the workings of our second law of qualitative structure, which asserts that physical-symbol systems solve problems by means of heuristic search. Beyond that, we have examined some subsidiary characteristics of heuristic search, in particular the threat that it always faces of exponential explosion of the search tree, and some of the means it uses to avert that threat. Opinions differ as to how effective heuristic search has been as a problem-solving mechanism—the opinions depending on what task domains are considered and what criterion of adequacy is adopted. Success can be guaranteed by setting aspiration levels low—or failure by setting them high. The evidence might be summed up about as follows: few programs are solving problems at 'expert' professional levels. Samuel's checker program and Feigenbaum and Lederberg's DENDRAL are perhaps the best-known exceptions, but one could point also to a number of heuristic search programs for such operations-research problem domains as scheduling and integer programming. In a number of domains, programs perform at the level of competent amateurs: chess, some theorem-proving domains, many kinds of games and puzzles. Human levels have not yet been nearly reached by programs that have a complex perceptual 'front end': visual scene recognizers, speech understanders, robots that have to manœuvre in real space and time. Nevertheless, impressive progress has been made, and a large body of experience assembled about these difficult tasks.

We do not have deep theoretical explanations for the particular pattern of performance that has emerged. On empirical grounds, however, we might draw two conclusions. First, from what has been learned about human expert performance in tasks like chess, it is likely that any system capable of matching that performance will have to have access, in its memories, to very large stores of semantic information. Second, some part of the human superiority in tasks with a large perceptual component can be attributed to the special-purpose built-in parallel-processing structure of the human eye and ear.

In any case, the quality of performance must necessarily depend on the characteristics both of the problem domains and of the symbol systems used to tackle them. For most real-life domains in which we are interested, the domain structure has so far not proved sufficiently simple to yield theorems about complexity, or to tell us, other than empirically, how large real-world problems are in relation to the abilities of our symbol systems to solve them. That situation may change, but until it does, we must rely upon empirical explorations, using the best problem solvers we know how to build, as a principal source of knowledge about the magnitude and characteristics of problem difficulty. Even in highly structured areas like linear programming, theory has been much more useful in strengthening the heuristics that underlie the most powerful solution algorithms than in providing a deep analysis of complexity.

Intelligence Without Much Search

Our analysis of intelligence equated it with ability to extract and use information about the structure of the problem space, so as to enable a problem solution to be generated as quickly and directly as possible. New directions for improving the problem-solving capabilities of symbol systems can be equated, then, with new ways of extracting and using information. At least three such ways can be identified.

Non-Local Use of Information.   First, it has been noted by several investigators that information gathered in the course of tree search is usually only used *locally,* to help make decisions at the specific node where the information was generated. Information about a chess position, obtained by dynamic analysis of a subtree of continuations, is usually used to evaluate just that position, not to evaluate other positions that may contain many of the same features. Hence, the same facts have to be rediscovered repeatedly at different nodes of the search tree. Simply to take the information out of the context in which it arose and use it generally does not solve the problem, for the information may be valid only in a limited range of contexts. In recent years, a few exploratory efforts have been made to transport information from its context of origin to other appropriate contexts. While it is still too early to evaluate the power of this idea, or even exactly how it is to be achieved, it shows considerable promise. An important line of investigation that Berliner (1975) has been pursuing is to use causal analysis to determine the range over which a particular piece of information is valid. Thus if a weakness in a chess position can be traced back to the move that made it, then the same weakness can be expected in other positions descendant from the same move.

The HEARSAY speech-understanding system has taken another approach to making information globally available. That system seeks to recognize speech strings by pursuing a parallel search at a number of different levels: phonemic, lexical, syntactic, and semantic. As each of these searches provides and evaluates hypotheses, it supplies the information it has gained to a common 'blackboard' that can be read by all the sources. This shared information can be used, for example, to eliminate hypotheses, or even whole classes of hypotheses, that would otherwise have to be searched by one of the processes. Thus increasing our ability to use tree-search information non-locally offers promise for raising the intelligence of problem-solving systems.

Semantic Recognition Systems. A second active possibility for raising intelligence is to supply the symbol system with a rich body of semantic information about the task-domain it is dealing with. For example, empirical research on the skill of chess masters shows that a major source of the master's skill is stored information that enables him to recognize a large number of specific features and patterns of features on a chess board, and information that uses this recognition to propose actions appropriate to the features recognized. This general idea has, of course, been incorporated in chess programs almost from the beginning. What is new is the realization of the number of such patterns and associated information that may have to be stored for master-level play: something on the order of 50,000.

The possibility of substituting recognition for search arises because a particular, and especially a rare, pattern can contain an enormous amount of information, provided that it is closely linked to the structure of the problem space. When that structure is 'irregular', and not subject to simple mathematical description, then knowledge of a large number of relevant patterns may be the key to intelligent behaviour. Whether this is so in any particular task-domain is a question more easily settled by empirical investigation than by theory. Our experience with symbol systems richly endowed with semantic information and pattern-recognizing capabilities for accessing it is still extremely limited.

The discussion above refers specifically to semantic information associated with a recognition system. Of course, there is also a whole large area of AI research on semantic information processing and the organization of semantic memories that falls outside the scope of the topics we are discussing in this paper.

Selecting Appropriate Representations. A third line of enquiry is concerned with the possibility that search can be reduced or avoided by selecting an appropriate problem space. A standard example that illustrates this possibility dramatically is the mutilated chequer-board problem. A standard 64-square chequer-board can be covered exactly with 32 tiles, each a 1 × 2 rectangle covering exactly two squares. Suppose, now, that we cut off squares at two diagonally opposite corners of the chequer-board, leaving a total of 62 squares. Can this mutilated board be covered exactly with 31 tiles? With (literally) heavenly patience, the impossibility of achieving such a covering can be demonstrated by trying all possible arrangements. The alternative, for those with less patience and more intelligence, is to observe that the two diagonally opposite corners of a chequer-board are of the same colour. Hence, the mutilated chequer-board has two fewer squares of one colour than of the other. But each tile covers one square of one colour and one square of the other, and any set of tiles must cover the same number of squares of each colour. Hence, there is no solution. How can a symbol system discover this simple inductive argument as an alternative to a hopeless attempt to solve the problem by search among all possible coverings? We would award a system that found a solution high marks for intelligence.

Perhaps, however, in posing these problems we are not escaping from search processes. We have simply displaced the search from a space of possible problem solutions to a space of possible representations. In any event, the whole process of moving from one representation to another, and of discovering and evaluating representations, is largely unexplored territory in the domain of problem-solving research. The laws of qualitative structure governing representations remain to be discovered. The search for them is almost sure to receive considerable attention in the coming decade.

Conclusion

That is our account of symbol systems and intelligence. It has been a long road from Plato's *Meno* to the present, but it is perhaps encouraging that most of the progress along that road has been made since the turn of the twentieth century, and a large fraction of it since the mid-point of the century. Thought was still wholly intangible and ineffable until modern formal logic interpreted it as the manipulation of formal tokens. And it seemed still to inhabit mainly the heaven of Platonic ideals, or the equally obscure spaces of the human mind, until computers taught us how symbols could be processed by machines. A. M. Turing made his great contributions at the mid-century crossroads of these developments that led from modern logic to the computer.

Physical-Symbol Systems. The study of logic and computers has revealed to us that intelligence resides in physical-symbol systems. This is computer science's most basic law of qualitative structure.

Symbol systems are collections of patterns and processes, the latter being capable of producing, destroying, and modifying the former. The most important properties of patterns are that they can designate objects, processes, or other patterns, and that when they designate processes, they can be interpreted. Interpretation means carrying out the designated process. The two most significant classes of symbol systems with which we are acquainted are human beings and computers.

Our present understanding of symbol systems grew, as indicated earlier, through a sequence of stages. Formal logic familiarized us with symbols, treated syntactically, as the raw material of thought, and with the idea of manipulating them according to carefully defined formal processes. The Turing machine made the syntactic processing of symbols truly machinelike, and affirmed the potential universality of strictly defined symbol systems. The stored-program concept for computers reaffirmed the interpretability of symbols, already implicit in the Turing machine. List-processing brought to the forefront the denotational capacities of symbols, and defined symbol-processing in ways that allowed independence from the fixed structure of the underlying physical machine. By 1956 all of these concepts were available, together with hardware for implementing them. The study of the intelligence of symbol systems, the subject of artificial intelligence, could begin.

Heuristic Search.   A second law of qualitative structure for AI is that symbol systems solve problems by generating potential solutions and testing them—that is, by searching. Solutions are usually sought by creating symbolic expressions and modifying them sequentially until they satisfy the conditions for a solution. Hence, symbol systems solve problems by searching. Since they have finite resources, the search cannot be carried out all at once, but must be sequential. It leaves behind it either a single path from starting-point to goal or, if correction and backup are necessary, a whole tree of such paths.

Symbol systems cannot appear intelligent when they are surrounded by pure chaos. They exercise intelligence by extracting information from a problem domain and using that information to guide their search, avoiding wrong turns and circuitous bypaths. The problem domain must contain information—that is, some degree of order and structure—for the method to work. The paradox of the *Meno* is solved by the observation that information may be remembered, but new information may also be extracted from the domain that the symbols designate. In both cases, the ultimate source of the information is the task-domain.

The Empirical Base.   Research on artificial intelligence is concerned with how symbol systems must be organized in order to behave intelligently. Twenty years of work in the area has accumulated a considerable body of knowledge, enough to fill several books (it already has), and most of it in the form of rather concrete experience about the behaviour of specific classes of symbol systems in specific task-domains. Out of this experience, however, there have also emerged some generalizations, cutting across task-domains and systems, about the general characteristics of intelligence and its methods of implementation.

We have tried to state some of these generalizations here. They are mostly qualitative rather than mathematical. They have more the flavour of geology or evolutionary biology than the flavour of theoretical physics. They are sufficiently strong to enable us today to design and build moderately intelligent systems for a considerable range of task domains, as well as to gain a rather deep understanding of how human intelligence works in many situations.

What Next?   In our account we have mentioned open questions as well as settled ones; there are many of both. We see no abatement of the excitement of exploration that has surrounded this field over the past quarter century. Two resource limits will determine the rate of progress over the next such period. One is the amount of computing power that will be available. The second, and probably the more important, is the number of talented young computer scientists who will be attracted to this area of research as the most challenging they can tackle.

A. M. Turing concluded his famous paper 'Computing Machinery and Intelligence' with the words: 'We can only see a short distance ahead, but we can see plenty there that needs to be done.'

Many of the things Turing saw in 1950 that needed to be done have been done, but the agenda is as full as ever. Perhaps we read too much into his simple statement above, but we like to think that in it Turing recognized the fundamental truth that all computer scientists instinctively know. For all physical-symbol systems, condemned as we are to serial search of the problem environment, the critical question is always: What to do next?

REFERENCES

Berliner, H. (1975). 'Chess as Problem Solving: The Development of a Tactics Analayzer.' Unpublished Ph.D. thesis. Carnegie-Mellon University.

McCarthy, J. (1960). 'Recursive Functions of Symbolic Expressions and their Computation by Machine.' *Commun. ACM* 3 (Apr.): 184–95.

McCulloch, W. S. (1961). 'What is a Number, that a Man may know it, and a Man that he may know a Number?' *General Semantics Bulletin* nos. 26–7: 7–18. Repr. in W. S. McCulloch, *Embodiments of Mind*, pp. 1–18. Cambridge, Mass.: MIT Press.

Nilsson, N. J. (1971). *Problem-Solving Methods in Artificial Intelligence*. New York: McGraw-Hill.

6

ARTIFICIAL INTELLIGENCE: A PERSONAL VIEW

DAVID C. MARR

Artificial Intelligence is the study of complex information-processing problems that often have their roots in some aspect of biological information-processing. The goal of the subject is to identify interesting and solvable information-processing problems, and solve them.

The solution to an information-processing problem divides naturally into two parts. In the first, the underlying nature of a particular computation is characterized, and its basis in the physical world is understood. One can think of this part as an abstract formulation of *what* is being computed and *why*, and I shall refer to it as the 'theory' of a computation. The second part consists of particular algorithms for implementing a computation, and so it specifies *how*. The choice of algorithm usually depends upon the hardware in which the process is to run, and there may be many algorithms that implement the same computation. The theory of a computation, on the other hand, depends only on the nature of the problem to which it is a solution. Jardine and Sibson (1971) decomposed the subject of cluster-analysis in precisely this way, using the term 'method' to denote what I call the theory of a computation.

To make the distinction clear, let us take the case of Fourier analysis. The (computational) theory of the Fourier transform is well understood, and is expressed independently of the particular way in which it is computed. There are however several algorithms for implementing a Fourier transform—the Fast Fourier transform (Cooley and Tukey 1965), which is a serial algorithm; and the parallel 'spatial' algorithms that are based on the mechanisms of coherent optics. All these algorithms carry out the same computation, and the choice of which one to use depends upon the available hardware. In passing, we also note that the distinction between serial and parallel resides at the algorithm level, and is not a deep property of a computation.

D. C. Marr, 'Artificial Intelligence—A Personal View' from *Artificial Intelligence* 9 (1977): 37–48. Reprinted by permission of Elsevier Science Publisher B.V.

Strictly speaking then, a *result* in Artificial Intelligence consists of the isolation of a particular information-processing problem, the formulation of a computational theory for it, the construction of an algorithm that implements it, and a practical demonstration that the algorithm is success-ful. The important point here, and it is what makes progress possible, is that once a computational theory has been established for a particular problem, it never has to be done again, and in this respect a result in AI behaves like a result in mathematics or any of the hard natural sciences. Some judgement has to be applied when deciding whether the computa-tional theory for a problem has been formulated adequately; the state-ment 'take the opponent's king' defines the goal of chess, but it is hardly an adequate characterization of the computational problem of doing it.[1] The kind of judgement that is needed seems to be rather similar to that which decides whether a result in mathematics amounts to a substantial new theorem, and I do not feel uncomfortable about having to leave the basis of such judgements unspecified.[2]

This view of what constitutes a result in AI is probably acceptable to most scientists. Chomsky's (1965) notion of a 'competence' theory for English syntax is precisely what I mean by a computational theory for that problem. Both have the quality of being little concerned with the gory details of algorithms that must be run to express the competence (i.e. to implement the computation). That is not to say that devising suitable algorithms will be easy, but it is to say that before one can devise them, one has to know what exactly it is that they are supposed to be doing, and this information is captured by the computational theory. When a problem decomposes in this way, I shall refer to it as having a *Type-1* theory.

The fly in the ointment is that while many problems of biological information-processing have a Type-1 theory, there is no reason why they should all have. This can happen when a problem is solved by the simul-taneous action of a considerable number of processes, *whose interaction is its own simplest description*, and I shall refer to such a situation as a *Type-2* theory.[3] One promising candidate for a Type-2 theory is the problem of

---

[1] One computational theory that in principle can solve chess is exhaustive search. The real interest lies however in formulating the pieces of computation that we apply to the game. One presumably wants a computational theory that has a rather general application, together with a demonstration that it happens to be applicable to some class of games of chess, and evidence that we play games in this class.

[2] New algorithms for implementing a known computational theory may subsequently be devised without throwing substantial new light upon the theory, just as Winograd's (1976) Very Fast Fourier Transform shed no new light on the nature of Fourier analysis.

[3] The underlying point here is that there is often a natural modularity in physics (e.g. under normal conditions, electrical interactions are independent of gravitational interactions), but

predicting how a protein will fold. A large number of influences act on a large polypeptide chain as it flaps and flails in a medium. At each moment only a few of the possible interactions will be important, but the importance of those few is decisive. Attempts to construct a simplified theory must ignore some interactions; but if most interactions are crucial at some stage during the folding, a simplified theory will prove inadequate. Interestingly, the most promising studies of protein folding are currently those that take a brute-force approach, setting up a rather detailed model of the amino acids, the geometry associated with their sequence, hydrophobic interactions with the circumambient fluid, random thermal perturbations etc., and letting the whole set of processes run until a stable configuration is achieved (Levitt and Warshel 1975).

The principal difficulty in AI is that one can never be quite sure whether a problem has a Type-1 theory. If one is found, well and good; but failure to find one does not mean that it does not exist. Most AI programs have hitherto amounted to Type-2 theories, and the danger with such theories is that they can bury crucial decisions, that in the end provide the key to the correct Type-1 decomposition of the problem, beneath the mound of small administrative decisions that are inevitable whenever a concrete program is designed. This phenomenon makes research in AI difficult to pursue and difficult to judge. If one shows that a given information-processing problem is solved by a particular, neatly circumscribed computational theory, then that is a secure result. If on the other hand one produces a large and clumsy set of processes that solves a problem, one cannot always be sure that there isn't a simple underlying computational theory for one or more related problems whose formulation has somehow been lost in the fog. With any candidate for a Type-2 theory, much greater importance is attached to the performance of the program. Since its only possible virtue might be that it works, it is interesting only if it does. Often, a piece of AI research has resulted in a large program without much of a theory, which commits it to a Type-2 result, but that program either performs too poorly to be impressive or (worse still) has not even been implemented. Such pieces of research have to be judged very harshly, because their lasting contribution is negligible.

Thus we see that as AI pursues its study of information-processing problems, two types of solution are liable to emerge. In one, there is a clean underlying theory in the traditional sense. Examples of this from vision are Horn's (1975) method for obtaining shape from shading, the

some processes involve several at the same time and with roughly equal importance, like protein folding. Thus the Type-1–Type-2 distinction is not a pure dichotomy, and there is a spectrum of possibilities between them.

notion of the primal sketch as a representation of the intensity changes and local geometry of an image (Marr 1976), Ullman's (1976) method for detecting light sources, Binford's (1971) generalized cylinder representation, on which Marr and Nishihara's [1978] theory of the internal representation and manipulation of 3-D structures was based, a recent theory of stereo vision (Marr 1974; Marr and Poggio 1976)[4] and Poggio and Reichardt's (1976) analysis of the visual orienting behaviour of the housefly. One characteristic of these results is that they often lie at a relatively low level in the overall canvas of intellectual functions, a level often dismissed with contempt by those who purport to study 'higher, more central' problems of intelligence. Our reply to such criticism is that low-level problems probably do represent the easier kind, but that is precisely the reason for studying them first. When we have solved a few more, the questions that arise in studying the deeper ones will be clearer to us.

But even relatively clean Type-1 theories such as these involve Type-2 theories as well. For example, Marr and Nishihara's 3-D representation theory asserts that the deep underlying structure is based on a distributed, object-centred co-ordinate system that can be thought of as a stick figure, and that this representation is explicitly manipulated during the analysis of an image. Such a theory would be little more than speculation unless it could also be shown that such a description may be computed from an image and can be manipulated in the required way. To do so involves several intermediate theories, for some of which there is hope of eventual Type-1 status, but others look intractably of Type-2. For example, a Type-1 theory now exists for part of the problem of determining the appropriate local co-ordinate system from the contours formed in an object's image (Marr [1977]), but it may be impossible to derive a Type-1 theory for the basic grouping processes that operate on the primal sketch to help separate figure from ground. The figure–ground 'problem' may not be a single problem, being instead a mixture of several subproblems which combine to achieve figural separation, just as the different molecular interactions combine to cause a protein to fold. There is in fact no reason why a solution to the figure–ground problem should be derivable from a single underlying theory. The reason is that it needs to contain a procedural representation of many facts about images that derive ultimately via evolution from the cohesion and continuity of matter in the

---

[4] The notion of co-operative computation, or relaxation labeling (Zucker 1976), is a notion at the algorithm level. It suggests a way of implementing certain computations, but does not address the problem of what should be implemented, which seems to be the real issue for vision no less than elsewhere.

physical world. Many kinds of knowledge and different techniques are involved; one just has to sort them out one by one. As each is added the performance of the whole improves, and the complexity of the images that can be handled increases.

We have already seen that to search for a Type-2 theory for a problem may be dangerous if in fact it has a Type-1 theory. This danger is most acute in premature assaults on a high-level problem, for which few or none of the concepts that underlie its eventual Type-1 theory have yet been developed, and the consequence is a complete failure to formulate correctly the problems that are in fact involved. But it is equally important to realize that the opposite danger exists lower down. For example, in our current theory of visual processing, the notion of the primal sketch seems respectable enough, but one might have doubts about the aesthetics of the grouping processes that decode it. There are many of them, their details are somewhat messy; and seemingly arbitrary preferences occur (e.g. for vertical or horizontal organizations). A clear example of a Type-2 theory is our assertion that texture-vision discriminations rest on these grouping processes and first-order discriminations applied to the information held in primal sketch of the image (Marr 1976). As such, it is less attractive than Julesz's (1975) clean (Type-1) theory that textured regions are discriminable only if there is a difference in the first- or second-order statistics of their intensity arrays. But as Julesz himself found, there exist patterns with different second-order statistics that are nevertheless indiscriminable; and one can in fact view our own work as attempting to define precisely what characteristics of the second-order statistical structure cause discriminability (see Schatz 1977, in preparation).

This inevitably forces us to relinquish the beauty of Julesz's concise theory, but I feel that one should not be too distressed by the need at this level of investigation to explore rather messy and untidy details. We already know that separate modules must exist for computing other aspects of visual information—motion, stereoscopy, fluorescence, colour—and there is no reason why they should all be based on a single theory. Indeed one would a priori expect the opposite; as evolution progressed, new modules came into existence that could cope with yet more aspects of the data, and as a result kept the animal alive in ever more widely ranging circumstances. The only important constraint is that the system as a whole should be roughly modular, so that new facilities can be added easily.

So, especially at the more peripheral stages of sensory information-processing, and perhaps also more centrally, one should not necessarily give up if one fails to find a Type-1 theory—there may not be one. More importantly even if there were, there would be no reason why that theory should bear much relation to the theory of more central phenomena. In vision, for example, the theory that says 3-D representations are based on stick-figure co-ordinate systems and shows how to manipulate them, is independent of the theory of the primal sketch, or for that matter of most other stages *en route* from the image to that representation. In particular, it is especially dangerous to suppose that an approximate theory of a peripheral process has any significance for higher-level operations. For example, because Julesz's second-order statistics idea is so clean and so neatly fits much data, one might be tempted to ask whether the idea of second-order interactions is in some way central to higher processes. In doing so one should bear in mind that the true explanation of visual texture discrimination may be quite different in nature even if the theory is very often a correct predictor of performance.

The reason for drawing this point out at such length is that it bears upon another issue, namely the type of theory that the grammar of natural language might have. The purpose of human language is presumably to transform a data-structure that is not inherently one-dimensional into one-dimensional form for transmission as a sequential utterance, thereafter to be retranslated into some rough copy of the original in the head of the listener. Viewed in this light, it becomes entirely possible that there may exist no Type-1 theory of English syntax of the type that transformational grammar attempts to define—that its constraints resemble wired-in conventions about useful ways of executing this tedious but vital operation, rather than deep principles about the nature of intelligence. An abstract theory of syntax may be an illusion, approximating what really happens only in the sense that Julesz's second-order statistics theory approximates the behaviour of the set of processes that implement texture vision and which, in the final analysis, are all the theory that there is. In other words, the grammar of natural language may have a theory of Type-2 rather than of Type-1.

Even if a biological information-processing problem has only a Type-2 theory, it may still be possible to infer more from a solution to it than the solution itself. This comes about because at some point in the implementation of a set of processes, design constraints attached to the machine in which they will run start to affect the structure of the implementation. This observation adds a different perspective to the two types of research carried out by linguists and by members of the artificial intelligence community. If the theory of syntax is really of Type-2, then any important implications about the CNS are likely to come from details of the way in which its constituent processes are implemented, and these are often explorable only by implementing them.

1. IMPLICATIONS OF THIS VIEW

If one accepts this view of AI research, one is led to judge its achievements according to rather clear criteria. What information-processing problem has been isolated? Has a clean theory been developed for solving it, and if so how good are the arguments that support it? If no clean theory has been given what is the evidence that favours a set-of-processes solution or suggests that no single clean theory exists for it, and how well does the proposed set of mechanisms work? For very advanced problems like story-understanding, current research is often purely exploratory. That is to say, in these areas our knowledge is so poor that we cannot even begin to formulate the appropriate questions, let alone solve them. It is important to realize that this is an inevitable phase of any human endeavour, personally risky (almost surely no exploring pioneer will himself succeed in finding a useful question), but a necessary precursor of eventual success.

Most of the history of AI (now fully 16 years old) has consisted of exploratory studies. Some of the best-known are Slagle's (1963) symbolic integration program, Weizenbaum's (1965) Eliza program, Evans's (1968) analogy program, Raphael's (1968) SIR, Quillian's (1968) semantic nets, and Winograd's (1972) Shrdlu. All of these programs have (in retrospect) the property that they are either too simple to be interesting Type-1 theories, or very complex yet perform too poorly to be taken seriously as a Type-2 theory. Perhaps the only really successful Type-2 theory to emerge in the early phase of AI was Waltz's (1975) program. And yet many things have been learnt from these experiences—mostly negative things (the first 20 obvious ideas about how intelligence might work are too simple or wrong) but including several positive things. The MACSYMA algebraic manipulation system (Moses 1974) is undeniably successful and useful, and it had its roots in programs like Slagle's. The mistakes made in the field lay not in having carried out such studies—they formed an essential part of its development—but consisted mainly in failures of judgement about their value, since it is now clear that few of the early studies themselves formulated any solvable problems. Part of the reason for these internal failures of judgement lay in external pressures for early results from the field, but this is not the place to discuss what in the end are political matters.

Yet, I submit, one would err to judge these failures of judgement too harshly. They are merely the inevitable consequence of a necessary enthusiasm, based on a view of the long-term importance of the field that seems to me correct. All important fields of human endeavour start with a personal commitment based on faith rather than on results. AI is just one more example. Only a sour, crabbed, and unadventurous spirit will hold it against us.

2. CURRENT TRENDS

Exploratory studies are important. Many people in the field expect that, deep in the heart of our understanding of intelligence, there will lie at least one and probably several important principles about how to organize and represent knowledge that in some sense captures what is important about the *general* nature of our intellectual abilities. An optimist might see the glimmer of such principles in programs like those of Sussman and Stallman (1975), of Marr and Nishihara [1978], in the overall attitudes to central problems set out by Minsky (1975), and possibly in some of Schank's (1973, 1975) work, although I sometimes feel that he failed to draw out the important points. While still somewhat cloudy, the ideas that seem to be emerging (and which owe much to the early exploratory studies) are:

1. That the 'chunks' of reasoning, language, memory, and perception ought to be larger than most recent theories in psychology have allowed (Minsky 1975). They must also be very flexible—at least as flexible as Marr and Nishihara's stick-figure 3-D models, and probably more. Straightforward mechanisms that are suggested by the terms 'frame' and 'terminal', are certainly too inflexible.

2. That the perception of an event or of an object must include the simultaneous computation of several different descriptions of it, that capture diverse aspects of the use, purpose, or circumstances of the event or object.

3. That the various descriptions described in (2) include coarse versions as well as fine ones. These coarse descriptions are a vital link in choosing the appropriate overall scenarios demanded by (1), and in establishing correctly the roles played by the objects and actions that caused those scenarios to be chosen.

An example will help to make these points clear. If one reads:

(A) The fly buzzed irritatingly on the window-pane.

(B) John picked up the newspaper.

the immediate inference is that John's intentions towards the fly are fundamentally malicious. If he had picked up the telephone, the inference would be less secure. It is generally agreed that an 'insect-damaging' scenario is somehow deployed during the reading of these sentences, being suggested in its coarsest form by the fly buzzing irritatingly. Such a scenario will contain a reference to something that can squash an insect on a brittle surface—a description which fits a newspaper but not a telephone. We might therefore conclude that when the newspaper is mentioned (or in the case of vision, seen) not only is it described internally as a newspaper, and some rough 3-D description of its shape and axes set up, but it is also described as a light, flexible object with area. Because sentence (B) might have continued 'and sat down to read', the newspaper may also be being described as reading-matter; similarly, as a combustible article, and so forth. Since one does not usually know in advance what aspect of an object or action is important, it follows that most of the time, a given object will give rise to several different coarse internal descriptions. Similarly for actions. It may be important to note that the description of fly-swatting or reading or fire-lighting does not have to be attached to the newspaper; merely that a description of the newspaper is available that will match its role in each scenario.

The important thing about Schank's 'primitive actions' seems to me not the fact that there happens to be a certain small number of them, nor the idea that every act is expressed solely by reduction to them (which I cannot believe at all), nor even the idea that the scenarios to which they are attached contain all the answers for the present situation (that is where the missing flexibility comes in). The importance of a primitive, coarse catalogue of events and objects lies in the role such coarse descriptions play in the ultimate access and construction of perhaps exquisitely tailored specific scenarios, rather in the way that a general 3-D animal model in Marr and Nishihara's theory can finish up as a very specific Cheshire Cat, after due interaction between the image and information stored in the primitive model. What after sentence (A) existed as little more than a malicious intent towards the innocent fly becomes, with the additional information about the newspaper, a very specific case of fly-squashing.

Marr and Nishihara have labelled the problem of providing multiple-descriptions for the newspaper its 'reference-window problem'. Exactly how it is best done, and exactly what descriptions should accompany different words or perceived objects, is not yet known. These insights are the result of exploratory studies, and the problems to which they lead have yet to be precisely formulated, let alone satisfactorily solved. But it is now certain that some problems of this kind do exist and are important; and it seems likely that a fairly respectable theory of them will eventually emerge.

3. MIMICRY VERSUS EXPLORATION

Finally, I would like to draw one more distinction that seems to be important when choosing a research problem, or when judging the value of completed work. The problem is that studies—particularly of natural-language understanding, problem-solving, or the structure of memory—can easily degenerate into the writing of programs that do no more than mimic in an unenlightening way some small aspect of human performance. Weizenbaum (1976) now judges his program Eliza to belong to this category, and I have never seen any reason to disagree. More controversially, I would also criticize on the same grounds Newell and Simon's work on production systems, and some of Norman and Rumelhart's (1974) work on long-term memory.

The reason is this. If one believes that the aim of information-processing studies is to formulate and understand particular information-processing problems, then it is the structure of those problems that is central, not the mechanisms through which they are implemented. Therefore, the first thing to do is to find problems that we can solve well, find out how to solve them, and examine our performance in the light of that understanding. The most fruitful source of such problems is operations that we perform well, fluently (and hence unconsciously) since it is difficult to see how reliability could be achieved if there were no sound underlying method. On the other hand, problem-solving research has tended to concentrate on problems that we understand well intellectually but perform poorly on, like mental arithmetic and criptarithmatic or on problems like geometry theorem-proving, or games like chess, in which human skills seem to rest on a huge base of knowledge and expertise. I argue that these are exceptionally good grounds for *not* studying how we carry out such tasks yet. I have no doubt that when we do mental arithmetic we are doing *something* well, but it is not arithmetic, and we seem far from understanding even one component of what that something is. Let us therefore concentrate on the simpler problems first, for there we have some hope of genuine advancement.

If one ignores this stricture, one is left in the end with unlikely looking mechanisms whose only recommendation is that they cannot do something we cannot do. Production systems seem to me to fit this description quite well. Even taken on their own terms as mechanisms, they leave a lot to be desired. As a programming-language they are poorly designed, and hard to use, and I cannot believe that the human brain could possibly be burdened with such poor implementation decisions at so basic a level.

A parallel may perhaps be drawn between production systems for students of problem-solving, and Fourier analysis for visual neuro-physiologists. Simple operations on a spatial frequence representation of an image can mimic several interesting visual phenomena that seem to be exhibited by our visual systems. These include the detection of repetition, certain visual illusions, the notion of separate linearly adding channels, separation of overall shape from fine local detail, and a simple expression of size invariance. The reason why the spatial frequency domain is ignored by image analysts is that it is virtually useless for the main job of vision—building up a description of what is there from the intensity array. The intuition that visual physiologists lack, and which is so important, is for how this may be done. A production system exhibits several interesting ideas—the absence of explicit subroutine calls, a blackboard communication channel, and some notion of a short-term memory. But just because production systems display these side-effects (as Fourier analysis 'displays' some visual illusions) does not mean that they have anything to do with what is really going on. My own guess would be, for example, that the fact that short-term memory can act as a storage register is probably the least important of its functions. I expect that there are several 'intellectual reflexes' that operate on items held there, about which nothing is yet known, and which will eventually be held to be the crucial things about it because they perform central functions like opening up an item's reference window. Studying our performance in close relation to production systems seems to me a waste of time, because it amounts to studying a mechanism not a problem, and can therefore lead to no Type-1 results. The mechanisms that such research is trying to penetrate will be unravelled by studying problems, just as vision research is progressing because it is the *problem* of vision that is being attacked, not neural visual mechanisms.

A reflection of the same criticism can be made of Norman and Rumelhart's work, where they studied the way information seems to be organized in long-term memory. Again, the danger is that questions are not asked in relation to a clear information-processing problem. Instead, they are asked and answers proposed in terms of a mechanism—in this case, it is called an 'active structural network' and it is so simple and general as to be devoid of theoretical substance. They may be able to say that such and such an 'association' seems to exist, but they cannot say of what the association consists, nor that it has to be so because to solve problem X (which we can solve) you *need* a memory organized in such-and-such way; and that if one has it, certain apparent 'associations' occur as side-effects. Experimental psychology can do a valuable job in discovering facts that need explaining, including those about long-term memory, and the work of Shepard (1975), Rosch (in press), and of Warrington (1975) (for example) seems to me very successful at this; but like experimental neurophysiology, experimental psychology will not be able to explain those facts unless information-processing research has identified and solved the appropriate problems X.[5] It seems to me that finding such problems X, and solving them, is what AI should be trying to do.[6]

[5] In the present state of the art, it seems wisest to concentrate on problems that probably have Type-1 solutions, rather than on those that are almost certainly of Type-2.

[6] Although I take full responsibility for the purely personal views set out here, any virtues that they may have are due in part to many conversations with Drew McDermott. This report describes work done at the Artificial Intelligence Laboratory of the Massachusetts Institute of Technology. Support for the Laboratory's artificial intelligence research is provided by the Advanced Research Projects Agency of the Department of Defense under Office of Naval Research contract number N00014–75–C–0643.

REFERENCES

Binford, T. O (1971). 'Visual Perception by Computer.' IEEE Conf. Systems and Control, Miami.

Chomsky, A. N. (1965). *Aspects of the Theory of Syntax*. Cambridge, Mass.: MIT Press.

Cooley, J. M., and Tukey, J. W. (1965). 'An Algorithm for the Machine Computation of Complex Fourier Series.' *Math. Comp.* 19: 297–301.

Evans, T. (1968). 'A Program for the Solution of Geometric-Analogy Intelligence Test Questions.' In M. Minsky (ed.), *Semantic Information Processing*, pp. 271–353. Cambridge, Mass.: MIT Press.

Horn, B. K. P. (1975). 'Obtaining Shape from Shading Information.' In P. H. Winston (ed.), *The Psychology of Computer Vision*, pp. 115–55. New York: McGraw-Hill.

Jardine, N., and Sibson, R. (1971). *Mathematical Taxonomy*. New York: Wiley.

Julesz, B. (1975). 'Experiments in the Visual Perception of Texture.' *Scientific American* 232: 34–43.

Levitt, M., and Warshel, A. (1975). 'Computer Simulation of Protein Folding.' *Nature* 253: 694–8.

Marr, D. (1974). 'A Note on the Computation of Binocular Disparity in a Symbolic, Low-Level Visual Processor.' MIT AI Lab. Memo 327.

—— (1976). 'Early Processing of Visual Information' *Phil. Trans. Roy. Soc. B* 275: 483–524.

—— [1977]. 'Analysis of Occluding Contour.' [*Proc. Roy. Soc. London, B* 197: 441–75.]

—— and Nishihara, H. K. [1978]. 'Representation and Recognition of the Spatial Organization of Three Dimensional Shapes.' [*Proc. Roy. Soc. London, B* 200: 269–94.]

—— and Poggio, T. (1976). 'Cooperative Computation of Stereo Disparity.' *Science* 194: 283–7.

Minsky, M. (1975). 'A Framework for Representing Knowledge.' In P. H. Winston (ed.), *The Psychology of Computer Vision*, pp. 211–77. New York: McGraw-Hill.

Moses, J. (1974), 'MACSYMA—The Fifth Year.' *SIGSAM Bull., ACM* 8: 105–10.

Norman, D. A., and Rumelhart, D. E. (1974). 'The Active Structural Network.' In D. A. Norman and D. E. Rumelhart (eds.), *Explorations in Cognition*, pp. 35–64. San Francisco: W. H. Freeman and Co.

Poggio, T., and Reichardt, W. (1976). 'Visual Control of the Orientation Behavior of the Fly: Towards the Underlying Neural Interactions.' *Quarterly Reviews of Biophysics* 9: 377–438.

Quillian, M. R. (1968). 'Semantic Memory.' In M. Minsky (ed.), *Semantic Information Processing*, pp. 227–70. Cambridge, Mass.: MIT Press.

Raphael, B. (1968). 'SIR: Semantic Information Retrieval.' In M. Minsky (ed.), *Semantic Information Processing*, pp. 33–145. Cambridge, Mass.: MIT Press.

Rosch, E. (in press). 'Classification of Real-World Objects: Origins and Representations in Cognition.' *Bulletin de psychologie*.

Schank, R. C. (1973). 'Identification of Conceptualizations Underlying Natural Language.' In R. C. Schank and K. M. Colby (eds.), *Computer Models of Thought and Language*. San Francisco: W. H. Freeman.

—— (1975). *Conceptual Information Processing*. Amsterdam: North-Holland.

Schatz, B. R. (1977). 'The Computation of Immediate Texture Discrimination.' MIT AI Lab. Memo 426.

Shepard, R. N. (1975). 'Form, Formation, and Transformation of Internal Representations.' In R. Solso (ed.), *Information Processing and Cognition: The Loyola Symposium*, pp. 87–122. Hillsdale, NJ: Erlbaum.

Slagle, J. R. (1963). 'A Heuristic Program that Solves Symbolic Integration Problems in Freshman Calculus.' In E. A. Feigenbaum and J. Feldman (eds.), *Computers and Thought*, pp. 191–203. New York: McGraw-Hill.

Sussman, G. J., and Stallman, R. M. (1975). 'Heuristic Techniques in Computer-Aided Circuit Analysis.' *IEEE Transactions on Circuits and Systems*, CAS-22: 857–65.

Ullman, S. (1976). 'On Visual Detection of Light Sources.' *Biol. Cybernetics* 21: 205–12.

Waltz, D. L. (1975). 'Understanding Line Drawings of Scenes with Shadows.' In P. H. Winston (ed.), *The Psychology of Computer Vision*, pp. 19–91. New York: McGraw-Hill.

Warrington, E. K. (1975). 'The Selective Impairment of Semantic Memory.' *Q. J. Exp. Psychol.* 27: 635–57.

Weizenbaum, J. (1965). 'ELIZA—A Computer Program for the Study of Natural Language Communication between Man and Machine.' *Commun. ACM* 9: 36–45.

Weizenbaum, J. (1976). *Computer Thought and Human Reason*. San Francisco: W. H. Freeman.

Winograd, S. (1976). 'Computing the Discrete Fourier Transform.' *Proc. Nat. Acad. Sci.* 73: 1005–6.

Winograd, T. (1972). *Understanding Natural Language*. New York: Academic Press.

Zucker, S. W. (1976). 'Relaxation Labelling and the Reduction of Local Ambiguities.' University of Maryland Computer Science TR-451.

7

COGNITIVE WHEELS:
THE FRAME PROBLEM OF AI

DANIEL C. DENNETT

Once upon a time there was a robot, named R_1 by its creators. Its only task was to fend for itself. One day its designers arranged for it to learn that its spare battery, its precious energy supply, was locked in a room with a time bomb set to go off soon. R_1 located the room, and the key to the door, and formulated a plan to rescue its battery. There was a wagon in the room, and the battery was on the wagon, and R_1 hypothesized that a certain action which it called PULLOUT (WAGON, ROOM) would result in the battery being removed from the room. Straightaway it acted, and did succeed in getting the battery out of the room before the bomb went off. Unfortunately, however, the bomb was also on the wagon. R_1 *knew* that the bomb was on the wagon in the room, but didn't realize that pulling the wagon would bring the bomb out along with the battery. Poor R_1 had missed that obvious implication of its planned act.

Back to the drawing board. 'The solution is obvious,' said the designers. 'Our next robot must be made to recognize not just the intended implications of its acts, but also the implications about their side-effects, by deducing these implications from the descriptions it uses in formulating its plans.' They called their next model, the robot-deducer, R_1D_1. They placed R_1D_1 in much the same predicament that R_1 had succumbed to, and as it too hit upon the idea of PULLOUT (WAGON, ROOM) it began, as designed, to consider the implications of such a course of action. It had just finished deducing that pulling the wagon out of the room would not change the colour of the room's walls, and was embarking on a proof of the further implication that pulling the wagon out would cause its wheels to turn more revolutions than there were wheels on the wagon—when the bomb exploded.

Daniel C. Dennett, 'Cognitive Wheels: The Frame Problem of AI', in C. Hookway (ed.), *Minds, Machines, and Evolution: Philosophical Studies* (1984), pp. 129–51. Reprinted by permission of Cambridge University Press.

Back to the drawing board. 'We must teach it the difference between relevant implications and irrelevant implications,' said the designers, 'and teach it to ignore the irrelevant ones.' So they developed a method of tagging implications as either relevant or irrelevant to the project at hand, and installed the method in their next model, the robot-relevant-deducer, or R_2D_1 for short. When they subjected R_2D_1 to the test that had so unequivocally selected its ancestors for extinction, they were surprised to see it sitting, Hamlet-like, outside the room containing the ticking bomb, the native hue of its resolution sicklied o'er with the pale cast of thought, as Shakespeare (and more recently Fodor) has aptly put it. 'Do something!' they yelled at it. 'I am,' it retorted. 'I'm busily ignoring some thousands of implications I have determined to be irrelevant. Just as soon as I find an irrelevant implication, I put it on the list of those I must ignore, and . . .' the bomb went off.

All these robots suffer from the *frame problem*.[1] If there is ever to be a robot with the fabled perspicacity and real-time adroitness of R_2D_2, robot-designers must solve the frame problem. It appears at first to be at best an annoying technical embarrassment in robotics, or merely a curious puzzle for the bemusement of people working in Artificial Intelligence (AI). I think, on the contrary, that it is a new, deep epistemological problem—accessible in principle but unnoticed by generations of philosophers—brought to light by the novel methods of AI, and still far from being solved. Many people in AI have come to have a similarly high regard for the seriousness of the frame problem. As one researcher has quipped, 'We have given up the goal of designing an intelligent robot, and turned to the task of designing a gun that will destroy any intelligent robot that anyone else designs!'

I will try here to present an elementary, non-technical, philosophical introduction to the frame problem, and show why it is so interesting. I have no solution to offer, or even any original suggestions for where a solution might lie. It is hard enough, I have discovered, just to say clearly what the frame problem is—and is not. In fact, there is less than perfect agreement in usage within the AI research community. McCarthy and

[1] The problem is introduced by John McCarthy and Patrick Hayes in their 1969 paper. The task in which the problem arises was first formulated in McCarthy 1960. I am grateful to John McCarthy, Pat Hayes, Bob Moore, Zenon Pylyshyn, John Haugeland, and Bo Dahlbom for the many hours they have spent trying to make me understand the frame problem. It is not their fault that so much of their instruction has still not taken.

I have also benefited greatly from reading an unpublished paper, 'Modelling Change: The Frame Problem', by Lars-Erik Janlert, Institute of Information Processing, University of Umea, Sweden. It is to be hoped that a subsequent version of that paper will soon find its way into print, since it is an invaluable vade-mecum for any neophyte, in addition to advancing several novel themes.

Hayes, who coined the term, use it to refer to a particular, narrowly conceived problem about representation that arises only for certain strategies for dealing with a broader problem about real-time planning systems. Others call this broader problem the frame problem—'the whole pudding,' as Hayes has called it (personal correspondence)—and this may not be mere terminological sloppiness. If 'solutions' to the narrowly conceived problem have the effect of driving a (deeper) difficulty into some other quarter of the broad problem, we might better reserve the title for this hard-to-corner difficulty. With apologies to McCarthy and Hayes for joining those who would appropriate their term, I am going to attempt an introduction to the whole pudding, calling *it* the frame problem. I will try in due course to describe the narrower version of the problem, 'the frame problem proper' if you like, and show something of its relation to the broader problem.

Since the frame problem, whatever it is, is certainly not solved yet (and may be, in its current guises, insoluble), the ideological foes of AI such as Hubert Dreyfus and John Searle are tempted to compose obituaries for the field, citing the frame problem as the cause of death. In *What Computers Can't do* (Dreyfus 1972), Dreyfus sought to show that AI was a fundamentally mistaken method for studying the mind, and in fact many of his somewhat impressionistic complaints about AI models and many of his declared insights into their intrinsic limitations can be seen to hover quite systematically in the neighbourhood of the frame problem. Dreyfus never explicitly mentions the frame problem,[2] but is it perhaps the smoking pistol he was looking for but didn't *quite* know how to describe? Yes, I think AI can be seen to be holding a smoking pistol, but at least in its 'whole pudding' guise it is everyone's problem, not just a problem for AI, which, like the good guy in many a mystery story, should be credited with a discovery, not accused of a crime.

One does not have to hope for a robot-filled future to be worried by the frame problem. It apparently arises from some very widely held and

[2] Dreyfus mentions McCarthy 1960: 213–14, but the theme of his discussion there is that McCarthy ignores the difference between a *physical state* description and a *situation* description, a theme that might be succinctly summarized: a house is not a home.

Similarly, he mentions *ceteris paribus* assumptions (in the introduction to the rev. ed., pp. 56 ff.), but only in announcing his allegiance to Wittgenstein's idea that 'whenever human behaviour is analyzed in terms of rules, these rules must always contain a *ceteris paribus* condition . . .' But this, even if true, misses the deeper point: the need for something like *ceteris paribus* assumptions confronts Robinson Crusoe just as ineluctably as it confronts any protagonist who finds himself in a situation involving human culture. The point is not, it seems, restricted to *Geisteswissenschaft* (as it is usually conceived); the 'intelligent' robot on an (otherwise?) uninhabited but hostile planet faces the frame problem as soon as it commences to plan its days.

innocuous-*seeming* assumptions about the nature of intelligence, the truth of the most undoctrinaire brand of physicalism, and the conviction that it must be possible to explain how we think. (The dualist evades the frame problem—but only because dualism draws the veil of mystery and obfuscation over all the tough how-questions; as we shall see, the problem arises when one takes seriously the task of answering certain how-questions. Dualists inexcusably excuse themselves from the frame problem.)

One utterly central—if not defining—feature of an intelligent being is that it can 'look before it leaps'. Better, it can *think* before it leaps. Intelligence is (at least partly) a matter of using well what you know—but for what? For improving the fidelity of your expectations about what is going to happen next, for planning, for considering courses of action, for framing further hypotheses with the aim of increasing the knowledge you will use in the future, so that you can preserve yourself, by letting your hypotheses die in your stead (as Sir Karl Popper once put it). The stupid—as opposed to ignorant—being is the one who lights the match to peer into the fuel tank,[3] who saws off the limb he is sitting on, who locks his keys in his car and then spends the next hour wondering how on earth to get his family out of the car.

But when we think before we leap, *how do we do it?* The answer seems obvious: an intelligent being learns from experience, and then uses what it has learned to guide expectation in the future. Hume explained this in terms of habits of expectation, in effect. *But how do the habits work?* Hume had a hand-waving answer—associationism—to the effect that certain transition paths between ideas grew more likely-to-be-followed as they became well worn, but since it was not *Hume's* job, surely, to explain in more detail the mechanics of these links, problems about how such paths could be put to good use—and not just turned into an impenetrable maze of untraversable alternatives—were not discovered.

Hume, like virtually all other philosophers and 'mentalistic' psychologists, was unable to see the frame problem because he operated at what I call a purely semantic level, or a *phenomenological* level. At the phenomenological level, all the items in view are *individuated by their meanings*. Their meanings are, if you like, 'given'—but this just means that the theorist helps himself to all the meanings he wants. In this way the semantic relation between one item and the next is typically plain to see, and one just assumes that the items behave as items with those meanings *ought* to behave. We can bring this out by concocting a Humean account of a bit of learning.

The example is from an important discussion of rationality by Christopher Cherniak [1983] in 'Rationality and the Structure of Memory'.

Suppose that there are two children, both of whom initially tend to grab cookies from the jar without asking. One child is allowed to do this unmolested but the other is spanked each time she tries. What is the result? The second child learns not to go for the cookies. Why? Because she has had experience of cookie-reaching followed swiftly by spanking. What good does that do? Well, the *idea* of cookie-reaching becomes connected by a habit path to the idea of spanking, which in turn is connected to the idea of pain . . . so *of course* the child refrains. Why? Well, that's just the effect of that idea on that sort of circumstance. But why? Well, what else ought the idea of pain to do on such an occasion? Well, it might cause the child to pirouette on her left foot, or recite poetry, or blink, or recall her fifth birthday. But given what the idea of pain *means*, any of those effects would be absurd. True; now *how* can ideas be designed so that their effects are what they ought to be, given what they mean? Designing some internal things — an idea, let's call it — so that it behaves *vis-à-vis* its brethren as if it meant *cookie* or *pain* is the only way of endowing that thing with that meaning; it couldn't mean a thing if it didn't have those internal behavioural dispositions.

That is the mechanical question the philosophers left to some dimly imagined future researcher. Such a division of labour might have been all right, but it is turning out that most of the truly difficult and deep puzzles of learning and intelligence get kicked downstairs by this move. It is rather as if philosophers were to proclaim themselves expert explainers of the methods of a stage magician, and then, when we ask them to explain how the magician does the sawing-the-lady-in-half trick, they explain that it is really quite obvious: the magician doesn't really saw her in half; he simply makes it appear that he does. 'But how does he do *that*?' we ask. 'Not our department', say the philosophers — and some of them add, sonorously: 'Explanation has to stop somewhere.'[4]

When one operates at the purely phenomenological or semantic level, where does one get one's data, and how does theorizing proceed? The term 'phenomenology' has traditionally been associated with an introspective method — an *examination* of what is presented or given to consciousness. A person's phenomenology just was by definition the contents of his or her consciousness. Although this has been the ideology all along, it has never been the practice. Locke, for instance, may have thought his 'historical, plain method' was a method of unbiased self-observation, but in fact it was largely a matter of disguised aprioristic reasoning about what ideas

---

[4] Note that on this unflattering portrayal, the philosophers might still be doing *some* valuable work; think of the wild goose chases one might avert for some investigator who had rashly concluded that the magician really did saw the lady in half and then miraculously reunite her. People have jumped to such silly conclusions, after all; many philosophers have done so, for instance.

and impressions *had to be* to do the jobs they 'obviously' did.[5] The myth that each of us can observe our mental activities has prolonged the illusion that major progress could be made on the theory of thinking by simply reflecting carefully on our own cases. For some time now we have known better: we have conscious access to only the upper surface, as it were, of the multi-level system of information-processing that occurs in us. Nevertheless, the myth still claims its victims.

So the analogy of the stage magician is particularly apt. One is not likely to make much progress in figuring out *how* the tricks are done by simply sitting attentively in the audience and watching like a hawk. Too much is going on out of sight. Better to face the fact that one must either rummage around backstage or in the wings, hoping to disrupt the performance in telling ways; or, from one's armchair, think aprioristically about how the tricks *must* be done, given whatever is manifest about the constraints. The frame problem is then rather like the unsettling but familiar 'discovery' that so far as armchair thought can determine, a certain trick we have just observed is flat impossible.

Here is an example of the trick. Making a midnight snack. How is it that I can get myself a midnight snack? What could be simpler? I suspect there is some leftover sliced turkey and mayonniase in the fridge, and bread in the breadbox—and a bottle of beer in the fridge as well. I realize I can put these elements together, so I concoct a childishly simple plan: I'll just go and check out the fridge, get out the requisite materials, and make myself a sandwich, to be washed down with a beer. I'll need a knife, a plate, and a glass for the beer. I forthwith put the plan into action and it works! Big deal.

Now of course I couldn't do this without knowing a good deal—about bread, spreading mayonniase, opening the fridge, the friction and inertia that will keep the turkey between the bread slices and the bread on the plate as I carry the plate over to the table beside my easy chair. I also need to know about how to get the beer out of the bottle into the glass.[6] Thanks to my previous accumulation of experience in the world, fortunately, I am equipped with all this worldly knowledge. Of course some of the knowledge I need *might* be innate. For instance, one trivial thing I have to know is that when the beer gets into the glass it is no longer in the bottle, and that if I'm holding the mayonnaise jar in my left hand I cannot also be spreading the mayonnaise with the knife in my left hand. Perhaps these

---

[5] See my 1982*a*, a commentary on Goodman 1982.

[6] This knowledge of physics is not what one learns in school, but in one's crib. See Hayes 1978, 1979.

are straightforward implications—instantiations—of some more fundamental things that I was in effect *born knowing* such as, perhaps, the fact
that if something is in one location it isn't also in another, different
location; or the fact that two things can't be in the same place at the same
time; or the fact that situations change as the result of actions. It is hard to
imagine just how one could learn these facts from experience.

Such utterly banal facts escape our notice as we act and plan, and it is
not surprising that philosophers, thinking phenomenologically *but introspectively*, should have overlooked them. But if one turns one's back on
introspection, and just thinks 'hetero-phenomenologically'[7] about the
purely informational demands of the task—what *must* be known by any
entity that can perform this task—these banal bits of knowledge rise to
our attention. We can easily satisfy ourselves that no agent that did not *in
some ways* have the benefit of the information (that beer in the bottle is
not in the glass, etc.) could perform such a simple task. It is one of
the chief methodological beauties of AI that it makes one be a
phenomenologist in this improved way. As a hetero-phenomenologist,
one reasons about what the agent must 'know' or figure out *unconsciously
or consciously* in order to perform in various ways.

The reason AI forces the banal information to the surface is that the
tasks set by AI start at zero: the computer to be programmed to simulate
the agent (or the brain of the robot, if we are actually going to operate in
the real, non-simulated world), initially knows nothing at all 'about the
world'. The computer is the fabled *tabula rasa* on which every required
item must somehow be impressed, either by the programmer at the outset
or via subsequent 'learning' by the system.

We can all agree, today, that there could be no learning at all by an
entity that faced the world at birth as a *tabula rasa*, but the dividing line
between what is innate and what develops maturationally and what is
actually learned is of less theoretical importance than one might have
thought. While some information has to be innate, there is hardly any
particular item that must be: an appreciation of *modus ponens*, perhaps,
and the law of the excluded middle, and some sense of causality. And
while some things we know must be learned—e.g. that Thanksgiving falls
on a Thursday, or that refrigerators keep food fresh—many other 'very
empirical' things could in principle be innately known—e.g. that smiles
mean happiness, or that unsuspended, unsupported things fall. (There is

---

[7] For elaborations of hetero-phenomenology, see Dennett 1978, ch. 10, 'Two Approaches to
Mental Images', and Dennett 1982b. See also Dennett 1982c.

some evidence, in fact, that there is an innate bias in favour of perceiving things to fall with gravitational acceleration.)[8]

Taking advantage of this advance in theoretical understanding (if that is what it is), people in AI can frankly ignore the problem of learning (it seems) and take the shortcut of *installing* all that an agent has to 'know' to solve a problem. After all, if God made Adam as an adult who could presumably solve the midnight snack problem *ab initio*, AI agent-creators can *in principle* make an 'adult' agent who is equipped with worldly knowledge *as if* it had laboriously learned all the things it needs to know. This may of course be a dangerous short cut.

The installation problem is then the problem of installing in one way or another all the information needed by an agent to plan in a changing world. It is a difficult problem because the information must be installed in a usable format. The problem can be broken down initially into the semantic problem and the syntactic problem. The semantic problem—called by Allen Newell the problem at the 'knowledge level' (Newell 1982)—is the problem of just what information (on what topics, to what effect) must be installed. The syntactic problem is what system, format, structure, or mechanism to use to put that information in.[9]

The division is clearly seen in the example of the midnight snack problem. I *listed* a few of the very many humdrum facts one needs to know to solve the snack problem, but I didn't mean to suggest that those facts are stored in me—or in any agent—piecemeal, in the form of a long list of sentences explicitly declaring each of these facts for the benefit of the agent. That is of course one possibility, officially: it is a preposterously extreme version of the 'language of thought' theory of mental representation, with each distinguishable 'proposition' separately inscribed in the system. No one subscribes to such a view; even an encyclopedia achieves

---

[8] Gunnar Johannsen has shown that animated films of 'falling' objects in which the moving spots drop with the normal acceleration of gravity are unmistakeably distinguished by the casual observer from 'artificial' motions. I do not know whether infants have been tested to see if they respond selectively to such displays.

[9] McCarthy and Hayes (1969) draw a different distinction between the 'epistemological' and the 'heuristic'. The difference is that they include the question 'In what kind of internal notation is the system's knowledge to be expressed?' in the epistemological problem (see p. 466), dividing off *that* syntactic (and hence somewhat mechanical) question from the procedural questions of the design of 'the mechanism that on the basis of the information solves the problem and decides what to do'.

One of the prime grounds for controversy about just which problem the frame problem is springs from this attempted division of the issue. For the answer to the syntactical aspects of the epistemological question makes a large difference to the nature of the heuristic problem. After all, if the syntax of the expression of the system's knowledge is sufficiently perverse, then in spite of the *accuracy* of the representation of that knowledge, the heuristic problem will be impossible. And some have suggested that the heuristic problem would virtually disappear if the world knowledge were felicitously couched in the first place.

important economies of explicit expression via its organization, and a walking encyclopedia—not a bad caricature of the envisaged AI agent—must use different systemic principles to achieve efficient representation and access. We know trillions of things; we know that mayonnaise doesn't dissolve knives on contact, that a slice of bread is smaller than Mount Everest, that opening the refrigerator doesn't cause a nuclear holocaust in the kitchen.

There must be in us—and in any intelligent agent—some highly efficient, partly generative or productive system of representing—storing for use—all the information needed. Somehow, then, we must store many 'facts' at once—where facts are presumed to line up more or less one-to-one with non-synonymous declarative sentences. Moreover, we cannot realistically hope for what one might call a Spinozistic solution—a *small* set of axioms and definitions from which all the rest of our knowledge is deducible on demand—since it is clear that there simply are no entailment relations between vast numbers of these facts. (When we rely, as we must, on experience to tell us how the world is, experience tells us things that do not at all follow from what we have heretofore known.)

The demand for an efficient system of information storage is in part a space limitation, since our brains are not all that large, but more importantly it is a time limitation, for stored information that is not reliably accessible for use in the short real-time spans typically available to agents in the world is of no use at all. A creature that can solve any problem given enough time—say a million years—is not in fact intelligent at all. We live in a time-pressured world and must be able to think quickly before we leap. (One doesn't have to view this as an a priori condition on intelligence. One can simply note that we do in fact think quickly, so there is an empirical question about how we manage to do it.)

The task facing the AI researcher appears to be designing a system that can plan by using well-selected elements from its store of knowledge about the world it operates in. 'Introspection' on how *we* plan yields the following description of a process: one envisages a certain situation (often very sketchily); one then imagines performing a certain act in that situation; one then 'sees' what the likely outcome of that envisaged act in that situation would be, and evaluates it. What happens backstage, as it were, to permit this 'seeing' (and render it as reliable as it is) is utterly inaccessible to introspection.

On relatively rare occasions we all experience such bouts of thought, unfolding in consciousness at the deliberate speed of pondering. These are occasions in which we are faced with some novel and relatively difficult problem, such as: How can I get the piano upstairs? or Is there any way to electrify the chandelier without cutting through the plaster ceiling? It would be quite odd to find that one had to think *that* way (consciously and slowly) in order to solve the midnight snack problem. But the suggestion is that even the trivial problems of planning and bodily guidance that are beneath our notice (though in some sense we 'face' them) are solved by similar processes. Why? I don't *observe* myself planning in such situations. This fact suffices to convince the traditional, introspective phenomenologist that no such planning is going on.[10] The hetero-phenomenologist, on the other hand, reasons that *one way or another* information about the objects in the situation, and about the intended effects and side-effects of the candidate actions, *must* be used (considered, attended to, applied, appreciated). Why? Because otherwise the 'smart' behaviour would be sheer luck or magic. (Do we have any model for how such unconscious information-appreciation might be accomplished? The only model we have *so far* is *conscious*, deliberate information-appreciation. Perhaps, AI suggests, this is a good model. If it isn't, we are all utterly in the dark for the time being.)

We assure ourselves of the intelligence of an agent by considering counterfactuals: if I had been told that the turkey was poisoned, or the beer explosive, or the plate dirty, or the knife too fragile to spread mayonnaise, would I have acted as I did? If I were a stupid 'automaton'—or like the *Sphex* wasp who 'mindlessly' repeats her stereotyped burrow-checking routine till she drops[11]—I might infelicitously 'go through the motions' of

[10] Such observations also convinced Gilbert Ryle, who was, in an important sense, an intro-spective phenomenologist (and not a 'behaviourist'). See Ryle 1949.
One can readily imagine Ryle's attack on AI: 'And *how many* inferences do I perform in the course of performing my sandwich? What syllogisms convince me that the beer will stay in the glass?' For a further discussion of Ryle's skeptical arguments and their relation to cognitive science, see my 'Styles of Mental Representation', Dennett 1983.
[11] 'When the time comes for egg laying the wasp *Sphex* builds a burrow for the purpose and seeks out a cricket which she stings in such a way as to paralyze but not kill it. She drags the cricket into her burrow, lays her eggs alongside, closes the burrow, then flies away, never to return. In due course, the eggs hatch and the wasp grubs feed off the paralyzed cricket, which has not decayed, having been kept in the wasp equivalent of deep freeze. To the human mind, such an elaborately organized and seemingly purposeful routine conveys a convincing flavour of logic and thoughtfulness—until more details are examined. For example, the wasp's routine is to bring the paralyzed cricket to the burrow, leave it on the threshold, go inside to see that all is well, emerge, and then drag the cricket in. If, while the wasp is inside making her preliminary inspection the cricket is moved a few inches away, the wasp, on emerging from the burrow, will bring the cricket back to the threshold, but not inside, and will then repeat the preparatory procedure of entering the burrow to see that everything is all right. If again the cricket is removed a few inches while the wasp is inside, once again the wasp will move the cricket up to the threshold and re-enter the burrow for a final check. The wasp never thinks of pulling the cricket straight in. On one occasion, this procedure was repeated forty times, always with the same result' (Wooldridge 1963).
This vivid example of a familiar phenomenon among insects is discussed by me in *Brainstorms*, and in Hofstadter 1982.

making a midnight snack oblivious to the recalcitrant features of the environment.[12] But in fact, my midnight-snack-making behaviour is multifariously sensitive to current and background information about the situation. The only way it could be so sensitive—runs the tacit hetero-phenomenological reasoning—is for it to examine, or test for, the information in question. The information manipulation may be unconscious and swift, and it need not (it *better* not) consist of hundreds or thousands of *seriatim* testing procedures, but it must occur somehow, and its benefits must appear in time to help me as I commit myself to action.

I may of course have a midnight snack routine, developed over the years, in which case I can partly rely on it to pilot my actions. Such a complicated 'habit' would have to be under the control of a mechanism of some complexity, since even a rigid sequence of steps would involve periodic testing to ensure that subgoals had been satisfied. And even if I am an infrequent snacker, I no doubt have routines for mayonnaise-spreading, sandwich-making, and getting-something-out-of-the-fridge, from which I could compose my somewhat novel activity. Would such ensembles of routines, nicely integrated, suffice to solve the frame problem for me, at least in my more 'mindless' endeavours? That is an open question to which I will return below.

It is important in any case to acknowledge at the outset, and remind oneself frequently, that even very intelligent people do make mistakes; we are not only not infallible planners; we are quite prone to overlooking large and retrospectively obvious flaws in our plans. This foible manifests itself in the familiar case of 'force of habit' errors (in which our stereotypical routines reveal themselves to be surprisingly insensitive to some portentous environmental changes while surprisingly sensitive to others). The same weakness also appears on occasion in cases where we have consciously deliberated with some care. How often have you embarked on a project of the piano-moving variety—in which you've thought through or even 'walked through' the whole operation in advance—only to discover that you must backtrack or abandon the project when some perfectly foreseeable but unforeseen obstacle or unintended side-effect loomed? If we smart folk seldom actually paint ourselves into corners, it may be not because we plan ahead so well as that we supplement our sloppy planning powers with a combination of recollected lore (about fools who paint themselves into corners, for instance) and frequent progress checks as we proceed. Even so, we must know enough to call up the right lore at the right time, and to recognize impending problems as such.

---

[12] See my 1982c: 58–9, on 'Robot Theater'.

158   DANIEL C. DENNETT

To summarise: we have been led by fairly obvious and compelling considerations to the conclusion that an intelligent agent must engage in swift
information-sensitive 'planning' which has the effect of producing reliable
but not foolproof expectations of the effects of its actions. That these
expectations are normally in force in intelligent creatures is testified to by
the startled reaction they exhibit when their expectations are thwarted.
This suggests a graphic way of characterizing the minimal goal that can
spawn the frame problem: we want a midnight-snack-making robot to be
'surprised' by the trick plate, the unspreadable concrete mayonnaise, the
fact that we've glued the beer glass to the shelf. To be surprised you have
to have expected something else, and in order to have expected the right
something else, you have to have *and use* a lot of information about the
things in the world.[13]

The central role of expectation has led some to conclude that the frame
problem is not a new problem at all, and has nothing particularly to do
with planning actions. It is, they think, simply the problem of having good
expectations about any future events, whether they are one's own actions,
the actions of another agent, or the mere happenings of nature. That is the
problem of induction—noted by Hume and intensified by Goodman
(Goodman 1965), but still not solved to anyone's satisfaction. We know
today that the problem of induction is a nasty one indeed. Theories of
subjective probability and belief fixation have not stabilized in reflective
equilibrium, so it is fair to say that no one has a good, principled answer to
the general question: given that I believe all *this* (have all this evidence),
what *ought* I to believe as well (about the future, or about unexamined
parts of the world)?

The reduction of one unsolved problem to another is some sort of
progress, unsatisfying though it may be, but it is not an option in this case.
The frame problem is not the problem of induction in disguise. For
suppose the problem of induction were solved. Suppose—perhaps
miraculously—that our agent has solved all its induction problems or had
them solved by fiat; it believes, then, all the right generalizations from its
evidence, and associates with all of them the appropriate probabilities and

[13] Hubert Dreyfus has pointed out that *not expecting x* does not imply *expecting y* (where x ≠
y), so one can be startled by something one didn't expect without it having to be the case that
one (unconsciously) expected something else. But this sense of *not expecting* will not suffice to
explain startle. What are the odds against your seeing an Alfa Romeo, a Buick, a Chevrolet,
and a Dodge parked in alphabetical order some time or other within the next five hours? Very
high, no doubt, all things considered, so I would not expect you to expect this; I also would not
expect you to be startled by seeing this unexpected sight—except in the sort of special case
where you had reason to expect something else at that time and place.

Startle reactions are powerful indicators of cognitive state—a fact long known by the police
(and writers of detective novels). *Only* someone who expected the refrigerator to contain
Smith's corpse (say) would be *startled* (as opposed to mildly interested) to find it to contain the
rather unlikely trio: a bottle of vintage Chablis, a can of cat food, and a dishrag.

conditional probabilities. This agent, *ex hypothesi*, believes just what it ought to believe about all empirical matters in its ken, including the probabilities of future events. It might still have a bad case of the frame problem, for that problem concerns how to represent (so it can be *used*) all that hard-won empirical information—a problem that arises independently of the truth value, probability, warranted assertability, or subjective certainty of any of it. Even if you have excellent *knowledge* (and not mere belief) about the changing world, how can this knowledge be represented so that it can be efficaciously brought to bear?

Recall poor R_1D_1, and suppose for the sake of argument that it had perfect empirical knowledge of the probabilities of all the effects of all its actions that would be detectable by it. Thus it believes that with probability 0.7864, executing PULLOUT (WAGON, ROOM) will cause the wagon wheels to make an audible noise; and with probability 0.5, the door to the room will open in rather than out; and with probability 0.999996, there will be no live elephants in the room, and with probability 0.997 the bomb will remain on the wagon when it is moved. How is R_1D_1 to find this last, relevant needle in its haystack of empirical knowledge? A walking encyclopedia will walk over a cliff, for all its knowledge of cliffs and the effects of gravity, unless it is designed in such a fashion that it can find the right bits of knowledge at the right times, so it can plan its engagements with the real world.

The earliest work on planning systems in AI took a deductive approach. Inspired by the development of Robinson's methods of resolution theorem proving, designers hoped to represent all the system's 'world knowledge' explicitly as axioms, and use ordinary logic—the predicate calculus—to deduce the effects of actions. Envisaging a certain situation S was modelled by having the system entertain a set of axioms describing the situation. Added to this were background axioms (the so-called 'frame axioms' that give the frame problem its name) which describe general conditions and the general effects of every action type defined for the system. To this set of axioms the system would apply an action—by postulating the occurrence of some action A in situation S—and then deduce the effect of A in S, producing a description of the outcome situation S'. While all this logical deduction looks like nothing at all in our conscious experience, research on the deductive approach could proceed on either or both of two enabling assumptions: the methodological assumption that psychological realism was a gratuitous bonus, not a goal, of 'pure' AI, or the substantive (if still vague) assumption that the deductive processes described would somehow model the backstage processes beyond conscious access. In other words, either we don't do our thinking deductively in the predicate calculus but a robot might; or we do (unconsciously) think deductively in the predicate calculus. Quite aside from doubts about its psychological realism, however, the deductive approach has not been made to work—the proof of the pudding for any robot— except for deliberately trivialized cases.

Consider some typical frame axioms associated with the action-type: *move x onto y.*

   (1) If $z \neq x$ and I move x onto y, then if z was on w before, then z is on w after.

   (2) If x is blue before, and I move x onto y, then x is blue after.

Note that (2), about being blue, is just one example of the many boring 'no-change' axioms we have to associate with this action-type. Worse still, note that a cousin of (2), also about being blue, would have to be associated with every other action-type—with *pick up x* and with *give x to y*, for instance. One cannot save this mindless repetition by postulating once and for all something like

   (3) If anything is blue, it stays blue,

for that is false, and in particular we will want to leave room for the introduction of such action-types as *paint x red.* Since virtually any aspect of a situation can change under some circumstance, this method requires introducing for each aspect (each predication in the description of S) an axiom to handle whether that aspect changes for each action-type.

This representational profligacy quickly gets out of hand, but for some 'toy' problems in AI, the frame problem can be overpowered to some extent by a mixture of the toyness of the environment and brute force. The early version of SHAKEY, the robot at SRI, operated in such a simplified and sterile world, with so few aspects it could worry about that it could get away with an exhaustive consideration of frame axioms.[14]

Attempts to circumvent this explosion of axioms began with the proposal that the system operate on the tacit assumption that nothing changes in a situation but what is explicitly asserted to change in the definition of the applied action (Fikes and Nilsson 1971). The problem here is that, as Garrett Hardin once noted, you don't do just one thing. This was R_1's problem, when it failed to notice that it would pull the bomb out with the wagon. In the explicit representation (a few pages back) of my midnight snack solution, I mentioned carrying the plate over to the table. On this proposal, my model of S' would leave the turkey back in the kitchen, for I didn't explicitly say the turkey would come along with the

---

[14] This early feature of SHAKEY was drawn to my attention by Pat Hayes. See also Dreyfus 1972: 26. SHAKEY is put to quite different use in Dennett 1982b.

plate. One can of course patch up the definition of 'bring' or 'plate' to handle this problem, but only at the cost of creating others. (Will a few more patches tame the problem? At what point should one abandon patches and seek an altogether new approach? Such are the methodological uncertainties regularly encountered in this field, and of course no one can responsibly claim in advance to have a good rule for dealing with them. Premature counsels of despair or calls for revolution are as clearly to be shunned as the dogged pursuit of hopeless avenues; small wonder the field is contentious.)

While one cannot get away with the tactic of supposing that one can do just one thing, it remains true that very little of what could (logically) happen in any situation does happen. Is there some way of fallibly marking the likely area of important side-effects, and assuming the rest of the situation to stay unchanged? Here is where relevance tests seem like a good idea, and they may well be, but not within the deductive approach. As Minsky notes:

> Even if we formulate relevancy restrictions, logistic systems have a problem using them. In any logistic system, all the axioms are necessarily 'permissive'—they all help to permit new inferences to be drawn. Each added axiom means more theorems; none can disappear. There simply is no direct way to add information to tell such a system about kinds of conclusions that should *not* be drawn! . . . If we try to change this by adding axioms about relevancy, we still produce all the unwanted theorems, plus annoying statements about their irrelevancy (Minsky 1981: 125).

What is needed is a system that genuinely *ignores* most of what it knows, and operates with a well-chosen portion of its knowledge at any moment. Well-chosen, but not chosen by exhaustive consideration. How, though, can you give a system *rules* for ignoring—or better, since explicit rule-following is not the problem, how can you design a system that reliably ignores what it ought to ignore under a wide variety of different circumstances in a complex action environment?

John McCarthy calls this the qualification problem, and vividly illustrates it via the famous puzzle of the missionaries and the cannibals.

> Three missionaries and three cannibals come to a river. A rowboat that seats two is available. If the cannibals ever outnumber the missionaries on either bank of the river, the missionaries will be eaten. How shall they cross the river?
> Obviously the puzzler is expected to devise a strategy of rowing the boat back and forth that gets them all across and avoids disaster . . .
> Imagine giving someone the problem, and after he puzzles for a while, he suggests going upstream half a mile and crossing on a bridge. 'What bridge?' you say. 'No bridge is mentioned in the statement of the problem.' And this dunce replies, 'Well, they don't say there isn't a bridge.' You look at the English and even at the translation of the English into first order logic, and you must admit that 'they don't

say' there is no bridge. So you modify the problem to exclude bridges and pose it again, and the dunce proposes a helicopter, and after you exclude that, he proposes a winged horse or that the others hang onto the outside of the boat while two row.

You now see that while a dunce, he is an inventive dunce. Despairing of getting him to accept the problem in the proper puzzler's spirit, you tell him the solution. To your further annoyance, he attacks your solution on the grounds that the boat might have a leak or lack oars. After you rectify that omission from the statement of the problem, he suggests that a sea monster may swim up the river and may swallow the boat. Again you are frustrated, and you look for a mode of reasoning that will settle his hash once and for all (McCarthy 1980: 29–30).

What a normal, intelligent human being does in such a situation is to engage in some form of *non-monotonic inference*. In a classical, monotonic logical system, *adding* premises never *diminishes* what can be proved from the premises. At Minsky noted, the axioms are essentially permissive, and once a theorem is permitted, adding more axioms will never invalidate the proofs of earlier theorems. But when we think about a puzzle or a real-life problem, we can achieve a solution (and even prove that it is a solution, or even the only solution to *that* problem), and then discover our solution invalidated by the addition of a new element to the posing of the problem; e.g. 'I forgot to tell you—there are no oars' or 'By the way, there's a perfectly good bridge upstream.'

What such late additions show us is that, contrary to our assumption, other things weren't equal. We had been reasoning with the aid of a *ceteris paribus* assumption, and now our reasoning has just been jeopardized by the discovery that something 'abnormal' is the case. (Note, by the way, that the abnormality in question is a much subtler notion than anything anyone has yet squeezed out of probability theory. As McCarthy notes, 'The whole situation involving cannibals with the postulated properties cannot be regarded as having a probability, so it is hard to take seriously the conditional probability of a bridge given the hypothesis' (ibid.).)

The beauty of a *ceteris paribus* clause in a bit of reasoning is that one does not have to say exactly what it means. 'What do you mean, "other things being equal"? Exactly which arrangements of which other things count as being equal?' If one had to answer such a question, invoking the *ceteris paribus* clause would be pointless, for it is precisely in order to evade that task that one uses it. If one could answer that question, one wouldn't need to invoke the clause in the first place. One way of viewing the frame problem, then, is as the attempt to get a computer to avail itself of this distinctively human style of mental operation. There are several quite different approaches to non-monotonic inference being pursued in AI today. They have in common only the goal of capturing the human talent for *ignoring* what should be ignored, while staying alert to relevant recalcitrance when it occurs.

One family of approaches, typified by the work of Marvin Minsky and Roger Schank (Minsky 1981; Schank and Abelson 1977), gets its ignoring-power from the attention-focusing power of stereotypes. The inspiring insight here is the idea that all of life's experiences, for all their variety, boil down to variations on a manageable number of stereotypic themes, paradigmatic scenarios—'frames' in Minsky's terms, 'scripts' in Schank's.

An artificial agent with a well-stocked compendium of frames or scripts, appropriately linked to each other and to the impingements of the world via its perceptual organs, would face the world with an elaborate system of what might be called habits of attention and benign tendencies to leap to particular sorts of conclusions in particular sorts of circumstances. It would.'automatically' pay attention to certain features in certain environments and assume that certain unexamined normal features of those environments were present. Concomitantly, it would be differentially alert to relevant divergences from the stereotypes it would always begin by 'expecting'.

Simulations of fragments of such an agent's encounters with its world reveal that in many situations it behaves quite felicitously and apparently naturally, and it is hard to say, of course, what the limits of this approach are. But there are strong grounds for skepticism. Most obviously, while such systems perform creditably when the world co-operates with their stereotypes, and even with *anticipated* variations on them, when their worlds turn perverse, such systems typically cannot recover gracefully from the misanalyses they are led into. In fact, their behaviour *in extremis* looks for all the world like the preposterously counter-productive activities of insects betrayed by their rigid tropisms and other genetically hard-wired behavioural routines.

When these embarrassing misadventures occur, the system designer can improve the design by adding provisions to deal with the particular cases. It is important to note that in these cases, the system does not redesign itself (or learn) but rather must wait for an external designer to select an improved design. This process of redesign recapitulates the process of natural selection in some regards; it favours minimal, piecemeal, *ad hoc* redesign which is tantamount to a wager on the likelihood of patterns in future events. So in some regards it is faithful to biological themes.[15] Nevertheless, until such a system is given a considerable capacity to learn from its errors without designer intervention, it will continue to respond in insectlike ways, and such behaviour is profoundly unrealistic as a model of

---

[15] In one important regard, however, it is dramatically unlike the process of natural selection, since the trial, error, and selection of the process is far from blind. But a case can be made that the impatient researcher does nothing more than telescope time by such foresighted interventions in the redesign process.

human reactivity to daily life. The short cuts and cheap methods provided by a reliance on stereotypes are evident enough in human ways of thought, but it is also evident that we have a deeper understanding to fall back on when our short cuts don't avail, and building some measure of this deeper understanding into a system appears to be a necessary condition of getting it to learn swiftly and gracefully.

In effect, the script or frame approach is an attempt to *pre-solve* the frame problems the particular agent is likely to encounter. While insects do seem saddled with such control systems, people, even when they do appear to be relying on stereotypes, have back-up systems of thought that can deal more powerfully with problems that arise. Moreover, when people do avail themselves of stereotypes, they are at least relying on stereotypes of their own devising, and to date no one has been able to present any workable ideas about how a person's frame-making or script-writing machinery might be guided by its previous experience.

Several different sophisticated attempts to provide the representational framework for this deeper understanding have emerged from the deductive tradition in recent years. Drew McDermott and Jon Doyle have developed a 'non-monotonic logic' (1980), Ray Reiter has a 'logic for default reasoning' (1980), and John McCarthy has developed a system of 'circumscription', a formalized 'rule of conjecture that can be used by a person or program for "jumping to conclusions"' (1980). None of these is, or is claimed to be, a complete solution to the problem of *ceteris paribus* reasoning, but they might be components of such a solution. More recently, McDermott has offered a 'temporal logic for reasoning about processes and plans' (McDermott 1982). I will not attempt to assay the formal strengths and weaknesses of these approaches. Instead I will concentrate on another worry. From one point of view, non-monotonic or default logic, circumscription, and temporal logic all appear to be radical improvements to the mindless and clanking deductive approach, but from a slightly different perspective they appear to be more of the same, and at least as unrealistic as frameworks for psychological models.

They appear in the former guise to be a step towards greater psychological realism, for they take seriously, and attempt to represent, the phenomenologically salient phenomenon of common sense *ceteris paribus* 'jumping to conclusions' reasoning. But do they really succeed in offering any plausible suggestions about how the backstage implementation of that conscious thinking is accomplished *in people*? Even if on some glorious future day a robot with debugged circumscription methods manœuvred well in a non-toy environment, would there be much likelihood that its constituent processes, *described at levels below the phenomeno-logical*, would bear informative relations to the unknown lower-level backstage processes in human beings? To bring out better what my worry is, I want to introduce the concept of a *cognitive wheel.*

We can understand what a cognitive wheel might be by reminding ourselves first about ordinary wheels. Wheels are wonderful, elegant triumphs of technology. The traditional veneration of the mythic inventor of the wheel is entirely justified. But if wheels are so wonderful, why are there no animals with wheels? Why are no wheels to be found (functioning as wheels) in nature? First, the presumption of that question must be qualified. A few years ago the astonishing discovery was made of several microscopic beasties (some bacteria and some unicellular eukaryotes) that have wheels of sorts. Their propulsive tails, long thought to be flexible flagella, turn out to be more or less rigid corkscrews, which rotate continuously, propelled by microscopic motors of sorts, complete with main bearings.[16] Better known, if less interesting for obvious reasons, are the tumbleweeds. So it is not quite true that there are no wheels (or wheeliform designs) in nature.

Still, macroscopic wheels — reptilian or mammalian or avian wheels — are not to be found. Why not? They would seem to be wonderful retractable landing gear for some birds, for instance. Once the question is posed, plausible reasons rush in to explain their absence. Most important, probably, are the considerations about the topological properties of the axle/bearing boundary that make the transmission of material or energy across it particularly difficult. How could the life-support traffic arteries of a living system maintain integrity across this boundary? But once that problem is posed, solutions suggest themselves; suppose the living wheel grows to mature form in a non-rotating, non-functional form, and is then hardened and sloughed off, like antlers or an outgrown shell, but not completely off: it then rotates freely on a lubricated fixed axle. Possible? It's hard to say. Useful? Also hard to say, especially since such a wheel would have to be free-wheeling. This is an interesting speculative exercise, but certainly not one that should inspire us to draw categorical, a priori conclusions. It would be foolhardy to declare wheels biologically impossible, but at the same time we can appreciate that they are at least very distant and unlikely solutions to *natural* problems of design.

Now a cognitive wheel is simply any design proposal in cognitive theory (at any level from the purest semantic level to the most concrete level of 'wiring diagrams' of the neurones) that is profoundly unbiological, however wizardly and elegant it is as a bit of technology.

Clearly this is a vaguely defined concept, useful only as a rhetorical abbreviation, as a gesture in the direction of real difficulties to be spelled out carefully. 'Beware of postulating cognitive wheels' masquerades as

[16] For more details, and further reflections on the issues discussed here, see Diamond 1983.

good advice to the cognitive scientist, while courting vacuity as a maxim to follow.[17] It occupies the same rhetorical position as the stockbroker's maxim: buy low and sell high. Still, the term is a good theme-fixer for discussion.

Many critics of AI have the conviction that *any* AI system is and must be nothing but a gearbox of cognitive wheels. This could of course turn out to be true, but the usual reason for believing it is based on a misunderstanding of the methodological assumptions of the field. When an AI model of some cognitive phenomenon is proposed, the model is describable at many different levels, from the most global, phenomenological level at which the behaviour is described (with some presumptuousness) in ordinary mentalistic terms, down through various levels of implementation all the way to the level of program code—and even further down, to the level of fundamental hardware operations if anyone cares. No one supposes that the model maps onto the process of psychology and biology *all the way down*. The claim is only that for some high level or levels of description below the phenomenological level (which merely *sets* the problem) there is a mapping of model features onto what is being modelled: the cognitive processes in living creatures, human or otherwise. It is understood that all the implementation details below the level of intended modelling will consist, no doubt, of cognitive wheels—bits of unbiological computer activity mimicking the gross effects of cognitive subcomponents by using methods utterly unlike the methods still to be discovered in the brain. Someone who failed to appreciate that a model composed microscopically of cognitive wheels could still achieve a fruitful isomorphism with biological or psychological processes at a higher level of aggregation would suppose there were good a priori reasons for generalized skepticism about AI.

But allowing for the possibility of valuable intermediate levels of modelling is not ensuring their existence. In a particular instance a model might descend directly from a phenomenologically recognizable level of psychological description to a cognitive wheels implementation without shedding any light at all on how we human beings manage to enjoy that phenomenology. I *suspect* that all current proposals in the field for dealing with the frame problem have that shortcoming. Perhaps one should

[17] I was interested to discover that at least one researcher in AI mistook the rhetorical intent of my new term on first hearing; he took 'cognitive wheels' to be an accolade. If one thinks of AI, as he does, not as a research method in psychology but as a branch of engineering attempting to extend human cognitive powers, then of course cognitive wheels are breakthroughs. The vast and virtually infallible memories of computers would be prime examples; others would be computers' arithmetical virtuosity and invulnerability to boredom and distraction. See Hofstadter (1982) for an insightful discussion of the relation of boredom to the structure of memory and the conditions of creativity.

dismiss the previous sentence as mere autobiography. I find it hard to imagine (for what that is worth) that any of the *procedural details* of the mechanization of McCarthy's circumscriptions, for instance, would have suitable counter-parts in the backstage story yet to be told about how human common-sense reasoning is accomplished. If these procedural details lack 'psychological reality' then there is nothing left in the proposal that might model psychological processes except the phenomenological-level description in terms of jumping to conclusions, ignoring, and the like—and we already know we do that.

There is an alternative defence of such theoretical explorations, however, and I think it is to be taken seriously. One can claim (and I take McCarthy to claim) that while formalizing common-sense reasoning in his fashion would not tell us anything *directly* about psychological processes of reasoning, it would clarify, sharpen, systematize the purely semantic-level characterization of the demands on any such implementation, bio-logical or not. Once one has taken the giant step forward of taking information-processing seriously as a real process in space and time, one can then take a small step back and explore the implications of that advance at a very abstract level. Even at this very formal level, the power of circumscription and the other versions of non-monotonic reasoning remains an open but eminently explorable question.[18]

Some have thought that the key to a more realistic solution to the frame problem (and indeed, in all likelihood, to any solution at all) must require a complete rethinking of the semantic-level setting, prior to concern with syntactic-level implementation. The more or less standard array of predic-ates and relations chosen to fill out the predicate-calculus format when representing the 'propositions believed' may embody a fundamentally inappropriate parsing of nature for this task. Typically, the interpretation of the formulae in these systems breaks the world down along the familiar lines of objects with properties at times and places. Knowledge of situations and events in the world is represented by what might be called sequences of verbal snapshots. State S, constitutively described by a list of sentences true at time t asserting various n-adic predicates true of various particulars, gives way to state S', a similar list of sentences true at t'. Would it perhaps be better to reconceive of the world of planning in terms of histories and processes?[19] Instead of trying to model the capacity to

---

[18] McDermott 1982 ('A Temporal Logic for Reasoning about Processes and Plans', §6, 'A Sketch of an Implementation',) shows strikingly how many *new* issues are raised once one turns to the question of implementation, and how indirect (but still useful) the purely formal consid-erations are.

[19] Patrick Hayes has been exploring this theme, and a preliminary account can be found in 'Naïve Physics 1: The Ontology of Liquids', (Hayes 1978).

keep track of things in terms of principles for passing through temporal cross-sections of knowledge expressed in terms of terms (*names* for *things*, in essence) and predicates, perhaps we could model keeping track of things more directly, and let all the cross-sectional information about what is deemed true moment by moment be merely implicit (and hard to extract—as it is for us) from the format. These are tempting suggestions, but so far as I know they are still in the realm of handwaving.[20]

Another, perhaps related, handwaving theme is that the current difficulties with the frame problem stem from the conceptual scheme engendered by the serial-processing von Neumann architecture of the computers used to date in AI. As large, fast parallel processors are developed, they will bring in their wake huge conceptual innovations which are now of course only dimly imaginable. Since brains are surely massive parallel processors, it is tempting to suppose that the concepts engendered by such new hardware will be more readily adaptable for realistic psychological modelling. But who can say? For the time being, most of the optimistic claims about the powers of the parallel-processing belong in the same camp with the facile observations often encountered in the work of neuroscientists, who postulate marvellous cognitive powers for various portions of the nervous system without a clue how they are realized.[21]

Filling in the details of the gap between the phenomenological magic show and the well-understood powers of small tracts of brain tissue is the immense research task that lies in the future for theorists of every persuasion. But before the problems can be solved they must be encountered, and to encounter the problems one must step resolutely into the gap and ask how-questions. What philosophers (and everyone else) have always known is that people—and no doubt all intelligent agents—can engage in swift, sensitive, risky-but-valuable *ceteris paribus* reasoning. How do we do it? AI may not yet have a good answer, but at least it has encountered the question.[22]

---

[20] Oliver Selfridge's forthcoming monograph, *Tracking and Trailing* (Bradford Books/MIT Press), promises to push back this frontier, I think, but I have not yet been able to assimilate its messages. There are also suggestive passages on this topic in Ruth Garrett Millikan's *Language, Thought, and Other Biological Categories*, also forthcoming from Bradford Books.

[21] To balance the 'top-down' theorists' foible of postulating cognitive wheels, there is the 'bottom-up' theorists' penchant for discovering *wonder tissue*. (Wonder tissue appears in many locales. J. J. Gibson's theory of perception, for instance, seems to treat the whole visual system as a hunk of wonder tissue, for instance, resonating with marvellous sensitivity to a host of sophisticated 'affordances'. See e.g. Gibson 1979.)

[22] One of the few philosophical articles I have uncovered that seem to contribute to the thinking about the frame problem—though not in those terms—is Ronald de Sousa's 'The Rationality of Emotions' (de Sousa 1979). In the section entitled 'What are Emotions For?' de Sousa suggests, with compelling considerations, that: the function of emotion is to fill gaps left by

[mere wanting plus] 'pure reason' in the determination of action and belief. Consider how Iago proceeds to make Othello jealous. His task is essentially to direct Othello's attention, to suggest questions to ask . . . Once attention is thus directed, inferences which, before on the same evidence, would not even have been,thought of, are experienced as compelling. In de Sousa's understanding, 'emotions are determinate patterns of salience among objects of attention, lines of inquiry, and inferential strategies' (p. 50) and they are not 'reducible' in any way to 'articulated propositions'. Suggestive as this is, it does not, of course, offer any concrete proposals for how to endow an inner (emotional) state with these interesting powers. Another suggestive — and overlooked — paper is Howard Darmstadter's 'Consistency of Belief' (Darmstadter 1971: 301–10). Darmstadter's exploration of *ceteris paribus* clauses and the relations that might exist between beliefs as psychological states and sentences believers may utter (or have uttered about them) contains a number of claims that deserve further scrutiny.

REFERENCES

Cherniak, C. [1983]. 'Rationality and the Structure of Memory.' *Synthese* [57: 163–86].
Darmstadter, H. (1971). 'Consistency of Belief.' *J. Philosophy* 68: 301–10.
Dennett, D. C. (1978). *Brainstorms*. Cambridge, Mass.: MIT Press/Bradford Books.
—— (1982a). 'Why Do We Think What We Do About Why We Think What We Do?' *Cognition* 12: 219–27.
—— (1982b). 'How to Study Consciousness Empirically; Or Nothing Comes to Mind.' *Synthese* 53: 159–80.
—— (1982c). 'Beyond Belief.' In A. Woodfield (ed.), *Thought and Object*, pp. 1–96. Oxford: Clarendon Press.
—— (1983). 'Styles of Mental Representation.' *Proc. Aristotelian Soc.* 83: 213–26.
Diamond, J. (1983). 'The Biology of the Wheel.' *Nature* 302: 572–3.
Dreyfus, H. L. (1972). *What Computers Can't Do*. New York: Harper & Row.
Fikes, R., and Nilsson, N. (1971). 'STRIPS: A New Approach to the Application of Theorem Proving to Problem Solving,' *Artificial Intelligence* 2: 189–208.
Gibson, J. J. (1979). *The Ecological Approach to Visual Perception*. Boston, Mass.: Houghton-Mifflin.
Goodman, N. (1965). *Fact, Fiction and Forecast*, 2nd edn. Indianapolis: Bobbs-Merrill.
—— (1982). 'Thoughts Without Words.' *Cognition* 12: 211–17.
Hayes, P. J. (1978). 'Naïve Physics I: The Ontology of Liquids.' Working Paper 35, Institute for Semantic and Cognitive Studies, Geneva.
—— (1979). 'The Naïve Physics Manifesto.' In D. Michie (ed.), *Expert Systems in the Micro-Electronic Age*, pp. 242–70. Edinburgh: Edinburgh University Press.
Hofstadter, D. (1982). 'Can Inspiration be Mechanized?' *Scientific American* 247: 18–34.
McCarthy, J. (1968). 'Programs with Common Sense.' *Proceedings of the Teddington Conference on the Mechanization of Thought Processes*, London. Repr. in M. Minsky (ed.)., *Semantic Information Processing*, pp. 403–18. Cambridge, Mass.: MIT Press.
—— (1980). 'Circumscription — A Form of Non-Monotonic Reasoning.' *Artificial Intelligence* 13: 27–39.
—— and Hayes, P. J. (1969). 'Some Philosophical Problems from the Standpoint of Artificial Intelligence.' In B. Meltzer and D. Michie (eds.), *Machine Intelligence 4*, pp. 463–502. Edinburgh: Edinburgh University Press.

McDermott, D. (1982). 'A Temporal Logic for Reasoning about Processes and Plans.' *Cognitive Science* 6: 101–55.

—— and Doyle, J. (1980). 'Non-Monotonic Logic.' *Artifical Intelligence* 13: 41–72.

Millikan, R. G. [1984]. *Language, Thought and Other Biological Categories*. Cambridge, Mass.: MIT Press/Bradford Books.

Minsky, M. (1981). 'A Framework for Representing Knowledge.' Originally published as MIT AI Lab. Memo 3306. Quotation drawn from excerpts repr. in J. Haugeland (ed.), *Mind Design*, pp. 95–128. Cambridge, Mass.: MIT Press/Bradford Books.

Newell, A. (1982). 'The Knowledge Level.' *Artificial Intelligence* 18: 87–127.

Reiter, R. (1980). 'A Logic for Default Reasoning.' *Artificial Intelligence* 13: 81–132.

Ryle, G. (1949). *The Concept of Mind*. London: Hutchinson.

Schank, R. C., and Abelson, R. P. (1977). *Scripts, Plans, Goals, and Understanding: An Inquiry into Human Knowledge*. Hillsdale, NJ: Erlbaum.

Selfridge, O. (forthcoming). *Tracking and Trailing*. Cambridge, Mass.: MIT Press/Bradford Books.

de Sousa, R. (1979). 'The Rationality of Emotions.' *Dialogue* 18: 41–63.

Wooldridge, D. (1963). *The Machinery of the Brain*. New York: McGraw-Hill.

8

THE NAÏVE PHYSICS MANIFESTO

PATRICK J. HAYES

The garden still offered its corners of weed, blackened cabbages, its stones and flowerstalks. And the house its areas of hot and cold, dark holes and talking boards, its districts of terror and blessed sanctuary; together with an infinite range of objects and ornaments that folded, fastened, creaked and sighed, opened and shut.

from 'Cider with Rosie'
by Laurie Lee

1. INTRODUCTION

Artificial Intelligence is full of 'toy problems': small, artificial axiomatizations or puzzles designed to exercise the talents of various problem-solving programs or representational languages or systems. The subject badly needs some non-toy worlds to experiment with. In this document I propose the construction of a formalization of a sizeable portion of common-sense knowledge about the everyday physical world: about objects, shape, space, movement, substances (solids and liquids), time, etc.

In what follows I will outline the proposal and distinguish it from some others, superficially similar; discuss some of the general issues which arise; argue that it *needs* to be done; that it *can* be done; and outline a way of *getting* it done. Along the way I will outline the theory of meaning which this proposal assumes, and criticize some others.

2. THE PROPOSAL: SUMMARY

The proposal is to construct a formalization of a large part of ordinary everyday knowledge of the physical world. Such a formalization could, for example, be a collection of assertions in a first-order logical formalism, or

Patrick J. Hayes, 'The Naïve Physics Manifesto', in D. Michie (ed.), Expert *Systems in the Micro-Electronic Age* (Edinburgh University Press, 1979), pp. 242–70.

a collection of KRL 'units', or a microplanner program, or one of a number of other things. The proposal is *not* to develop a new formalism to express this knowledge in. Although we recognize that formalism-hacking may be necessary in due course, we believe that existing, well-understood formalisms have many, as yet unexplored, possibilities. The proposal is *not* to write a program which can solve problems, or plan actions, or whatever, in the formalism. Although it is important to bear control and search issues—in short, *computational issues*—in mind, we propose to deliberately postpone detailed consideration of implementation. All too often, serious work on representational issues in AI has been diverted or totally thwarted by premature concern for computational issues.

The formalism we propose should have the following characteristics (we elaborate on these later):

1. *Thoroughness*. It should cover the whole range of everyday physical phenomena: not just the blocks-world, for example. Since in some important sense the world (even the everyday world) is infinitely rich in possible phenomena, this thoroughness will never be perfect. Nevertheless, we should *try* to fill in all the major holes, or at least identify them.

2. *Fidelity*. It should be reasonably detailed. For example, such aspects of a block in a blocks-world as shape, material, weight, rigidity, and surface texture should be available as concepts in a blocks-world description, as well as *support* relationships. Again, since the world is infinitely detailed, perfect fidelity is impossible: but we should try to do better than the very low fidelity of the common 'toy problem' axiomatizations, in which, for example, the relationship of one block being 'above' another is merely a partial ordering, so that the integers are a possible model of the axioms.

3. *Density*. The ratio of facts to concepts needs to be fairly high. Put another way: the units have to have *lots* of slots. Low-density formalizations are in some sense trivial: they fail to say enough about the concepts they contain to pin down the meaning at all precisely. Sometimes, for special purposes, as for example in foundational studies, this can be an advantage: but not for us.

4. *Uniformity*. There should be a common formal framework (language, systems, etc.) for the whole formalization, so that the inferential connections between the different parts (axioms, frames, . . .) can be clearly seen, and divisions into subformalizations are not prejudged by deciding to use one formalism for one area and a different one for a different area.

(As I shall emphasize later, I also think it is methodologically important to allow the use of a variety of formalisms: often a particular subarea can be neatly expressed in some idiosyncratic way. But there is no contradiction: for we will also insist that such idiosyncratic formalisms are systematically reducible to the basic formalism: they will be regarded as 'semantic sugar'. This is important: for although *computational* properties of a representation may depend crucially upon the use of such idiosyncratic formalisms, there must be a common *representational* framework within which the meaning-content of any piece of representation can be related to that of any other.)

I believe that a formalization of naïve physics with these properties can be constructed within a reasonable time-scale. The reasons for such optimism are explained later. It is important however to clearly distinguish this proposal from some others with which it may be confused, because some of these seem to be far less tractable.

3. WHAT THE PROPOSAL ISN'T

(a) It is *not* proposed to make a computer program which can 'use' the formalism in some sense. For example, a problem-solving program, or a natural-language comprehension system with the representation as target. It is tempting to make such demonstrations from time to time. (They impress people; and it is satisfying to have actually *made* something which works, like building model railways; and one's students can get Ph.D.'s that way.) But they divert attention from the main goal. In fact, I believe they have several more dangerous effects. It is perilously easy to conclude that, because one has a program which *works* (in some sense), its representation of its knowledge must be more or less *correct* (in some sense). Regrettably, the little compromises and simplifications needed in order to get the program to work in a reasonable space or in reasonable time, can often make the representation even less satisfactory than it might have been.

This is not to say that computational questions should be *ignored* in constructing the proposed formalization. For example, the question of the *length of derivations* of ordinary common-sense inferences is important, and our notion of 'density' has direct computational consequences, for example, for storage and retrieval strategies. But the construction of 'demonstration' programs seems to serve no really useful purpose (McDermott 1977 argues similarly).

I emphasize this point because there is a prevailing attitude in AI that research which does not result fairly quickly in a working program of some kind is somehow useless, or at least, highly suspicious. This may be partly to blame for the dearth of really serious efforts in the representational direction, and the proliferation of programs and techniques which work well (or sometimes badly) in trivially small domains, but which are wholly limited by scale factors, and which therefore tell us nothing about thinking about realistically complicated worlds. (Backtracking search and the STRIPS representation of actions by add and delete lists are two good examples. I suspect that production systems are another.)

Ideally, one should in principle be able to get a working program from the formalization by assuming a particular inference mechanism and adding *further* information at the meta-level, which 'controls' the inferences this mechanism performs (Hayes 1973; Kowalski 1977; Pratt 1977; Bundy 1978). Looked at this way, a formalization can be thought of as a 'core' of inferential abilities, whose appropriate deployment at any moment for a particular task has to be further specified. (However, this pleasant picture has no doubt to be modified to account for idiosyncratic representations which have especially desirable computational properties.)

The decision to postpone details of implementation can be seen as an implicit claim that the representational content of a large formalization can be separated fairly clearly from the implementational decisions: this is by no means absolutely obvious, although I believe it to be substantially true.

(*b*) It is *not* proposed to develop a new formalism or language to write down all this knowledge in. In fact, I propose (as my friends will have already guessed) that first-order logic is a suitable basic vehicle for representation. However, let me at once qualify this.

I have no particular brief for the usual *syntax* of first-order logic. Personally I find it agreeable: but if someone likes to write it all out in KRL, or semantic networks, or 'fancy' semantic networks of one sort or another, or what have you; well, that's fine. The important point is that one *knows what it means*: that the formalism has a clear *interpretation* (I avoid the word 's*m*nt*cs' deliberately). At the level of interpretation, there is little to choose between any of these, and most are strictly weaker than predicate calculus, which also has the advantage of a clear, explicit model theory, and a well-understood proof theory (Hayes 1977, 1978*a*).

Secondly, let me emphasize again that idiosyncratic notations may sometimes be useful for idiosyncratic subtheories. For example, in sketching an axiomatic theory of fluids (Hayes 1978*b*), I found it useful to think of the possible physical states of fluids as being essentially states of a finite-state machine. This summarizes a whole lot of lengthy, and rather clumsy first-order axioms into one neat diagram. Still, it *means* the same as the axioms: first-order is still, as it were, the reference language. Other examples are the 'evaluable' predicates and functions sometimes included in theorem-proving programs where for example the term '(plus 2 3)' is *evaluated* to the constant '5', no axioms being provided for arithmetic expressions. But this can always[1] be regarded as a computationally efficient way of representing the same *meaning* as would be represented by the (infinite) collection of axioms '(plus 2 3) = 5', '(plus 2 2) = 4', etc.

Thirdly, first-order logic as it stands is almost certainly not rich enough and will need to be extended. I have already found two extensions which I think are necessary; quotation, so that the formalism can describe its own formulae; and a sort of non-unique Skolem function, similar to Hilbert's ε-symbol. The notion of the default may be another (although I have not felt any particular need for this concept so far). I expect that such extensions will arise naturally through difficulties in using the formalism; and as I think it is dangerous to try to predict such difficulties ahead of time, I will not try.

4. THE AXIOM-CONCEPT GRAPH: CLUSTERS AND DENSITY

Let us imagine that a naïve physics formalization exists (I tend to believe that it does, in fact, inside my head), and try to analyse its structure. It consists, fundamentally, of a large number of assertions, involving a large number of (non-logical) symbols: relation symbols, function and constant symbols (*or*: frame headers, slot names, etc.; *or*: node and arc labels, etc.). In future I will not bother to re-emphasize these obvious parallels, but will assume the reader is aware of them.) For a neutral word, let us call these symbols *tokens*.

The *meaning* of tokens is defined by the structure of the formalization, by the pattern of inferential connections between the assertions. This structure can be very complex, but we can make some fundamental points by treating it in an essentially qualitative way.

Let us say that a formalization is *dense* if, for each token, there are many axioms involving it. A dense formalization has many links between the separate concepts expressed by tokens in the formalization. Density is clearly a matter of degree. Formalizations which are not dense in this way (*sparse* formalizations) are unsatisfactory, since they do not pin down exactly enough the meanings of the tokens they contain. If all a formalization says about the relation *above* (in the blocks-world) is that it is the transitive closure of *on*; and all it says about *on* is that if nothing is *on* a

---

[1] This remark slurs over a minefield of technical difficulties. It is not appropriate to go into these here, however.

block then you can pick it up; and all it says about *picking* up is that after *picking*, the block is *held*, etc. (in the familiar way); if this is *all* that it says with these tokens, then one can hardly say that the concepts of *on*, *above*, etc., are represented in the formalization at all. For, these concepts have connections to many other concepts as well, in our heads. If one thing is above another, there are all sorts of consequences. Maybe the first will fall on the second, if its supports fail: there are consequences for the relative appearance of the two objects: the top one might provide a shelter for the bottom one; if the bottom one supports the top one, then the bottom one will be under some strain due to the weight of the top one; and so on. We need to try to capture this *richness* of conceptual linking.

(It is worth emphasizing that the view of meaning espoused here differs profoundly from the view which holds that tokens in a formalization are essentially words in a natural language (Wilks 1977). According to this latter view, the tokens *do* represent the concept intended, by *fiat*: they are 'semantic primitives', out of which all other meanings are composed. I will return to this idea shortly.)

Any formalization which hopes to approach the richness of our own conceptual apparatus must be dense. Of course, density is not a *sufficient* condition for success: it is not difficult to invent wholly useless axiomatizations of arbitrarily high density.

It is useful to consider a simplified model of a formalization. Imagine a graph whose nodes are tokens of the formalization, and whose arcs correspond to axioms: an arc links two nodes if the corresponding axiom contains those two tokens. (Strictly speaking, this has to be a polygraph (Landin 1970), since axioms may well contain more than two tokens. However, we will be making only heuristic use of the idea, in any case, so technicalities are inappropriate.) We will call this the axiom-concept (a-c) graph. The formalization is dense if the a-c graph is highly connected, sparse if the graph is sparse. However, we cannot expect density to be uniform: there will be more dense clusters of concepts with many relationships between them, less tightly linked to the rest of the formalization.

Identifying these clusters is both one of the most important and one of the most difficult methodological tasks in developing a naïve physics. I think that several serious mistakes have been made in the past: here, for example, causality is, I now tend to think, *not* a cluster: there is no useful, more-or-less self-contained theory of causality. 'Causality' is a word for what happens when other things happen, and what happens, depends on circumstances. If there is liquid around, for example, things will often happen very differently from when everything is nice and dry. What happens with liquids, however, is part of the *liquids* cluster, not part of some theory of 'what-happens-when'. Mistakes like this are hard to overcome, since a large conceptual structure *can* be entered anywhere. The symptom of having got it wrong is that it seems hard to say anything very useful about the concepts one has proposed (because one has entered the graph at a locally sparse place, rather than somewhere in a cluster). But this can also be because of having chosen one's concepts badly, lack of imagination, or any of several other reasons. It is easier, fortunately, to recognize when one is in a cluster: assertions suggest themselves faster than one can write them down.

There is also a useful notion of a supercluster: a cluster which is related to a large number of other clusters. I think that the collection of concepts to do with 3-dimensional shape and orientation are a supercluster in our own mental conceptual structures: concepts such as above, below, tall, fat, wide, behind, touching, resting on, angle of slope, edge (of a surface), surface (of a volume), side, vertical, top, bottom, . . . These obviously have many internal relationships: they form a cluster. They also must appear significantly in whatever conceptual frameworks underlie visual perception and locomotion in space; they are crucial in describing assemblies; in the theory of liquids (Hayes 1978*b*); also, I believe, in the descriptions of physical actions and events (Hayes 1978*c*); and so on.

Superclusters can be recognized by the fact that they crop up, in this way, in a variety of other clusters. Other plausible candidates for superclusters include a theory of measuring scales (which should provide such notions as accuracy, vagueness, utility for various kinds of tasks), a theory of time-measurement, and the collection of notions concerned with *inside*, *outside*, containment and *ways through* from one place to another.

This suggestive terminology of clustering should not be taken literally: I do not mean to suggest that there are many sharply isolated parts of our conceptual structure which can be developed in total isolation from all others. And the 'a-c graph' model of an axiom system is itself over-simplified in several gross respects, in any case. Nevertheless, I think the basic idea, that there are collections of concepts which have close connec-tions between one another, is substantially correct and quite important.

5. THE A/C RATIO AND REDUCTIONIST FORMALIZATIONS

Let me turn now to an even cruder model of a formalization: the *ratio* of axioms to concepts (the *a/c* ratio). For a dense axiomatization, *a/c* will be large. Any interesting axiomatization will have *a/c* greater than one; but there *are* interesting axiomatizations in which *a/c* will be very close to unity.

Consider an axiomatic set theory. The idea, for foundational research, is to have a *small* axiomatic theory (e.g. Zermelo–Fraenkel set theory which has $c = 2$, viz. 'ε' and '*set*', and $a = 8$, so $a/c = 4$) which enables one to *define* a large number of mathematical concepts (e.g. the integers can be defined as sets of any one of several special kinds; the rationals are sets of pairs of integers; the reals are sets of infinite sets of rationals . . .) in such a way that the desired properties of these concepts (e.g. the principle of induction for integers: the continuity of the real line) *follow from* the structure of these definitions, and the axioms of the basic theory. It is important to realize that these properties of the defined concepts are *theorems* of the axiomatization which consists of set theory together with the definitions of the concepts. They are not axiomatic assumptions themselves, needed to pin down the meaning of the introduced concepts: the definitions pin down the concept as completely as it can be: all else then follows. Mathematics is reduced to a series of *lemmas* to set theory: or at least, that is the idea. (When one thinks of it this way, it seems almost incredible that such an audacious program should have so nearly succeeded.)

I want to emphasize how different this approach to capturing meaning in a formalism is, from the axiomatic approach to naïve physics which I am proposing. Set theory is reductionist in the extreme: I am urging a richly connected formalization, with many interactions between assumptions. The reductionist approach leads, indeed, to axiomatic theories, but they are extraordinarily *sparse*.

Consider the effect of adding a definition of a new concept *to a formalization*. This increases both a and c by one. If a/c is large, this will decrease it appreciably. (Indeed, one will have to add approximately (a/c) many axioms to bring the ratio back to what it was.

This emphasizes what is intuitively clear, that definitions which do not pay their keep by introducing concepts which are going to be of some general use, are probably a mistake: they dilute the formalization.) Suppose however that a and c are both small, say $a = 8$, $c = 2$. Then adding one definition reduces a/c from 4 to 3. Adding another reduces it to 2.5: adding 1,000 definitions reduces a/c to 1.059. Clearly, as the number of definitions in the formalization increases, a/c tends asymptotically to unity. The a-c graph of such a formalization has one very small cluster at the centre, surrounded by a cloud of nodes each linked radially to a few nodes closer to the centre. It is almost as sparse a connected graph as one could have, containing the concept tokens that it does. It has quite a different 'shape' from the connected, clustered graph of a dense axiomatic theory.

The existence of such a reductionist theory for mathematics is a remarkable fact, and it would indeed be astonishing if one could find an analogous reductionist theory for common-sense reasoning: a smallish collection of concepts, and axioms connecting them, such that all other concepts (e.g. all those expressed by English words) could be defined in terms of these few. In fact, this would be *so* astonishing that I feel confident in asserting that no such small theory exists. And yet, many approaches to the formal representation of meaning, in the AI literature, make such an assumption. These are the 'semantic primitive' approaches, exemplified by the work of Wilks (1975, for example) and Schank (1975, for example). Here, members of a smallish collection of tokens (Wilks 90 or so, Schank $14 + n$ for some n, as yet unknown) are taken as *primitive*. The *meaning* of an English word is then some formal expression built out of these primitive tokens, using some formal apparatus (which in Schank's case is usually presented graphically, although this is not essential). In our terms, the formalization consists mostly of definitions: its a/c ratio tends to unity, just like axiomatic set theory. In the work of Schank and his students one can clearly see that the axiomatic structure of the 'core' theory—cf. Rieger's 'inference molecules' associated with the 14 primitive action-tokens—is intended to serve the same sort of central organizing role that the set axioms play in the development of set theory. That is, desired properties of non-primitive concepts (such as *buying* or *giving*) follow from their definitions, and the meaning given to the primitives by the core theory. In Wilks's work, there does not seem to *be* any core formalization at all: we are merely presented (cf. Wilks 1977) with a list of tokens and a brief description of the concept they are supposed to mean. To think that a *formal* symbol has any meaning other than that specified by the structure of the formalization in which it occurs, that it has any *intrinsic* meaning, is to make a particularly unfortunate mistake. Wilks, however, takes the view that his semantic primitives are, in fact, *words*, like English words, not mere formal tokens, thus neatly escaping this objection, and explaining why it is not necessary to give any formalization in which they occur. I confess to still being puzzled as to how it is that his *program* knows what they mean.

This reductionist, semantic-primitives approach to meaning is essentially bound to low-fidelity, low-density representations. Such representations have their uses—they may be adequate for information-retrieval or machine translation applications, for example—and when they *do* work, have some very useful computational properties. But at some point we will have to face up to the problem of representing *detailed* knowledge of the world. This will require the abandoning of the 'definition' view of meaning of tokens. As Wilks says, 'No representation in primitives could be expected to distinguish by its structure the senses of *hammer, mallet* and *axe* . . .'. Perhaps not: but naïve physics should be able to.

6. MEANINGS, MODEL THEORY, AND FIDELITY

It might be asked: if the meanings of tokens are not specified by definitions, how *are* they specified? In one sense there is no answer to such a question. One cannot point to a particular structure and say, *that* is the meaning of a token. One can only say that a token *means* a concept to the extent that *the formalization taken as a whole* enables a sufficient number of inferences to be made whose conclusions contain the token, i.e. which mention the concept. This operational definition of meaning can, if the formalization has an adequate model theory, be recast in an extensional way: a token means a concept if, in every possible model of *the formalization taken as a whole*, that token denotes an entity which one would agree was a satisfactory instantiation of the concept in the possible state of affairs represented by the model.

Now, to do this requires that it be possible to think of a model as a 'state of affairs'. And since we want to formalize the common-sense world of physical reality, this means, for us, that a model of the formalization must be recognizable as a facsimile of physical reality: one in which the concept we are interested in can be recognized.

A model for a first-order axiomatization is a set—the set of entities which exist in the 'state of affairs' represented by the model—and a particular mapping from the tokens of the axiomatization into this set and the sets of relations and functions, of appropriate arity, over it. This is usually presented, in textbooks of elementary logic, in a rather formal, mathematical way: and this fact may have given rise to the curious but widespread delusion that a first-order model is merely another formal description of the world, just like the axiomatization of which it is a model; and that the Tarskian truth-recursion is a kind of translation from the latter to the former: a translation from one formal system to another (e.g. Wilks 1977). This is quite wrong. For a start, the relationship between an axiomatization and its models (or, dually, between a model and the set of axiomatizations which are true of it) is quite different from a translation. It is many-many rather than one-one, for example. Moreover, it has the algebraic character called a Galois connection, which is to say, roughly, that as the axiomatization is increased in size (as axioms are added), the collection of models—possible states of affairs—*decreases* in size. It is quite possible for a large, complex axiomatization to have small, simple models, and vice versa. In particular, a model can always be gratuitously complex (e.g. contain entities which aren't mentioned at all in the axiomatization). But the deeper mistake in this way of thinking is to confuse a *formal description* of a model — found in the textbooks which are developing a mathematical approach to the meta-theory of logic — with *the actual model*. This is like confusing a mathematical description of Sydney Harbour Bridge in a textbook of structural engineering with the actual bridge. A Tarskian model can actually *be* a piece of reality. If I have a blocks-world axiomatization which refers to three blocks, called 'A', 'B', and 'C' (i.e. these are the tokens used in the axiomatization to mention the blocks), and if I have a (real, physical) table in front of me, with three (real, physical) wooden blocks on it, then the set of those three blocks can *be* the set of entities of a model of the axiomatization (provided, that is, that I can go on to interpret the relations and functions of the axiomatization as physical operations on the wooden blocks, or whatever, in such a way that the assertions made by the axiomatization about the wooden blocks, when so interpreted, are *in fact* true). There is nothing in the Tarskian model theory of first-order logic which a priori prevents the real world being a model of an axiom system.

On the other hand, it is also true that many axiomatizations have models which do not contain solid physical objects, but in which tokens denote, say, integers or other symbols. In fact, any first-order axiomatization which has any model at all (i.e. which is consistent) also has a model in which only symbols exist — this is the 'Herbrand interpretation' in which we let tokens denote themselves. Thus, Tarskian model theory does not guarantee that axiomatizations are 'about' any particular world. It is always possible to consistently believe that the only things which exist are the symbols of the formalization itself.[2] This might be called the 'solipsist' interpretation: denying the existence of the external world, while at the same time having an elaborate theory of it.

Given then, that any axiomatization will have several (usually infinitely many) models, we cannot justify an axiomatization as an adequate description by merely exhibiting a model of it which is somehow similar to reality. For it may have a very much simpler model than that, and if it has such a simpler model, then the tokens occurring in it mean no more than they mean in that simpler model. This is exactly what I mean by 'fidelity'. A low-fidelity formalization of, say, the blocks-world will admit models

---

[2] Although in the usual formalization of first-order logic, this belief cannot be expressed: and in at least one extension in which it *can* be expressed, its negation — that some things exist which are not symbols — can also be expressed.

which are very much simpler than the intended one: a model, for example, in which the 'blocks' are integers and *above* means greater than, etc.; or perhaps, a model in which 'blocks' are points in a discrete 2-dimensional space, or whatever. An adequate formalization of a blocks-world—a high-fidelity formalization—will be such that any model of it must have an essentially 3-dimensional structure. Thus, SHRDLU's blocks-world axiomatization (Winograd 1972) uses 3-dimensional Cartesian co-ordinates. (It would be interesting to find a useful but less quantitative way of describing 3-dimensional structure: for example, that a rigid attachment needs *three* points of contact: two-legged stools fall over.)

A good guide to the fidelity of a formalization is how closely its *simplest* model resembles the *intended* model. This is, I think, an excellent argument for a representational language's having a model theory: it gives us a way of testing the fidelity of a representation. It is perilously easy to *think* that one's formalization has captured a concept (because one has used a convincing-sounding token to stand for it, for example), when in fact, for all the formalization knows, the token might denote something altogether more elementary, in a very simple model.

(This criterion suggests that we should pay attention to features of representational languages which can be used in a formalization to insist upon the complexity of the simplest model. In the case of first-order logic, these include functions (the use of which claims that there is a value of the function, applied to any suitable argument), explicit existential assertions (especially 'comprehension axioms' of various kinds—more on this below), the use of equality, and the use of a highly sorted logic (or, in terms more familiar to AI, the presence of an *isa* hierarchy related to the quantifiers: although it can be much more complex than a simple hierarchy, cf. the sorted logic used in Hayes (1971). We might expect, therefore, that these features will be heavily used in developing a naïve physics.)

On this account of the meaning of a token, it depends upon the *entire* formalization of which the token is part. Thus, an alteration to any part of the formalization can, in principle, change the meaning of every other part of it. And, I think, this is essentially correct: what it means, in introspective terms, is that learning a new fact or acquiring a new concept, is liable to have far-reaching consequences for the ways in which one understands the meanings of other concepts. It also means that people with different formalizations in their heads may understand the same token in different ways. What I mean by 'water' may not be exactly what you mean by 'water': it may be possible to find a substance and a set of circumstances such that I would call it water and you would not (for example, if you have never seen opaque water, you might deny that this opaque liquid which I have in a glass on my desk, was water). And yet, *we might both be right*, since our theories of 'water' may not be identical. And this may even be possible when our beliefs about water (in the direct sense of: all the assertions which actually contain the token 'water') are identical. Each of us might agree to everything the other said about water, and yet our concepts might be subtly different. The difference may lie in some related concept (such as viscosity, or drinkability) which we understand differently. It may not even be possible to say exactly which tokens we differ on: just that we have somewhat different theories of them. It follows from this that there is no *single* 'meaning' of a token: or at least, that to assume that there is, is to assume that people's cognitive formalizations are identical in structure. Much confusion follows if this is not borne clearly in mind. For example, Wilks has argued (Anderson *et al.* 1972) that since there are people who have never seen ice, the fact that water freezes *cannot* be part of the meaning of 'water': for if it were, one would have to say that those people did not understand the meaning of 'water': and that is ridiculous. One can cut through this tortuous piece of reasoning by observing that the word evidently means more to some people than to others: and to those who *do* know about ice, the fact that ice is frozen water can well be part of the meaning of 'water'.

However, in order that communication be possible at all, it should obviously be that people's cognitive structures are *similar*: and, as a working hypothesis, we will make such an assumption in developing naïve physics. One of the good reasons for choosing naïve *physics* to tackle first is that there seems to be a greater measure of interpersonal agreement here than in many fields.

There seems to be a notion of 'distance' in a formalization, such that the effect of an alteration on the meaning of a token is less, the further away the token is from the alteration. It is not clear to me whether or how this suggestive intuition can be made to stand up. It is tempting to identify this distance with shortest-path distance in the axiom-concept graph, and although this is not really adequate since it ignores the structure of the axioms, it is the best I can do at present.

Thanks to this distance-dilution effect, it seems a reasonable strategy to work on clusters more or less independently at first: the meaning of the tokens in a cluster is more tightly constrained by the structure of a cluster than by the links to other clusters. It seems reasonable therefore to introduce concepts, which occur definitively in some other cluster, fairly freely, assuming that their meaning is, or will be, reasonably tightly specified by that other cluster. For example, in considering liquids, I needed to be able to talk about volumetric shape: assuming—and, I now claim, reasonably—that a shape cluster would specify these for me. Of course, their occurrence in the liquids cluster *does* alter their meaning: our concept of a horizontal surface would hardly be complete if we had never seen a large, still body of water—but the assumption of a *fairly autonomous* theory of shape still seems reasonable.

What I am claiming here is, at bottom, that although the 'definitions' view of meaning is wrong, we can—indeed, must—act *as though* it were correct, in order to make progress. It is good methodological scaffolding.

One last point on the model-theoretic view of meaning. As I have said, any consistent first-order axiomatization has a model in which there are only symbols. For all that the simplest such model may be very complex, one might feel that if all it contains are *symbols*, it hardly can be said to be like the real physical world, even if it is in some sense similar in 'abstract' structure.

To answer this objection, we have to talk about the body and sensory input. Imagine the naïve-physics formalization has a (physical) body with sense-organs. The way in which a formalization can be attached to the physical world is by taking a 'realist' view of the data supplied to it by its perception. Thus, part of naïve physics should be a theory of *appearance*: such a theory is now being developed by work in visual perception (especially the work of Marr and Horn, which consciously attempts to relate appearance to physical structure). While it may be that a detailed, hi-fi theory of wooden blocks could all be a dream: block-tokens might denote (say) integers in some models, for example: this would not be true if the theory also specified that if (to over-simplify) one directed one's gaze at a block with *such* a kind of surface, and *such* an orientation and in *such* a kind of lighting conditions, then one would see *such* a kind of image. For, an integer, or a symbol, is not the sort of thing one can direct one's gaze towards. (And, even if it were, in some *outré* sense, it certainly wouldn't look like a brick.) It might be objected: but you are *assuming* that the tokens for 'directing one's gaze' must denote the physical action of so doing, which begs the question: for there may be an interpretation of these tokens in a model of the integers, say, which also satisfies the axiomatization. And yes, I am exactly begging the question: I assume that 'motor tokens'—symbols which describe bodily movements—are *directly* related to the body. They constitute a *body image*[3] which has a very special relationship to the (actual) body (I imagine it to be similar to that between a graphic data-structure and the physical picture on the screen).

What this assumption means, then, is that a naïve physics can be 'connected' to the real physical world because it has a physical body, equipped with sense organs: the notion of *directing* one's gaze (or of *feeling*, or *pushing*, etc.) is *essentially* physical, and has this character

---

[3] I am indebted to Sylvia Weir for introducing me to this notion.

because it has a fixed interpretation in the body's sensorimotor system. It is this fixed, physical interpretation of some of the tokens in an axiomatization which attaches the axiomatization as a whole must contain real, physical entities and relationships. In some sense, therefore, one would expect that most of the naïve physics was 'close' to the visual-appearance (or touching-and-feeling or smelling or hearing) cluster(s) of concepts: that no part of naïve physics is *very* remote from sensory evidence. And indeed, when elaborate theories are involved in real physics, containing concepts which *are* inferentially remote from the evidence, there is usually a general feeling of mild disquiet. One hears talk of *theoretical entities*, for example (electrons are a good example) in the philosophy-of-science literature. What these discussions seem to me often to lack is a strong enough sense of the way in which even everyday, mundane ideas such as a *piece of wood* or *being wet* are constructs just as theoretical, albeit in a different and more naïve theory of the world.

A moral of this for naïve physics is that we should be always ready to seize on a chance to relate concepts to sensory or sensorimotor concepts. In working with liquids, for example, I found a notion of movement-in-space very useful: and it has obvious utility in other areas as well. This can be related directly to a visual-perception theory. If you look at a space in which there is movement, then the movement can be *seen* . Ullman (1977) explains in delightful detail how to see it. Again, much of the richness of texture of our introspective common-sense world comes, I think from our knowledge of *what it feels like when we do things* like pushing, pulling, lifting: from, ultimately, the proprioceptive sensors in our body joints and muscles. I am *not* optimistic that we can capture this richness in a formalization in the forseeable future. (It requires the construction of a suitable physical body, equipped with the necessary sense.) But I think we *can* annotate the formalization by noting the concepts which would be 'attached' to bodily-movement-concepts—and that would be a useful and interesting exercise.

This whole area of psychosomatic relationship is one which deserves deeper study, and which I believe AI concepts can do much to clarify.

7. THOROUGHNESS AND CLOSURE

One way to have a high a/c ratio, it might seem, would be to keep c small, and to say a lot about a few concepts. And if this could be done, it would indeed be very useful and encouraging: if we could find some small, self-contained groups of concepts which could be formalized in total isolation to a reasonable degree of fidelity.

But there don't seem to be many of these. (Geometrical shape may be one.) The typical situation one finds is that, having chosen one's concepts to start on, one quickly needs to introduce tokens for others one had not contemplated: and in order to pin down *their* meanings, yet more concepts need to be introduced, and the proliferation of tokens seems to be getting out of hand. If one thinks of this as exploring the a-c graph of our conceptual structures, this phenomenon is of course, hardly surprising (especially if we assume, as we must, that the graph is very dense). One needs a sense of direction, to stay within the current cluster while recognizing paths into others. But even with such a sense (which will only be developed with experience), the proliferation of necessary concepts seems almost frightening at first.

But this proliferation *must* slow down eventually: for the formalization is finite. The point of the 'thoroughness' requirement is to *go on until it does slow down*: until one finds that the collection of concepts has closed upon itself, so that all things one wants to say in the formalization can be said using the tokens which have already been introduced. In the graphical analogy, until we have *spanned* the entire graph, and need only to add new arcs, filling out the graph until its density is sufficient to capture the meanings of its tokens.

This idea of closure is familiar to anyone who has built toy-world axiomatizations. One finds, suddenly, that there are enough concepts around so that one can say 'enough' about them all: enough, that is to enable the inferences that one had had in mind all along to be made. Closure can be achieved in very small formalizations : but if a formalization is closed *and* has high fidelity (so, high density), then it must, I believe, also be thorough: its scope must cover *all* major concepts of common-sense reasoning. This amount to claiming that the a-c graph is fairly strongly connected: there are no really isolated subgraphs.

This correlation between thoroughness and fidelity is a matter of degree. To achieve greater fidelity, one will need greater thoroughness. To *really* capture the notion of 'above', it is probably not enough to stay even within naïve physics: one would have to go into the various analogies to do with interpersonal status, for example. (Judges' seats are raised: Heaven is high, Hell is low: to express submission, lower yourself, etc.) Only a very *broad* theory can muster the power (*via* the Galois connection of model theory) to so constrain the meaning of the token 'above' that it fits to our concept *this* exactly. (Imagine a world in which the 'status' analogy was reversed, so that to be *below* someone was to be dominant and/or superior to them. That would be a possible model of naïve physics, but not of the larger theory of common sense: and it would be a very different world from ours.) A formalization must be deep without being broad, and must be deep to be dense: so a dense formalization must be deep and broad.

Clusters are exactly partial closures in this sense. A cluster contains a group of concepts which close in on one another to *some* extent: one does need other concepts, but within the cluster there are a lot of things to be said about the cluster's own concepts. Clustering is also, therefore, a matter of degree, and depends upon the fidelity, the level of detail.

The whole programme of tackling naïve physics in isolation from other parts of common sense is based on the view that there is a level of detail at which naïve physics forms a reasonably close cluster in the larger conceptual structure, and that this is a rich but tractable level of detail: it represents, I believe, an order of magnitude more thoroughness than has yet been achieved; but not, say, ten orders of magnitude.

8. SOME LIKELY CLUSTERS AND THEIR CONCEPTS

In this section I sketch some concrete ideas for concept-clusters. These are only sketches, and vague ones at that: fuller accounts will appear elsewhere, eventually. I do not want to suggest that this is in any way an exhaustive list.

It is quite likely that many of these are not clusters in any real sense. They may, for example, split into pieces on closer investigation; or, new links may be revealed which blur the edges of the clusters. Nevertheless, they seem to me to be quite good places to start exploring.

Measuring Scales

We need to be able to express *quantities* such as size, extent, amount (of a liquid or powder), weight, viscosity, etc. These seem to be properties of things. But we can measure weight, for example, in pounds or in kilograms, so we have to introduce the notion of various measuring scales which measure the same physical quantity. We could have various functions from objects to (say) the rationals, called weight-in-lbs and weight-in-kilos, etc.: but this is awkward, unnatural, and does not support a very dense collection of axioms. I think we should introduce a notion of an 'abstract space' of *weights* (*sizes*, *amounts*), so that *weights* is a function from objects to weights, and *lbs* (etc.) are functions from rationals to weights, and we can write:

weight (Fred) = pounds (150.32) = kilos (68.25)

These *measure spaces* of weights, sizes, etc. have, I think, a theory of their own. They probably have the structure of a tolerance space (Zeeman 1962), i.e. they have a finite 'grain'. They are notions of approximation, nearness, 'typical' measures of various kinds (normal-sized for an elephant), of inequalities, and of other related matters: and I guess much of this is independent of the particular quantity being measured.

One remark which may be apposite here is this. It is often argued that 'common sense' requires a different, fuzzy logic. The examples which are cited to support this view invariably involve fuzzy measuring scales or measure spaces. This, I believe, is where fuzziness may have a place: but that is *no* argument for fuzzy truth-values.

Shape, Orientation, and Dimension

Physical 3-dimensional shape. This cluster is not much investigated, it seems to me, in spite of the considerable work in robot manipulator languages (cf. Bolles 1976). It is also related to topics in visual perception, and as there is more work here, this may be a good way into it. I wish there were more I could say about shape, but I can make only a few loose remarks.

For naïve physics, vertical gravity is a constant fact of life, so vertical dimensions should be treated differently from horizontal dimensions: 'tall' and 'long' are different concepts. An object's shape is also often described differently when it is against a rigid surface (such as a wall) than when it is free-standing (width and length; or depth—from the wall—and width *or* length along the wall: width if one thinks of the object as being *put against* the wall, length if one thinks of it as *running along* the wall). I suspect— the details have not been worked out—that these differing collections of concepts arise from the reconciliation of various co-ordinate systems. A wall, for example, defines a natural co-ordinate system with a semi-axis along its normal.

An important aspect of shape is the relationship of surfaces to solids and edges to surfaces. The different names available for special cases indicates the richness of this cluster: top, bottom, side, rim, edge, lip, front, back, outline, end. Roget's thesaurus (class two, § two) supplies hundreds more. Again, these are *not* invariant under change of orientation, especially with respect to the gravity vertical. Such boundary concepts are also crucial in describing the shape of space, and are the basis of homology theory and differential geometry.

Inside and Outside

Consider the following collection of concepts: (inside), (outside), (door, portal, window, gate, way in, way out), (wall, boundary, container), (obstacle, barrier), (way past, way through).

I think these words hint at a cluster of related concepts which are of fundamental importance to naïve physics. Ths cluster concerns the dividing up of 3-space into *pieces* which have physical boundaries, and the ways in which these pieces of space can be connected to one another; and how objects, people, and liquids can get from one such *place* to another.

There are several reasons why I think this cluster is important. One is merely that it seems so, introspectively. Another is that these ideas, especially the idea of a *way through* and the things that can go wrong with it, seem widespread themes in folklore and legend, and support many common analogies. Another is that these ideas have cropped up fairly frequently in looking at other clusters, especially liquids and histories (see below). Another is that they are at the root of some important mathematics, viz. homotopy theory. But the main reason is that *containment limits causality*. One of the main reasons for being in a room is to isolate oneself from causal influences which are operating outside, or to prevent those inside the room from leaking out (respectively: to get out of the rain, to discuss a conspiracy). A good grasp of what kind of barriers are effective against what kinds of influence seem to be a centrally useful talent needed to be able to solve the 'frame problem'.

It is interesting to contrast these ideas with ideas of shape. Here, we are concerned with space as a place to be in: *room to move,* as it were; whereas in describing shape, space is *space occupied* by a substance. There are many concepts useful in both areas, however.

Histories: Describing Happenings

The now-classical approach to describing actions and change, pioneered by John McCarthy, is to use the concept of *state* or *situation*. This is thought of as a snapshot of the world at a given moment: actions and events are then functions from states to states. This framework of ideas is used even by many who deny that their formalism contains state variables, and has been consciously incorporated into several AI programming-languages. I now think however, that it is a mistake or a least a gross over-simplification.

Consider the following example (which Rod Burstall showed me many years ago, but which I did not appreciate at the time). Two people in New York agree to meet a week later in London. Then they go their separate ways: one to Edinburgh, one to San Francisco. Each of them leads an eventful week, independently of the other, and they duly meet as arranged. In order to describe this using situations, we have to say what happens to each of them after every event that happens to the other: for each situation, being conceptually a state of the whole world, encompasses them both.

What we need is a notion of a state of an event which has a restricted spatial extent. By a *history* I mean such an object, viz. a connected piece of space-time, typically bounded on all four dimensions, in which 'something happens' (where I intend this to include the special case of nothing happening).

A 3-dimensional spatial cross-section of a history is a place at a certain moment, i.e. a state of that place. Places can be larger or smaller, and can be nested inside one another: a room, a hotel, a street (understood as being the space inside all the buildings which open onto the street), and a city can all be places. A typical history will be, for example, the inside of a certain room from 1.00 p.m. until 4.00 p.m. on a certain afternoon. Space is, conceptually, made up of places, and space-time is made up of histories, fitted together in a jigsaw pattern. One can also think of a history as *the extension of* (an occurrence of) a process.

Any well-defined object or piece of space can be trivially extended into a history by multiplying it (in the algebraic direct-product sense) by a time-interval, but there are also somewhat less simple histories, such as trajectories, which 'slope' in space-time.

It is very useful to be able to refer to the *shape* of a history as well as of an object or piece of space. For example, a column of falling water (pouring from a jug, for example) defines a history, the shape of which is a vertical cylinder. The top and bottom of this cylinder are of some importance. in relating this history to others in which liquid is moving.

Histories can be related to one another in various ways. There are *adjacency* relationships, both spatial (e.g. vertically-above and touching, as in a column of water falling onto a table-top) and temporal (e.g. immediately-following, as in touching a switch, thus starting a motor), and hybrid (as in a collision between aircraft, which is the intersection of two trajectories). There are shape properties and relative-position relationships between places and hence between histories. There are the spatial containment relationships similarly inherited from places and objects, and also temporal containment relationships ('while'). And there are relationships between histories and various global co-ordinate systems, both of space and of time: what we might call the *address* of a history. There are many possible addressing systems, not all being metric co-ordinate frames, e.g. room-numbering systems within a building. All of these define what might be called a naïve geometry of space-time.

Not every patch of space counts as a well-defined place, and not every patch of space-time as a history. There have to be 'natural' boundaries defining the edges. What counts as a natural boundary is (deliberately) open-ended, but physical barriers are obvious examples, such as the walls of a room.

Since places can be nested (indeed, perhaps every place is *inside* some other place), every event is contained in many (perhaps infinitely many) histories. But for each type of event, there is a smallest which strictly contains it, viz. the smallest one which is spatially bounded by barriers which are opaque to the causal consequences of that sort of event. (*Roughly,* this place is the natural answer to the question 'where did (the event) happen?' (possible answers—in front of the desk; in the living room; in that house: in London)). The importance of this idea has already been mentioned: such barriers limit the extent to which causal consequences of events have to be pursued, and hence make prediction easier. One can predict, from a *static* description of the barrier-geometry, that various kinds of event can affect only a few histories. For example, only a restricted class of very unusual events taking place in a closed room (large explosions, flooding, large fires), can directly affect histories outside the room.

There are several other important increases in expressive and predictive power which histories give us, compared to the classical situations/actions ontology, but it would take too long to go into more detail here. A fuller account is in preparation.

Energy and Effort

In making predictions, there is a distinction which seems crucial between events which can 'just happen' (such as fallings) and events which require some effort or expenditure of energy (such as rocks flying through the air). The point being, of course, that if there is no effort being made in a given history, then the latter kind of event is ruled out.

Such a distinction runs counter to the law of conservation of energy, and I think quite correctly so for naïve physics (or we could say merely that the intuitive concept of 'effort' does not exactly correspond to the physics notion of 'work'). There are many everyday situations where energy is expended with little apparent result (hitting a nail into a brick, for example).

I am not sure what else can be said about the concept of effort: perhaps that sources of effort have a finite capacity (they wear out or get tired). It may be that is not so much a cluster as a concept which loosely links together a number of other clusters.

Agents can be sources of energy, but the two concepts are distinct, since some actions require no energy (speaking, for example), and there are sources of energy which have no volition. However, it may be that the two are equivalent notions within naïve *physics,* and can only be distinguished by the use of 'psychological' concepts such as volition.

Assemblies

Many solid physical objects are made from parts assembled together in some way: while others are simply a piece of (some kind of) stuff, such as a block of wood. There are a number of concepts which are connected with this idea of an assembly: notions such as being a component of, being a part of (for example, one's hand is a part of one, but not a component: one's liver is (arguably) a component: it is, as it were, detachable), attachment point, assembling and taking apart; glueing, nailing, screwing, etc. . . . There are also notions to do with the ways in which assembled parts can move relative to one another (shafts, pulleys, keyways, hinges), and these connect to the spatial-geometry concepts of shape and movement. And there are concepts to do with the mechanical properties of different kinds of stuff: rigidity, hardness, flexibility, being able to cut easily, being able to be glued, etc.

Support

Objects (or liquids) fall, if left to themselves. To stop them falling they must be supported. I think we can make a short list of all the ways in which a thing can be supported, as follows:

1. Something is underneath it, holding it up. Of course, this must be supported too, and so on; but the *ground* does not require support: *it* is the bottom of all support relationships (from which it follows that the ground is not an object).

2. Something is above it, and it is *hanging from* that thing.

3. Something is alongside it, and it is *attached to* that thing (cf. assemblies).

4. It is *floating* on some contained liquid.

5. It is *flying,* i.e. holding itself up in some way, without touching anything solid. This requires great effort on the part of the flying object, so that inanimate, 'passive' objects cannot fly (although kites may be an exception).

Of these, (1) is by far the safest: in all the others, a failure of some component (respectively: a breaking (e.g. of string), a detaching, a leak, a cessation of effort: all of these are histories of one kind or another) which can mean the ending of the support and hence the sudden beginning of a falling history, and falling histories have dangerous endings (typically). Hence, there is a mini-cluster of concepts around the idea of 'support-from-below': notions such as pile, tower, wall, etc.: stability and the ways it can fail (falling over, crumbling, coming apart, sliding, subsidence).

Substances and Physical States

There are different kinds of *stuff*: iron, water, wood, meat, stone, sand, etc. And these exist in different kinds of *physical state*: solid, liquid, powder, paste, jelly (jello for US readers), slime, paperlike, etc. Each kind of stuff has a *usual* state: iron is solid, water is liquid, sand is powder, etc., but this can sometimes be changed. For example, many stuffs will melt if you make them hot enough (which for some things is *very very* hot, i.e. *in practice* they can't be melted, e.g. sand; and others will *burn* when heated, e.g. wood or flour). Any liquid will freeze if you make it cold enough. Any solid *can* be powdered if you pulverise it with enough effort and determination, etc. There is no obvious standard way of changing a powder into a solid (but wetting it to get a paste, then drying the paste carefully, sometimes works).

Sometimes we have a separate concept for the same substance in two different states. Sand and rock are a good example. This is, I think, worthwhile when it is (1) extremely difficult to convert one to another, and (2) both occur in nature. (Contrast iron filings, which satisfies (1) but not (2).)

Some substances, left to themselves, *decompose*, i.e. change slowly into some other (useless) substance; or *mature*, i.e. change slowly into some other (useful) substance. Rusting and wet rot are examples of decomposition, cheese-making an example of maturation.

Every physical object which is not an assembly must be made of some stuff, and many of the properties of the object are in fact properties of the stuff of which it is made (rigidity, colour—unless it has been painted—hardness, etc.). It is, I think, important to separate these properties of a thing from those which are essentially connected with the shape or structure of the thing. Some objects are essentially defined by one kind of property (a lump of lead), others by other (a building block). Properties such as *weight*, which involve both size and material, have different implications depending upon the kind of object: a heavy lump of lead is a big lump of lead, whereas a heavy building brick must be made of some especially dense material. Solid objects must be made of materials whose physical state is solid, for only solids can be said to have a shape. It follows that if a solid object is heated to the melting-point of the material of which it is made, it must cease to exist as an object, since the requisite state of its substance no longer obtains. This is, I think, a very convincing account of melting.

Cookery would seem to be a good area in which to explore the possible transitions between physical states of various semi-solid substances, and manufacturing processes (moulding, casting, forging) another. I think the various methods of measuring amount and quantity might also be a useful area to explore. Wood and metal, for example, are wholesaled by weight or volume (basically), but retailed in various different systems depending upon the shape of the pieces (striplike or surfacelike or solid).

Forces and Movement

Naïve physics is pre-Galilean. I can still vividly remember the intellectual shock of being taught Newtonian 'laws of motion' at the age of 11: how could something be moving if there was no force acting on it? It is interesting to read Galileo's 'Dialogue Concerning the Principal Systems of the World' (1632), where he argues very convincingly, from everyday experiences, that Newton's first law must hold. But it takes a great deal of careful argument, and relies on the reader having some experience of smooth, polished surfaces and near perfect spheres. Another non-Newtonian intuition which every child has is that a released slingshot travels radially outwards, rather than tangentially. Other examples can be found.

If an object is moving, there are only five possibilities. It may be *falling*; or it may be being *pulled or pushed* by something; or it may be moving *itself* along, expending effort as it does so (and therefore cannot be a passive object); or it may be *sliding* (by which I mean to include sliding on a slippery surface or down a slippery slope); or it may be *rolling,* in which case it must either be or have rollers or wheels. The last two cases can keep on going for a while after all effort has ceased. (We would call this phenomenon *coasting*: a gesture to Galileo.) This list does not include rotary or oscillatory motions: it is meant to cover all movements in the sense of *change of position*.

I believe there may be actually two distinct ways of conceptualizing motion: as a displacement or as a trajectory. Displacement motion requires effort, and when the effort stops the motion stops: and it is characteristically under constant servo-control relative to position, i.e. it is conceptualized as *change of position*. Trajectory motion has inertia, keeps going unless stopped (when there is an impact), is characterized as smooth motion *along a path*. It requires effort to start up, to stop, or to alter, but less effort, or no effort, to maintain. Examples are a thrown projectile, a car, sliding on ice, jumping. Displacement motion is Greek, trajectory motion is Galilean. Concepts such as aiming, impact, velocity (as a measure space), acceleration, are connected with the latter: concepts such as going, coming, dodging, avoiding, movement towards, away, are connected with the former. Both displacements and trajectories are histories, but the former are, essentially, merely transitions from their beginnings to their ends, which are positions of an object, whereas the latter have a definite *shape*: they can be extrapolated in time, for example, whence the concept of aiming. Falling, slidings, rollings, and jumpings are examples of trajectories.

Forces can be transmitted in various ways. A rigid body can transmit a push, a string can transmit a pull. Luger and Bundy (1977) consider this mini-cluster in some detail.

Liquids

Liquid substances pose special problems, since, unlike pieces of solid stuff, 'pieces' of liquid are typically individuated *not* by being a particular piece of liquid, but by being in a particular *place* (a lake), or in some special relationship to a solid object (inside a cup). In Hayes (1978b) I enlarge on an approach to solving these problems.

9. SOME STRUCTURED FORMALIZATION TECHNIQUES

Constructing axiomatic formalizations which are 'heuristically adequate' (McCarthy and Hayes 1969) is an art, rather as writing good programs is an art. It is not yet very well developed (and, indeed, one of the main aims of tackling naïve physics is to gain skill in this relatively unexplored area), but a few stylistic points seem to be emerging.

One is the importance of *taxonomies*: finite exhaustive lists of the various types or categories of a kind of thing or of the possible states of a thing. We have seen these emerging already: ways of being supported, kinds of physical states, possible states of a fluid (there are six: contained, flowing, spreading, wetting, falling, or flying). In each case we have a group of axioms with the following form:

$$\varphi(x) \; = \; \varphi_1(x) \vee \ldots \vee \varphi_n(x)$$
$$\varphi_1(x) \supset T_1$$
$$\vdots$$
$$\varphi_n(x) \supset T_n$$

where the T_i are theories of what these particular cases are like. Such exhaustive lists can be very useful in making inferences, by a generalization of the graphical consistency-checking computations widely used in vision (cf. Mackworth 1977). Intuitively, the iff ($=$) means that if all but one of the disjuncts can be ruled out, then the remaining one must be the case. If the collection (T_i) of subtheories is appropriately structured, then this can be a powerful technique for obtaining short proofs. For example, in the case of *support,* we can rapidly infer that if a passive thing is held up by a string (and that's all), and the string breaks, then it must fall: for there is nothing underneath (case 1), there is no liquid for it to float on (case 3), it isn't attached to anything (case 2), and it can't fly: so it must be supported. So (by the basic *support* axiom), it must fall (i.e. this moment is the beginning of a falling history). A similar way of arguing can be used to show that water which flows to the edge of a table will fall (rather than, say, keep going horizontally, or pile up at the edge).

It may be significant that such taxonomies have the syntactic form of a definition (of φ in terms of the φ_i), but do not serve the role of definitions since the 'defined' token already occurs elsewhere in the axiomatization.

A second stylistic point concerns existence and comprehension axioms. As we pointed out earlier, axioms which establish the existence of entities are vital in a formalization which is to have non-trivial models. Examples we have met include the spaces (places) defined by physical boundaries (rooms, insides of cups), or by various (metric *and* non-metric) co-ordinate systems, and the histories which ensue when various states obtain, e.g. the falling which inevitably follows a state in which an object is unsupported. In these, and no doubt other, cases, there will be *comprehension axioms* which assert the existence of the required entity, and its relationship to the entities already established (the space *between* the walls: or *behind* the door: the falling *after* (and below) the moment (and place) where the object loses its support, and so on).

The point I want to make here, however, is that these are all *restricted* comprehension axioms. We cannot take arbitrary pieces of 3-space or 4-space-time and treat them as individuals: only ones which are related in a describable way to entities we already have confidence in, as it were. This selectiveness in ontological commitment is one of the characteristic differences, I believe, between common-sense reasoning and 'hard' scientific or philosophical reasoning. Common sense's ontology is prolix — entities of all sorts, concrete and abstract (objects, materials, colours, spaces, times, histories, events, . . .) are used with scant concern for philosophical tidiness, little appeal to an underlying ontological simplicity (compare subatomic physics, for example, or even Chemistry's periodic table): and yet it is also very controlled: contrast Goodman's (1966) nominalism, or axiomatic set theory, or the comprehension axiom scheme of the typed λ-calculus. The effect of both of these differences is to give a far more richly structured collection of entities: *fewer* than in these 'uniform' formalizations, but of far more *sorts*, and with a much richer collection of *kinds of relationship* between them.

One further point here: the use of global metric co-ordinate frameworks essentially restores the unrestricted comprehension by the back door. For, by using suitable co-ordinates, we can describe *any* piece of 3-space (air traffic corridors, for example, which have no physical boundaries at all), or *any* piece of space-time or *any* piece of fluid (e.g. the cubic centimetre whose top north-east corner was 5 cm below the surface at such-and-such a place in a certain river, at 19:30:06.8 hours on the 24 May 1962). The resulting ontological *freedom* and uniformity is probably one reason why co-ordinate systems are so useful in (real) science.

10. WHY IT NEEDS TO BE DONE

I take it as obvious that the construction of a program which can be said to have common sense must involve, ultimately, the formalization and common-sense knowledge such as naïve physics in some way or other (and also, of course, naïve psychology, naïve epistemology, etc.). While there are those who would disagree with this — those, for example, who believe that a simple uniform learning procedure might eventually exhibit intelligence — their pre-theoretical assumptions differ so much from those made by most AI workers that it is better, I think, to regard such work as belonging to an essentially different field. I shall, in any case, not discuss this particular issue further here.

There is real room for disagreement, however, on methodology. The most dominant view within AI seems to be that it is necessary to construct a 'complete' program in order to demonstrate that one's ideas on representation are feasible. Working systems which exhibit an impressive total behaviour are taken to be the ultimate criterion of success. So strong is this requirement, indeed, that in many centres it is difficult for a student to obtain a Ph.D. unless he has implemented an impressive working program. The naïve physics proposal, as outlined here, deliberately avoids the construction of such a complete program. Our aim is to construct a formalization which defines a *heuristically adequate* search space of possible inferences. Questions of how, exactly, to search this space; of controlling an interpreter; of information retrieval and relevancy—what might be called computational questions—as well as questions of the choice of data-structures; how to implement fast searches; the choice of programming-language—what might be called implementational questions—all will be deliberately ignored.

Relatively few workers in AI are adopting a similar methodological position, and yet I believe that it is vital to adopt it, in order to make a substantial progress on representational issues. McCarthy (1977) makes some similar arguments.

It is not just a question of diverting resources, although this is an important issue. The more fundamental reason is, that quick success in constructing a complete working program seems necessarily to involve making simplifications and restrictions which make it impossible to tackle the essential representation problems. This happens in two ways.

Firstly, to achieve success in making an AI performance program, one must choose the domain of the program very carefully indeed. To be tractable, it must be restricted in some way, usually fairly drastically. A standard style of restriction is to limit the scope of the program: a restricted subject-matter for a reasoning program, a restricted vocabulary for a micro-world for a natural-language program, a restricted range of seeable objects for a vision system, etc. It follows, then, that the representation needed by the program is one which is not particularly *thorough*, in the sense used earlier. Moreover, the representation often can (and often does) rely on this restricted scope in using techniques which work for small 'toy' worlds, but which are simply inapplicable to more thorough uses. I have noted several examples above.

Secondly, computationally effective representations tend to be of rather low density. There is a good reason for this: a dense representation necessarily defines a large and explosively expanding search space of possible inferences. *If* the only heuristic devices available for controlling an inferential search are weak and general (numerical) heuristics (depth-first search as in MICROPLANNER, local procedure invocation as in KRL-0, etc.), then effective computational behavior cannot be achieved with such a search space. But these weak, general methods are essentially the only ones we know: hence, to be computationally effective, we must have sparse representations.

These two pressures, then, taken together, encourage the construction of sparse representations of limited scope which are tailored carefully to the particular desired behavioural repertoire of the task-domain chosen for the performance of the program. But, as I have argued, thoroughness and density are essential properties of a representation which can be said to capture the meaning of common-sense knowledge adequately.

This is a methodological point, but there is a closely related point concerned with adequacy. A popular view in AI is that only the adequate performance of a program can be a criterion of success of an AI theory. (Indeed, I suggest elsewhere (Hayes 1978*d*) that this criterion is exactly what distinguishes AI from 'information-processing psychology'.) Accepting this, it is only a small step to accepting a sort of behaviourist criterion of adequacy for a representation: viz, that it support an adequate behaviour in some performance program. This criterion would accept a sparse, limited representation over a dense, thorough one, given the present state of the implementer's art. If the program *works*, so the argument would go, then its representation must adequately capture the intended meanings: for that is what we *mean* by 'adequate'.

The problem with this position, at least in its simpler versions, is that it takes no account of scale effects. One cannot make a program which behaves adequately in a large 'world' by any simple process of adding together programs which perform adequately in a number of smaller subworlds of that world. At least, this is so far the everyday world of common sense. The world just doesn't split up that neatly: one needs the interaction between the parts as well. Thus, a representation of, say, the blocks-world which is adequate for reasoning merely about blocks, on this criterion, will be less adequate for reasoning about blocks in the context of liquids, strings, rods, friction, pulleys, etc. And the pressures just noted, towards tailoring representations of limited scope, militate against what might be called upward compatibility of formalizations. Thus, even accepting the criterion as an ultimate test of an AI theory, which I do, I would still argue that applying it too rigorously too soon (typically, after 3 years work) is self-defeating. *We are never going to get an adequate formalization of common sense by making short forays into small areas, no matter how many of them we make.*

It might be objected that if all this is true, then a dense, thorough formalization *cannot* be part of an effective AI program. But this is false. I have argued that weak, general methods of control are unable to handle dense, thorough formalizations: clearly, we need more powerful control methods. There is one idea which I introduced in Hayes (1973) — see also Kowalski (1977), Pratt (1977), McDermott (1976), Davis (1976) — which suggests how to achieve the kind of power needed: to regard control not as a problem of defining a mechanism, but as itself a representational problem. We need to formalize the knowledge of *how* to make inferences, as well as the knowledge of the world which makes inferences possible. This meta-information can itself take part in the reasoning process, but also has a different and special relationship to the deductive interpreter: it describes its activity, rather than merely being grist for its mill. The development of formalisms, and associated interpreters, for expressing such meta-knowledge seems to me to be one of the most important tasks facing AI. But — and this is the crucial point for the present argument — the structures of both the formalisms and the meta-formalizations expressed in it, depend upon those of the world-knowledge formalism and the formalizations expressed in it. We cannot develop meta-formalizations in a vacuum (unless they be merely formalizations of the weak, general heuristics we already have); we must first have some realistically complex examples of common-sense formalizations, whose deductive properties will be described in the meta-formalization.

11. WHY IT CAN BE DONE

A different objection to the naïve physics proposal is that it is impossibly ambitious: that we don't know enough about formalizations to embark on such a large representational task; that it would take centuries, etc. Ultimately the only answer to such objections is to make the attempt and succeed, so all I can do here is to convey my reasons for feeling optimistic. There are four.

The first is based on my recent experiences in tackling the 'liquids' problem, which I have long believed was one of the most difficult problems in 'representation theory' (Hayes 1975). The idea of quantifying over pieces of space (defined by physical boundaries) rather than pieces of liquid, enables the major problems to be solved quite quickly, to my surprise. The key point here was finding the correct procedure for *individuating* a liquid object: the criterion by which one could refer to such a thing. I believe a similar concern for individuating criteria may well lead to progress in other clusters as well. McCarthy (private communication 1977) has, for example, begun a new approach to epistemic formalizations based on the individuation of 'concepts', i.e. thoughts in people's heads.

The second reason for optimism is the idea of histories outlined earlier. I believe that formalizations of the physical world have been hampered for years by an inadequate ontology for change and action, and that histories will provide a way round this major obstacle. The third reason is based on the no-programming methodology already discussed. To put it bluntly: hardly anybody has *tried* to build a large, heuristically adequate formalization. We may find that, when we are freed from the necessity to implement performance programs, it is easier than we think.

The fourth reason is that there is an obvious methodology for getting it done, and this methodology has, in recent years, proved very successful in a number of areas.

12. HOW TO GET IT DONE

There is a tried and true way of getting knowledge out of people's heads and into a formalization. Within AI, it has been called 'knowledge engineering' by Feigenbaum (1977); but essentially the same technique is used by linguists. It works as follows. In consultation with an 'expert' (i.e. a human being whose head contains knowledge: one knows it does because he is able to do the task one is interested in), one builds a preliminary formalization, based upon his introspective account of what the knowledge in his head is. This formalization then *performs* in a particular way, and its performance is compared with that of the expert. Typically it performs rather badly. The expert, observing this performance of the formalization in detail, is often able to pinpoint more exactly the inadequacies in his first introspective account, and can offer a more detailed and corrected version. This is formalized, criticized, and corrected: and so on. Typically, the expert, continually confronted with the formal consequences of his introspections, becomes better at detailed introspection, as time goes by.

In 'knowledge engineering', the expert is a specialist of some kind, and the formalization is, typically, a collection of condition-action rules which can be run on a suitable interpreter: a very modular program, in a sense. In linguistics, the formalization is a grammar of some sort which assigns syntactic structures to sentences, and the expert is a native speaker: indeed, the expert is usually the linguist himself. In both areas, the technique has proved extremely successful.

I believe this process of formalization, confrontation against intuition, and correction, can also be used to develop naïve physics. Here is a domain in which we are all experts, in the required sense. The *performance* of a formalization is, here, the pattern of inferences which it supports. Performance is adequate when the 'experts' agree that all and only the immediate, plausible consequences follow from the axioms of the formalization. (In fact, there is a weak notion of adequacy: the stronger notion would be that the *derivations* of the plausible consequences were also plausible. Attempting to use this stronger notion gives rise to severe methodological problems, since it requires one to have '2nd-order' introspections. Linguistics has an exactly analogous notion of strong adequacy for a grammatical theory, and suffers exactly similar methodological difficulties.) It seems to be sound to have several 'experts' involved, as it is easy to miss some obvious distinctions when working alone.

The ideal way to make progress is to have a committee. Each member is assigned what seems to be a cluster, and has to try to formalize it. They tell one another what they require from the other clusters: thus the 'histories' cluster will need some 'shape' concepts, and the 'assemblies' cluster will need some 'histories' concepts, and so on. Fairly frequently, the fragmentary formalizations are put together at a group meeting, criticized by other members (in their common-sense 'expert' role), and tested for adequacy. I will anticipate that some clusters will dissolve, and new ones will emerge, during these assembly meetings.

Initially, the formalizations need to be little more than carefully-worded English sentences. One can make considerable progress on ontological issues, for example, without actually *formalizing* anything. Fairly soon, however, it will be necessary to express the intuitions formally. Here, I think one should be liberal in allowing a free choice of formal language. Many people find framelike notations agreeable: others like semantic networks, etc. There is no reason why such superficial variants of first-order logic, or even more exotic formalisms, should be banned: the only important requirement is that the inferential relationships between the various formalisms should be made explicit. In practice, this means that they should all be translatable into predicate calculus: but this is no problem, since they all are. A more serious point is that particular clusters may suggest special *ad-hoc* representations. Shapes may be represented diagrammatically, for example. One can imagine a cluster, represented in some idiosyncratic way, whose internal inferential relationships were inaccessible from outside, but which was interfaced to the rest of the formalization by a defined translation of part of itself into the reference formalization (first-order logic): say statements of relative position and orientation. It will be difficult to prevent such things happening, and maybe one should not try. But there are grave dangers, since this way of proceeding prejudges the possible interactions in the formalization as a whole, and this may encapsulate a serious mistake in a way which will be hard to detect and even harder to rectify.

There are several other ways to find concept-clusters. For example: looking in a thesaurus; choosing a particular domain (cookery, volumetric measurements of various substances) and attempting to describe it; analysing some everyday act in detail (e.g. spreading a sheet over a bed by holding two corners and flicking: why does that work?). I expect these, and others, will be useful starting-points.

13. IS THIS SCIENCE?

It will be objected that to attempt to formalize knowledge is the abstract, i.e. divorced from a particular sensory modality or task-domain, is unscientific because there are no clear criteria of success or failure. What would it be like to fail? If this question cannot be answered, then naïve physics is mere literary criticism.

I think this objection is well taken and needs a more adequate reply than I am currently able to give. The point is, that one can always get *somewhere* with a formalization: who is to say that where one has got to is not far enough? One can only say, I think, that people's common intuition is the guide. If there are some 'obvious' physical facts which cannot be made to follow 'easily' or 'naturally' from the axioms, then more work has to be done. One can apply the usual scientific judgements of 'elegance', 'economy', etc. to compare rival formalizations. (All the quoted words cry out for further discussion, which I will not attempt here.) It is worth remarking that linguistics is in *exactly* the same position, and regularly squirms on the methodological hook: judgements of physical plausibility and elementary causality are certainly as reliable as judgements of grammaticality by native speakers; and indeed are largely independent of cultural and linguistic boundaries, so are probably rather more reliable as source data. (Which may suggest that we should borrow the competence/performance distinction to protect ourselves from behavioural refutation: but there are deep problems indeed about trying *that* trick.)

It would be nice if naïve physics could be linked more closely with the mass of data now available in Piagetian psychology on the development of physical concepts during childhood (although none of this is uncontroversial, it seems). Certainly, *compatibility* with this data should be a constraint on naïve-physics formalization. It is however a very weak constraint, since the data is compatible with many different developmental theories, and is not usually sufficiently detailed to distinguish one from another (see Prazdny 1978 for some comments in this direction). I would hope that the construction of a naïve physics might show up some new mechanism of developmental change in conceptual frameworks. To some extent this is happening already, since building a formalization is often a matter of *developing* (in exactly the right sense) the partial formalizations one already has.[4]

[4] This paper was written during a sabbatical visit to the Institut pour les études sémantiques et cognitives, Geneva. I am grateful to the Directrice, Mme M. King, for inviting me there; and to all members of the Thursday seminar, especially Maghi King, Giussepe Trautteur, and Henri Wermus. Conversations with Mimi Sinclair on Piagetian research were also of great help. My wife, Jackie, typed several drafts of the manuscript and was an unfailing source of reliable, sound, common-sense intuitions.

REFERENCES

Anderson, B., *et al.* (1972). 'Beyond Leibnitz.' *Memo AIM*, Stanford AI Project.

Binford, T. O., *et al.* (1976). 'Computer Integrated Assembly Systems.' *Memo AIM-285*, Stanford AI Project.

Bolles, R. C. (1976). 'Verification Vision within a Programmable Assembly System.' Stanford AI Memo No. 295.

Bundy, A. M. (1978). 'Exploiting the Properties of Functions to Control Search.' (To appear.)

Davis, R. (1976). 'Applications of Meta-level Knowledge to the Construction, Maintenance and Use of Large Knowledge Bases.' *HPP Memo 76-7*, Stanford University.

Feigenbaum, E. A. (1977). 'Themes and Case Studies of Knowledge Engineering.' *Proc. 5th IJCAI Conference*, pp. 1014–29, MIT.

Goodman, N. (1966). *The Structure of Appearance.* New York: Bobbs-Merrill Co.

Hayes, P. J. (1971). 'A Logic of Actions.' *Machine Intelligence 6.* Edinburgh: Edinburgh University Press.

—— (1973). 'Computation and Deduction.' *Proc. 2nd MFCS Symposium*, Czechoslovakian Academy of Sciences.

—— (1975). 'Problems and Non-Problems in Representation Theory.' *Proc. 1st AISB Conference*, pp. 63–79, Sussex University.

—— (1977). 'In Defence of Logic.' *Proc. 5th IJCAI Conference*, pp. 559–65, MIT.

—— (1978a). 'The Logic of Frames.' [In D. Meitzing (ed.), *Frame Conceptions and Text Understanding*, pp. 46–61. Berlin: Walter de Gruyter, 1979.]

—— (1978b). 'Naïve Physics I: Ontology of Liquids.' Working Paper 35, Institute for Semantic and Cognitive Studies, Geneva.

—— (1978c). 'Naïve Physics II: Histories.' (In preparation.)

—— (1978d). 'On the Difference between Psychology and Artificial Intelligence.' *AISB Bulletin*. (To appear.)

Kowalski, R. A. (1977). 'Algorithm = Logic + Control.' Memorandum, Imperial College, London.

Landin, P. J. (1970). 'A Program-Machine Symmetric Automata Theory.' *Machine Intelligence 5*, pp. 99–119, Edinburgh: Edinburgh University Press.

Luger, G., and Bundy, A. M. (1977). 'Representing Semantic Information in Pulley Problems.' *Proc. 5th IJCAI Conference*, p. 500, MIT.

McCarthy, J. (1977). 'Epistemological Problems of Artificial Intelligence.' *Proc. 5th IJCAI Conference*, pp. 1038–44, MIT.

—— and Hayes, P. J. (1969). 'Some Philosophical Problems from the Standpoint of Artificial Intelligence.' *Machine Intelligence 4*, pp. 463–502. Edinburgh: Edinburgh University Press.

McDermott, D. V. (1976). 'Flexibility and Efficiency in a Computer Program for Designing Circuits.' Ph.D. thesis, MIT AI Lab.

—— (1977). 'Artificial Intelligence and Natural Stupidity.' *SIGART Newsletter*, pp. 4–9. Repr. in J. Haugeland (ed.), *Mind Design*, pp. 143–60. Cambridge, Mass.: MIT Press.

Mackworth, A. K. (1977). 'Consistency in Networks of Relations.' *Artificial Intelligence* 8: 99–118.

Pratt, V. (1977). 'The Competence-Performance Distinction in Programming.' *Proc. 4th ACM Symposium on Principles of Programming Languages,* Los Angeles.

Prazdny, K. (1978). 'Stage Two of the Object Concept Development: A Computational Study.' Memorandum, Essex University.

Schank, R. C. (1975). *Conceptual Information Processing*. Amsterdam: North-Holland.

Ullman, S. (1977). 'The Interpretation of Visual Motion.' Ph.D. thesis, MIT.

Wilks, Y. A. (1975). 'A Preferential, Pattern-Matching Semantics for Natural Language Understanding.' *Artificial Intelligence* 6: 53–74.

—— (1977). 'Good and Bad Arguments about Semantic Primitives.' Memo 42, AI Dept., Edinburgh University.

Winograd, T. (1972). *Understanding Natural Language*. Edinburgh: Edinburgh University Press.

Zeeman, W. P. C. (1962). 'The Topology of the Brain and Visual Perception.' In K. Fort (ed.), *Topology of 3-Manifolds*. Englewood Cliffs, NJ: Prentice-Hall.

9

A CRITIQUE OF PURE REASON

DREW McDERMOTT

In 1978 Patrick Hayes promulgated the *Naïve Physics Manifesto*. (It finally appeared as an 'official' publication in Hobbs and Moore 1985.) In this paper, he proposed that an all-out effort be mounted to formalize common-sense knowledge, using first-order logic as a notation. This effort had its roots in earlier research, especially the work of John McCarthy, but the scope of Hayes's proposal was new and ambitious. He suggested that the use of Tarskian semantics could allow us to study a large volume of knowledge-representation problems free from the confines of computer programs. The suggestion inspired a small community of people to actually try to write down all (or most) of common-sense knowledge in predicate calculus. He launched the effort with his own paper on 'Liquids' (also in Hobbs and Moore 1985), a fascinating attempt to fix ontology and notation for a realistic domain. Since then several papers in this vein have appeared (Allen 1984; Hobbs 1986; Shoham 1985). I myself have been an enthusiastic advocate of the movement, having written general boosting papers (1978) as well as attempts to actually get on with the work (1982, 1985). I even co-authored a textbook orientated around Hayes's idea (Charniak and McDermott 1985).

It is therefore with special pain that I produce this report, which draws mostly negative conclusions about progress on Hayes's project so far, and the progress we can expect. In a nutshell, I will argue that the skimpy progress observed so far is no accident, that in fact it is going to be very difficult to do much better in the future. The reason is that the unspoken premiss in Hayes's arguments, that a lot of reasoning can be analysed as deductive or approximately deductive, is erroneous.

Drew McDermott, 'A Critique of Pure Reason' from *Computational Intelligence* 3 (1987): 151–60. Reprinted by permission of the National Research Council Canada.

This paper is an amplification of a talk given at the AI Society of New England meeting, November 1985. The paper contains a few masculine pronouns that some might object to; bear in mind that the use of such a pronoun to refer to an anonymous person does not preclude his being female.

I don't want what I say in this paper to be taken as a criticism of Pat Hayes, for the simple reason that he is not solely to blame for the position I am criticizing. I will therefore refer to it as the 'logicist' position in what follows. It is really the joint work of several people, including John McCarthy, Robert Moore, James Allen, Jerry Hobbs, Patrick Hayes, and me, of whom Hayes is simply the most eloquent.

1. THE LOGICIST ARGUMENT

I should first outline the logicist position. It starts from a premiss that almost everyone in AI would accept, that programs must be based on a lot of knowledge. Even a program that learns must start out knowing many more facts than it will ever learn. The next step is to assume that this knowledge must be represented somehow in the program. Almost everyone in AI would accept this, too, but with subtle reservations, which I will come back to later in the paper.

The next step is to argue that we can and should write down the knowledge that programs must have before we write the programs themselves. We know what this knowledge is; it's what everybody knows, about physics, about time and space, about human relationships and behaviour. If we attempt to write the programs first, experience shows that the knowledge will be short-changed. The tendency will be to over-simplify what people actually know in order to get a program that works. On the other hand, if we free ourselves from the exigencies of hacking, then we can focus on the actual knowledge in all its complexity. Once we have a rich theory of the common-sense world, we can try to embody it in programs. This theory will become an indispensible aid to writing those programs.

The next step is to argue that mathematical logic is a good notation for writing the knowledge down. It is in some sense the *only* notation. The notation we use must be understandable to those using it and reading it; so it must have a semantics; so it must have a Tarskian semantics, because there is no other candidate. The syntax of the notation is unimportant, so let's use a traditional logical notation, because we already know how to extend it as far as we're likely to go. The supposed advantages of newer notations (e.g. semantic networks) are based on confusion and fantasy — confusion about implementation versus content, and fantasy about what we sometimes wish our notations would mean.

It is possible to misunderstand the role that logic is supposed to play in the programs that we will ultimately write. Because we started by saying that knowledge ought to be represented, one might conclude that the axiomatic theories we construct will eventually appear explicitly in the programs, accompanied by an interpreter of some kind that reads the axioms in order to decide what to do. This model may in fact be realized, or it may not be. It is possible, for instance, that a transitivity axiom will not appear explicitly in a program, but may be embodied in some kind of graph traverser. The logicists' argument is that we ought to forget about the grubby details of programming for a while (for a generation, let's say) and instead write down what people know (or coherently believe) about everyday life. If we can actually come up with a formal theory of informal knowledge, then we will have something that future programmers will have to reckon with—literally. From this point of view, it is all the clearer why logicists favour classical logic over newer notations like associative networks. The purported differences between logic and the other notations is that the latter organize knowledge in various ways for use by programs, and that is precisely what the logicists are not interested in. All they want are the facts, ma'am, facts like 'liquids leave tilted containers.' Unfortunately, this is not the only motive for favouring logic. There is an unspoken premiss in the argument that a significant amount of thought is deductive. Without this premiss, the idea that you can write down what people know without regard for how they will use this knowledge is without foundation.

How are we supposed to know when we have made progress in formalizing knowledge? Suppose we write down a bunch of axioms. How will we know when we've written down most of what people know about the subject? Well, when we can't think of anything else to say. But people aren't that great at writing down everything they know about a subject; they tend to leave things out. How will we know when we're really getting there? I think the logicists take it for granted that we'll be done when all the straightforward inferences follow from the axioms that have been written down. If something is obvious to people, Hayes says somewhere, then it must have a short proof. For this dictum to make sense, a substantial portion of the inferences people make must be deductive. Water flows out of a tilted cup; this cup is tilted; *ergo*, water will flow out of it. If most inferences fit the deductive pattern, then the notion of logical proof provides an idealized model of inference. We don't have to have any particular process model in mind, because *every* process model we devise will be an approximation to this ideal inference engine, and will have to conform to the inferences that it licenses. When we design an inference mechanism, it can in fact record the reasons for its inferences in terms of the deductive support that they receive. This is the idea behind data dependencies (Doyle 1979): a program may use any method to arrive at a conclusion, but it ought to be able to list all the actual premises that justify it, so that if any of those are erased, the conclusion will be erased, too. The idealized inference engine justifies the practical inference engine.

2. DEFENDING DEDUCTION

But many inferences are not deductive. If I come upon an empty cup of soda pop that was full a while ago, I may infer that my wife drank it, and that's not a deduction (except in Sherlock Holmes's sense), but an inference to the best explanation. (The only way to mistake this for a deduction is to mistake logic programming for logic; more on this below.) If almost all inferences fall into this or some other non-deductive category, the logicist program will be in serious trouble. It must be the case that a significant portion of the inferences we want are deductions, or it will simply be irrelevant how many theorems follow deductively from a given axiom set. Whatever follows deductively would in that case be *ipso facto* trivial.

Unfortunately, the more you attempt to push the logicist project, the less deduction you find. What you find instead is that many inferences which seem so straightforward that they *must* be deductions turn out to have non-deductive components. We can begin with two examples from Hayes's (1985*b*) Liquids paper:

Suppose an open container with no leaks is empty, but at time *t* a falling history begins whose *bottom* is the free top of the container: for example, you turn on the bath tap with the plug in. By axiom (46), this leaving has an arriving on its other side, which is an inward-directed face of the inside of the bath. By axiom (59), there must be a filling inside the bath, so the *amount* of water increases: axiom (61). So long as the tap keeps running, it will go on increasing. Let us suppose that eventually the bath is full, i.e., it contains its capacity. [So the bath will overflow.] (Notice that if the container were closed—a tank being filled along a pipe, say—then the same line of reasoning would insist on there being a leaving which could not possibly occur . . . One can conclude from this contradiction . . . that the arriving must cease to exist at that time, and hence that the flowing . . . along the supply pipe . . . must cease also . . .)

This seems like a beautiful pair of arguments, a perfect illustration of Hayes's desideratum that obvious inferences have short proofs. Unfortunately, though they may be short, they are not proofs. Suppose we grant Hayes's analysis of the second case; we made the assumption that the filling lasted a certain amount of time, got a contradiction, and concluded it would not, after all, last that long. But then the first case, if we are to follow anything like uniform rules, must be a case of making an assumption and *not* getting a contradiction. All right, but you are not allowed in deduction to make an argument of the form 'Assume *P*; no contradiction?; okay, conclude *P*.' Something else is going on.

Another example is the treatment of the planning problem by logicists, such as Rosenschein (1981). They cast the problem thus: given some axioms about the effects of actions in the world, an initial state of affairs, and a desired state of affairs, find a sequence of actions that can be proven to transform the world from the initial state to the desired state. This is an interesting problem, but it has nothing to do with planning as practised by corporate managers, ordinary people, or robots. Think of the last time you made a plan, and ask yourself if you could have *proven* the plan would work. Chances are you could easily cite ten plausible circumstances under which the plan would *not* work, but you went ahead and adopted it anyway. In fact, all of the hard parts of planning—especially replanning during execution—are incompatible with the view that the object is to prove a plan correct.

This informal survey is borne out by the meagre results that the logicists, including me, have had. In case after case, what can actually be written down as axioms is fairly puny. (I forbear from citing other people's work in this context; one example of mine is McDermott 1985.) On the other side, one finds non-logicist researchers like Forbus who concentrate on writing algorithms to draw inferences, but let themselves be intimidated by the logicists into thinking they really should be able to express as axioms the content of the knowledge in those algorithms. The results (e.g. the axioms in Forbus 1984) are silly, and fall way short of expressing what they are supposed to. I used to think the failure was Forbus's, but I would now exonerate him, and blame the task, seemingly so feasible but actually impossible.[1]

The obstacles I am describing are not news to logicists. From the beginning, it has been clear that the logicist project had to be qualified, or specially interpreted. In what follows, I will describe all the known defences of logicism, and argue that they all fail. These defences are not mutually exclusive; each compensates for a different set of ailments, and most logicists have probably believed in most of them most of the time. Here is the list:

1. The 'idealization' defence: view deductive formulations of problems as idealizations. The deductive planning problem, for instance, can be seen as an idealized version of the 'real' planning problem, carried on,

---

[1] When I say Forbus was 'intimidated', I mean it literally. I refereed his paper, and asked him to try to be more logicist. *Mea culpa.*

perhaps, in a world theory that is an idealized version of the real-world theory.

2. The 'vocabulary' defence: emphasize that we can pick whatever predicates we like. It may be true that we cannot deduce that a particular plan will work, but if we change the problem to one of deducing should-do (*agent, plan*), then progress will be easier.

3. The 'queen of the sciences' defence: find fertile bonds between deduction and non-deductive inference. For instance, inference to the best explanation can be seen as finding premisses from which an observed conclusion follows deductively. In this way deduction comes to be the centre-piece of a grand theory of reasoning, surrounded by interesting variants of deduction.

4. The 'metatheory' defence: posit the existence of deductive 'metatheories', theories about how to find and edit the conclusions of the original, defective 'object level' theories.

5. The 'deducto-technology' defence: argue from the existence of logic programming that many realistic inference problems can be seen as essentially deductive.

6. The 'non-monotonic' defence: argue that by extending classical logic to allow defeasible conclusions we can capture a significantly larger set of inferences.

I will refute each of these in order, starting with the idealization defence.

Idealizations are not always bad; they are often essential. For instance, it may be a useful idealization to prove that a certain plan will win a game of chess, even though the proof neglects the possibility that someone might suddenly offer the planner a million dollars to throw the game. One is entirely justified in simply leaving that kind of possibility out of the axioms. However, what I am arguing against is a mentality that would assume in all chess situations that the goal is to find a provably winning strategy, or that would overlook more normal situations in favour of situations where such a proof was possible. The fact is that realistic chess programs (and human players) do nothing remotely resembling proving that a plan will work. Of course, for any given algorithm, say, game-tree search, there is a way of viewing what it does as deducing *something* (e.g. the minimax value of a tree), but this claim is of no interest to us.

I am afraid that in many cases where a deductive problem is claimed as an approximation to a realistic problem, it is actually an *analogue* to a realistic problem, the best deductive mimic of the real thing. In many cases, this fact may not be insuperable. The attempt to write down facts in the analogue domain may yield insights into the actual domain. The resulting ontology and axioms may be useful for eventually writing programs. What we cannot expect from an idealization is the *coverage* the logicists are expecting. Many concepts from the real domain will just not be found in the idealization. Contrariwise, there is the danger that too many concepts from the ideal domain will not be found in the real one, and the idealization will be so askew as to be useless. Still, as a strategy the use of idealizations does seem worthwhile, and I will come back to this idea at the end of the paper.

Next, the 'vocabulary' defence. The point made here is certainly one I would embrace. If one is designing a program to think about mathematics, being committed to a deductive approach does not entail confining the vocabulary of the program to Zermelo–Fraenkel set theory. Instead, one would want whatever predicates a human mathematician would use, such as interesting concept (C), appears provable (*theorem*), and so forth. For instance, one might take all the predicates used implicitly by Lenat in AM (Lenat 1982) and try to write a program that deduced that a concept was interesting or a theorem was probably provable.

The problem with this defence is that so far it hasn't helped. By broadening the range of problems that can be cast as deduction, we have in many cases simply added to the list of problems we can't solve using deduction. There are good reasons why AM was not a deductive program.

Another problem with the 'vocabulary' defence is that it allows us to replace a hard problem with a trivial one. For instance, in medical diagnosis, if we run into trouble deducing diagnosis (*patient, disease*), switching to possible diagnosis (*patient, disease*) is not going to help. The new problem in this case is too easy; all of the action is in differential diagnosis and weighing evidence, which will now be neglected, or passed off to some non-deductive module.

The 'queen of the sciences' defence may be elaborated thus: consider 'abduction', C. S. Peirce's term for explanatory hypothesis generation. This process is non-deductive, but we can think of it as a sort of 'inverse deduction'. For instance, to explain q, look for an implication of the form (if p q) that you already know, and propose p as an explanation. Put more generally, to explain q, find premises that combined with what you already know will entail q. If this model is correct, then even though abduction is not a kind of deduction, still it is justified by deduction. (This view is endorsed, with reservations, by Charniak and McDermott (1985).)

This account of explanation is known among philosophers as the *Deductive-Nomological Theory*. It is most commonly associated with the name of C. G. Hempel (Hempel and Oppenheim 1948; Hempel 1965). Unfortunately, it is believed by almost no one else. It has several bugs as a model of scientific explanation, which is what it was devised for, and seems hopeless as a model of explaining the behaviour of individual humans or physical systems. The problem is that a deductive chain between *explanans* and *explanandum* is neither necessary nor sufficient.

One reason it is not necessary is that we are content if the explanation merely makes the observed facts probable. Hempel allowed for this case, and so do all the diagnostic expert systems like Mycin (Shortliffe 1976) and Prospector (Duda *et al.* 1980).

But there are more devious examples. Suppose you read in the paper that Selma McGillicuddy, of Secaucus, just won the New Jersey lottery for the second time in the last two months, for more than a hundred thousand dollars each time. There is no reason to infer corruption, so you arrive at the explanation, *It's a fair lottery; occasionally someone will win twice in succession.* This is a satisfying explanation, but you cannot infer from it, 'Selma McGillicuddy wins the lottery twice in two months.' (Wesley Salmon first pointed out this class of explanation; see Salmon (1967, 1975).)

The reason why a deduction of the data is not sufficient is that the requirement is too easy to meet. There will in general be millions of deductions leading to the observed conclusion, almost all of which are absurd as explanations. For example, one day I noticed that my clock radio was two minutes fast. Since I am compulsive about accurate clocks, I was bothered, and sought an explanation. It occurred to me that there had been a power failure lasting two hours recently. One would therefore expect that the clock would be two hours slow, but I remembered that it had a battery backup clock. Hence, the proper explanation was that the battery-powered clock was inaccurate, and gained about a minute an hour.

Let us assume that this explanation could be turned into a deductive argument, with the conclusion, 'The clock is two minutes fast' So what? There are plenty of other deductions with the same conclusion. ('A visitor to our house maliciously set the clock ahead.' 'A burst of cosmic rays hit the clock just right.') One can argue that these are obviously inferior explanations, and all we could hope for from the 'Queen of the Sciences' picture is a characterization of an *adequate* explanation, but this is a sterile position. The condition of adequacy is just too trivial.

If we are not careful, it can become even more trivial. A premiss like 'Every clock in the room is two minutes fast' will explain 'This clock is two minutes fast,' if this clock is the only one in the room. Hempel sought to avoid this problem by requiring the premisses and conclusion to be 'lawlike'. It is not really clear what this property amounts to, but it is intended to rule out 'The number of planets is the least odd square of a prime number' as an explanation for 'There are nine planets.' In fact, it eliminates from consideration any explanation of a particular fact, and makes the theory into a theory of explaining laws.

In a way, the 'Queen' defence is a version of the 'idealization' defence, with similar weaknesses. There must be *some* link between the hypothesis and the evidence to be explained, but it is merely dogmatic to surmise that the link is deductive. In general, about all we can say about it is that a good hypothesis is one that satisfies a typical human inquirer. I will come back to this topic later. For now, we can conclude that deduction cannot be the centre-piece of a theory of abductive inference.

The 'meta-theory' defence argues that the problems with a deductive inference engine can be fixed via the intervention of a deductive 'meta-engine' that steps in and edits its output; or alters its premises; or turns it off to allow a competing theory to take over. For instance, in the two tank examples from Hayes, we can imagine this meta-engine doing *belief revision*, introducing premises about the persistence of flows, and retracting them when awkward questions arise.

The problem with this defence is its vacuity. The subject matter of the deductive meta-theory must presumably be 'legal interventions in object theories.' But there are no constraints on such a theory, from human intuition or anywhere else. There is certainly no constraint that the interventions preserve deductive soundness. If there were, this defence would not accomplish the required strengthening of deduction. So it is difficult to see how to rule out a theory like 'Believe all statements with an odd number of symbols on weekends; believe all statements with an even number on weekdays.' If the enterprise becomes one of crafting meta-theories of such arbitrary power, then we might as well admit we are programming after all.

There is nothing to say *in general* about the meta-theory idea; and for any given case there is too much to say. Let's take the idea of *belief revision*, which I bandied about a couple of paragraphs back. If you start studying this seriously, you eventually wind up studying non-monotonic logic (about which much more below). This study will dwarf the meta-theory framework. You will have to construct a very complex and detailed model to make any progress, and long before you are done it will be clear that it is completely irrelevant whether it is targeted for implementation as some kind of deductive meta-theory or instead as a LISP program. The meta-theory framework contributes nothing, unless you just prefer PROLOG to LISP.

This brings me to the fifth defence, 'deducto-technology'. One reason it is easy to overestimate the power of deduction is because of the existence of a powerful set of tools, such as backward chaining and unification, which are derived from automatic-theorem-proving research, but have found a wider popularity in systems like PROLOG (Clocksin and Mellish 1981) and MRS (Genesereth 1983). These tools turn out to provide an elegant model of computation, just as powerful as, and in some cases prettier than, traditional models. Because you can use these tools to do any computation, and because of their genesis in theorem provers, it is natural to draw the conclusion that any computation is in some sense deduction. It is difficult to refute the argument leading to this conclusion, because there is no argument as such, just vague associations among concepts. (The fallacy is certainly not hindered by the use of phrases like 'logical inferences per second' by the logic-programming community to refer to something as trivial as list-processing operations.) Serious researchers are not consciously taken in by the fallacy, but even they can get carried away by the cleverness of deducto-technology.

To take one example of the sort of woolly thinking we are up against here, consider the way in which values are computed in PROLOG-type systems. A goal containing variables is interpreted as a request to find values for the variables. The goal append ([a, b], [c, d], X) means, 'Find an X that is the result of appending [a, b] and [c, d].' If the axioms are written right values will be found; in this case, X will get bound to [a, b, c, d]. Contrast this goal with append ([a, b], [c, d], [a, b, c, d]), where the goal is to verify that [a, b, c, d] is the result. It is a property of reasonably well-behaved PROLOG programs that whenever they can find a value they can verify it. (The opposite property is much harder to achieve (Shoham and McDermott 1984).)

The problem is that while the idea of verifying conclusions carries over to deduction in general (since it's just the idea of proving something), the idea of calculating values does not. From the point of view of logic, append ([a, b], [c, d], X) is just a Skolemized version of (not(exists(X), append ([a, b], [c, d], X))). (See any textbook for an explanation of Skolemization, and of the 'not'.) It is essentially a useful accident that backward chaining verifies this conclusion by finding a value of X. If we try to generalize beyond backward chaining, the idea falls apart. Luckham and Nilsson (1971) give a variant that works for any resolution proof, but not every resolution proof generates a single value per variable. More important, once logic is extended beyond finitely axiomatizable first-order theories (and it often is in the representation-of-knowledge business), the whole idea of resolution and Skolemization becomes irrelevant.

Even when the idea works, logic does not provide a general theory of answer construction. Consider Robert Moore's 'Bomb in the Toilet' problem: you receive two indistinguishable ticking objects in the mail, plus an anonymous phone call warning that exactly one of them is a bomb. From old movies, you know that putting a bomb in the toilet is a sure-fire way to disarm it. What should you do? The answer is, put both objects in the toilet. (Perhaps a bathtub would be better.) But if we pose the problem logically as

    effect(Plan, and(disarmed(object1),
                       disarmed(object2)))

we might get back (using Luckham and Nilsson's procedure)

    Plan = put(object1, toilet)

or

    Plan = put(object2, toilet)

That is, the theorem prover will have cheerfully verified that there is a workable plan, without actually constructing one. Of course, we can't really ask any more. Deduction just doesn't provide a theory of computing arbitrary things. All it aspires to is a theory of verifying arbitrary things.

3. THE NON-MONOTONIC DEFENCE

Finally, we come to the most potent defence, the appeal to 'non-monotonic logic', the name given to a system of logic in which conclusions can be *defeasible*, that is, subject to withdrawal given more premisses. This sort of logic looks tailormade for examples like the two involving tanks described above. We want to infer that the water is still flowing into the tank so long as we have no reason to believe otherwise; when a contradiction materializes, the conclusion will be withdrawn.

Non-monotonicity is almost by definition incompatible with deduction. Hence, as Israel (1980) has pointed out, 'non-monotonic logic' is somewhat oxymoronic. It is as if to compensate for some deficiency of prime numbers we were to propose studying 'composite primes'. In practice, what is meant by 'a non-monotonic logic' is an inference system that provides a simple, general extension to ordinary logic that captures obvious defeasible inferences. We don't expect such a system to do inference to the best explanation, but we do expect it to infer that your car is still where you parked it last.

Since there might be many alternative 'simple, general' extensions to ordinary logic, we cannot draw any final conclusions about the prospects for non-monotonic logic. We can, however, survey what has been accomplished and evaluate its promise for the future. There are two main methods that have been employed, the *default* approach and the *circumscriptive*. The default approach attempts to formalize the 'negation as failure' idea of PLANNER (Hewitt 1969) and PROLOG (Clocksin and Mellish 1981). We extend ordinary logic by allowing inference rules of the form 'From premiss p and the inability to infer q, infer r.' The idea is that r is the default conclusion from p in the absence of special overriding information q. an example would be

$$\frac{(\text{bird } a) \; \text{Consistent}(\text{not}(\text{abnormal } a))}{(\text{can\_fly } a)}$$

where 'Consistent *formula*' means that *formula* is consistent with all the inferences in the system. For any given bird, we can then normally infer that it can fly, but if there are axioms for inferring abormality, then we can use them as 'gates' to turn this rule off. Systems of roughly this form have been studied by Reiter (1980), McDermott and Doyle (1980), Clark (1978), and others.

The circumscriptive approach, developed by McCarthy (1980) and his colleagues (Lifschitz 1985; Lifschitz, unpublished manuscript[2]), avoids adding new inference rules, and instead augments a first-order theory with an axiom that expresses the goal to 'minimize' some predicate. For instance, given an axiom

(forall(x)(if(and(bird x)(not(abnormal x)))
           (can_fly x)))

one would want to minimize the abnormal predicate, so that as before we can normally infer for any given bird that it can fly. To achieve this, we add to the original theory a second-order axiom. Let A(abnormal;bird) be the conjunction of all the axioms we already have. (It had better be finite.) In the parenthesis following the A we write the names of predicates we intend to substitute for. The semicolon separates the to-be-minimized predicate (abnormal) from the 'variable' predicates (bird); there may in general be one or more to be minimized, zero or more variable. So A(foo;baz) would be the same set of axioms with abnormal replaced by foo and bird replaced by baz. Given this notation, the new axiom is

(forall(p b)
   (if(and)(A(p;b))
          (forall(x)(if(p x)(abnormal x)))
      (forall(x)(if(abnormal x)(p x)))))

That is, if p is any predicate that satisfies A (after A has been weakened by changing bird) and is as strong as abnormal, then abnormal is as strong as p. Another step is now required, and that is to plug values in for p and b.

[2] Lifschitz, V. (1986) 'Pointwise circumscription.' Unpublished draft, 16 Jan. 1986.

Suppose that the only bird we know that can't fly is Clyde. Then A will include the axiom about normal bird flight, plus (bird Clyde) and (not (can_fly Clyde)). If we plug in p = (lambda(y)(= y Clyde)) and b = (lambda(y)(= y Clyde)), then, because A(p;b) becomes

```
(and(forall(x)(if(and(= x Clyde)(not(= x Clyde)))
                 (can_fly x )))
    (= Clyde Clyde)
    (not(can_fly Clyde)))
```

the instance we get of the circumscription axiom is

```
(if(and(forall(x)(if(and(= x Clyde)(not(= x Clyde)))
                    (can_fly x)))
       (= Clyde Clyde)
       (not(can_fly Clyde))
       (forall(x)(if(= x Clyde)(abnormal X))))
   (forall(x)(if(abnormal x)(= x Clyde)))))
```

But the antecedent of this implication follows from A(abnormal;bird), so we can conclude the consequent, that Clyde is the only abnormal object. Hence any other bird (if we can show him unequal to Clyde) will be judged able to fly.

Note how circumscription achieves non-monotonicity. When a new axiom is added to A, the circumscriptive axiom changes, and usually some theorem goes away.

There are two problems with all known varieties of non-monotonic logic. The first is that it is often not clear without considerable effort what the consequences of a set of rules are. The second is that they often fail to achieve the proper 'amplification'; that is, the rules will have overly weak consequences. I will label these two problems with the phrases 'You can't find out', and 'You don't want to know'.

In default logics, the 'You can't find out' problem arises because it is in general undecidable whether a formula is consistent with a theory. In fact, it is even hard to define exactly what is meant by the phrase 'consistent with a theory', when the theory in question is the one containing the default rules.' You can't tell what isn't inferrable until you've inferred everything, and so we are led to the idea of a 'stable extension' or 'fixed point' of a default theory. Such a fixed point is a set of formulas, intuitively a 'stable set of beliefs', which is characterized by a set of non-concluded formulas', such that (a) everything in the fixed point follows from the original theory plus the non-concluded formulas via default inference rules; and (b) no non-concluded formula follows. If Clyde is known to be a bird, then (not(abnormal Clyde)) will be one of the non-concluded formulas, and hence (can_fly Clyde) will be an element of the fixed point; unless (abnormal Clyde) is also deducible, in which case (can_fly Clyde) will not be in the fixed point. Unfortunately, the fixed points and the sets of non-concluded formulas are infinite, and in general hard to describe or find.

Circumscription is also hard to use. What all known versions of circumscription have in common is this procedure for arriving at conclusions:

- add second-order axiom to original theory
- guess predicate constant to plug in to the axiom
- simplify

This is the kind of procedure we followed in the example. The problem with it is that the information added is usually about the same size as the conclusions you ultimately want to draw. In fact, it usually looks about the same as those conclusions, with a few extra lambdas.

In principle, circumscription could be used mechanically; you could turn a crank and all the conclusions would come out. In practice, there is no way to enumerate useful instances of the second-order axiom, so circumscription has been used only on small examples for which the desired conclusions are already known. (In special cases, you can show that circumscription and default logics both reduce to computable algorithms, but these special cases are of no interest to us here.) Paradoxically, the hopelessly undecidable default logics suggest a practical algorithm that actually gets somewhere: To verify that p is consistent, try to prove its negation and fail. When this procedure halts, it is often a good heuristic approximation (Clark 1978).

The intractability of non-monotonic logic has led to a curious phenomenon. Logicists go ahead and use non-monotonic constructs, and state in the accompanying text what conclusions they hope will follow, without really knowing if they will. At this point, it is no longer clear in what sense the reasoning they are describing is justified by a formal system. This wishful thinking wouldn't matter much if the wishes came true. Unfortunately, this brings us to the 'You don't want to know' problem: When a non-monotonic system is studied carefully, it often happens that the conclusions the formal system actually allows are different from, typically weaker than, what was expected. In default formulations, the problem arises because the fixed points described above are often non-unique. Some of them are reasonable, but many correspond to sets of beliefs that would be rejected by the person writing the original rules. (There is no way to eliminate such fixed points by adding more default rules; that can only make matters worse.) If a theory has several alternative fixed points, what actually can be said to be a theorem of the theory? Either theories like this don't have theorems, in which case they can't serve as the idealized inference engine we are seeking; or we are stuck with a weak notion of theorem, in which a theorem is something that is inferred in all fixed points. Typically this alternative gives us disjunctive theorems, where some of the disjuncts are counter-intuitive intruders from unwanted fixed points. We want the conclusion p, but we wind up with p or q, where q is off the wall.

Such overweak disjunctions pop up in the circumscriptive versions, too. The phenomenon is somewhat different for circumscription, because the notion of fixed point doesn't play the same proof-theoretic role. But we do have a homologous idea, the *minimal model*, defined as follows. One model is 'smaller' than another with respect to some predicate P if it agrees on all other (non-variable) predicates and its P is a subset of the other model's. A minimal model is one with no model smaller. It can be shown that a formula is true in all minimal models of $A(P; V)$ if it follows from the $A(P; V)$ plus the circumscriptive second-order axiom given above.

The overweak disjunction problem now appears in the following form. Typically there will be minimal models that differ in important ways, such that some of the models are 'obviously wrong' to a human observer. On the syntactic side, circumscription will yield a disjunction, such that each disjunct characterizes a class of minimal models. Hence the situation is not really that different from the default-logic case, except that the disjunctions come about as a consequence of the basic machinery, rather than being tossed in as a kludgy way of defining the notion of theorem.

In a recent paper, Hanks and McDermott (1985, 1986) explored one instance of this phenomenon in detail. We studied a simplified version of the temporal logic of McDermott (1982), which was somewhat more complex than the previous non-monotonic systems that had been studied. We were hoping to show that the conclusions we wanted from the formal system really did follow. We expected that the multiple-fixed-point problem would defeat the default logics, but we expected circumscription to work. We were surprised to discover that circumscription had the same problem as the default formulations, although in retrospect the similarities among the various systems seem so overwhelming that the surprise is lessened.

The problem for all the logics is that concepts like 'minimization' and 'stable sets of beliefs' are just inappropriate for the temporal domain. The non-monotonic rule we wanted was (to put it informally) 'states of the world tend to remain undisturbed.' All the logics drew conclusions that minimized disturbances, but that's not what we really wanted. Instead, we wanted to avoid disturbances with unknown causes.

What we were trying to state was that a 'history' continues unless it is explicitly 'clipped' by subsequent events. Consider the following event sequence:

1. Fred is born   Fred starts to be ALIVE
2. A gun is loaded   Gun starts to be LOADED
3. Fred is shot with it   Fred becomes DEAD

We ought to be able to conclude that Fred is now dead (sorry for the violence). But another scenario would minimize disturbance equally well. In this one, the gun ceases to be loaded before event 3, for no particular reason except to avoid disturbing Fred's being alive.

Since that paper, Vladimir Lifschitz of McCarthy's group has shown[3] that a new idea, 'pointwise circumscription', will solve a simplified version of the Hanks–McDermott problem. No one knows if it solves the more complex version, let alone a realistic set of axioms about physics. No one knows what other problems are still out there. But what's really bothersome about this 'solution' is that it is even more top-heavy than previous versions of circumscription. We will have to know the answer, in which case circumscription will verify it for us. In addition, predicates are allowed to be in the class 'to-be-minimized' on parts of their domains, and 'variable' on other parts, and you have to supply the information about which part is which in the form of an extra relation.

This kind of solution destroys circumscription in order to save it. As with all forms of circumscription, we start with the conclusions that we want to augment our deductive theory with, and we find a second-order axiom that will give us those conclusions. If the first axiom we pick doesn't work, we find a different axiom. Once the exercise is carried out, we throw the axiom away; no one knows how to extract any other consequences than the ones we were verifying. Under these circumstances, what is the axiom doing for us? In what sense is it justifying the conclusions, rather than the desired conclusions justifying it? In practice, it would be just as easy to simply add those conclusions to the theory directly. This procedure would be every bit as non-monotonic (just change the added ingredients when the theory changes), and every bit as magic.

The original goal, of a simple, general extension of classical logic that would grind out 'obviously correct' conclusions, has eluded us. In the case of default formulations, that's because the lures yield non-recursively

[3] See no. 2 above.

enumerable theorems. In the case of circumscription, it's because we have to put the answer in before we can get it out. In both cases, the answers, when available, are often too weak, although with circumscription we often have the option of switching to a different circumscriptive axiom.

It is important to realize that this crisis does not affect programs that reason non-monotonically. Almost all computerized inference is non-monotonic and hence non-deductive. That's the problem we started with. What the crisis does affect is our attempt to extend deduction slightly to cover 'obvious' cases. As things now stand, there is no non-monotonic system that justifies the non-monotonic inferences our programs do. On the contrary, what ends up happening is that we have to expend a lot of effort contorting the formal systems to duplicate simple procedural reasoning. And the effort is a side-show or afterthought to the development of the program; it doesn't contribute anything.

As I said above, the situation may improve. Someone may discover tomorrow the kind of non-monotonic system we are looking for. But for now we must conclude that there is no appeal to non-monotonicity as a way out of some of the problems of deduction.

4. DOING WITHOUT DEDUCTION

Let us try to summarize the argument so far. I laid out the logicists' project, to express common-sense knowledge in the form of logical axioms. I sketched the justification for the project, and pointed out an implicit premiss, that a lot of inference is deductive. I have argued that this premiss is wrong, even if logic is extended in various ways.

With this premiss knocked out, how does the original argument fare? We can now see that no matter how many axioms you write down about a domain, most of the inferences you want will not follow from them. For that to happen, you must also supply a *program*. In other words, in most cases there is no way to develop a 'content theory' without a 'process model'. (These terms are due to Larry Birnbaum.) A content theory is supposed to be a theory of what people know, how they carve up the world, what their 'ontology' is. A process model explains how they use this knowledge. A content theory is at Newell's (1981) 'knowledge level', supposedly independent of how the facts it expresses are to be manipulated. What we can now conclude is that content theories are of limited usefulness, in the case where the contemplated inferences are non-deductive. You cannot just start listing facts people know, expressed in logic or any other notation, without saying something about how you assume they will be used by a program, and hence what class of inferences you are trying to account for. The only occasion when you can neglect that chore is when you can point to an important class of purely deductive inferences involving the knowledge. In that case, you do know enough about every candidate process model that you need say no more. But such classes, it now seems, are rare.

By the way, this point should apply just as much to Lenat *et al.*'s (1986) CYC project as to the logicist project. His group has availed themselves of a broader range of tools, and forsworn the discipline of logic, but the same objection presents itself: how will they know when they are making progress?

This argument against free-standing content theories has unfortunate repercussions on the original argument in favour of Tarskian semantics. When there was no program, then denotational semantics was the only way to specify the meanings of our notations. But there is a competing tradition about knowledge representation, which says that a knowledge-representation system is in essence a special-purpose high-level programming-language. This point of view is explicit in descriptions of systems like OPS5 (Brownston *et al.* 1985) and PROLOG (Clocksin and Mellish 1981), but it applies to many associative nets, too, which are often devices for organizing chunks of LISP code. Actually, OPS5 and PROLOG aren't such great examples, since they are general-purpose programming-languages. A better example might be a parser-rule notation like that of Marcus (1980). The notation expresses 'knowledge' about the syntax of a language, but it has no denotational semantics. Its semantics are *procedural*; a set of rules is correct if it makes the parser do the right thing.

The competing procedural tradition, in other words, is that a knowledge-representation system does not actually represent anything. This position makes the typical logicist's hair stand on end, because it means acknowledging that the represented knowledge is essentially to be used in just one way. It is hard to count 'ways', but picture the 'same fact', as needed by two different modules, each with its own special-purpose programming notation. The fact would have to be represented twice. Surely this is not a pleasant requirement to impose on an intelligent program.

It would be nice if a notation could have both denotational and procedural semantics. Nothing prevents this; any logic-based notation that is actually used by a program does *ipso facto* have such a dual semantics. (Pure PROLOG programs are an example.) One is tempted to conjecture the converse, that any procedural notation can be translated into an equivalent denotational notation. Isn't it just a matter of cleaning up a few inconsistencies, and making up some ontology? Unfortunately, this optimistic assessment is based on a misconceived notion of how devices like associative networks are actually used. In the minds of some researchers, the notation is supposed to have a formal semantics of some kind, and there is not much doubt that there are equivalent notations that look more like traditional logic. But in the minds of most users, the system is a collection of features—demons and whatnot—just like a standard programming environment (except, they hope, more exotic). Any way of using the features that achieves the immediate programming goal is legitimate. There is nothing shady about this. For every researcher whose system is misused, there are ten who would *encourage* such 'creative' use of their system. The chances of being able to find a denotational semantics for any such system are slim.

Still, this deplorable standard of practice cannot by itself deter us from seeking notations that have both denotational and procedural semantics. It's just that this pursuit now seems to lack any rationale. Some people insist that their notations have denotational semantics; others (rather more) can't stand that constraint. In spite of what I am arguing here, I still find myself temperamentally in the first group. If a student comes to me with a denotationless representation, it bothers me. Formerly I thought I had an argument to convince him to rethink, but now all I have is indigestion. The student can always point to his program and claim that it doesn't draw absurd conclusions from his absurd notation. The fact that *I* might draw an absurd conclusion is my problem.

To take one of my favourite examples, consider a simple fact like 'The Russians have warships deployed off the US coast.' Unless we are willing to resort to 'computerdeutsch' predicates like currently_deployed_off_US_coast, a proper representation of something like this will have to express explicitly what the US coast is, roughly how many ships there are and in what distribution they are found, what period of time is implied, and so on and on. But who says that's 'proper'? Any particular application program can probably get by with computerdeutsch. And many eager-beaver notation designers will resort to 'computerenglisch', such as

(have Russians (deployed warships (off (coast US))))

which seems even worse to me. But why? If the program works, what's wrong with it?

Hence, in the original logicist argument, there is a flaw in the second step, the claim that knowledge must be represented. Although most AI people would assent to this claim, we now see that most of them don't mean it. What they are thinking is roughly: We will have to write a lot of programs to get the knowledge in, and we will need special high-level notations to do it.

The logicist can take comfort in the fact that his opponents have a hard time distinguishing the 'high-level' programs that constitute representations from any old programs. If the distinction cannot be made, then all programs could be taken to 'represent knowledge', which I take to be the proceduralist position in the old procedural-declarative controversy. This controversy died because no one was really interested in this sense of 'represent', by which, for instance, a vision program could be said to represent knowledge about the physics of image formation. There seems to be a stronger sense in which AI programs manipulate explicit representations of objects and facts; denotational semantics provides one answer about what that sense is, but we now see how unattractive this answer is to many AI researchers.

5. DEFENDING PROCEDURES

It's not that the logicist never planned to write programs. He just expected that by the time they were written they would be seen as optimized versions of theorem provers. All that would be required to justify those programs would be to show that they were faithful to the axioms that underlay them.

Now that we have rejected this picture, we need new ways of justifying inferential programs. AI programs are notorious for being impenetrably complex. Sometimes this feature is painted as a virtue, as if the mystery of intelligence ought to be preserved in computational models of it. But a model that we don't understand is not a model at all, especially if it works on only a handful of examples (Marr 1977; Birnbaum 1986).

It is probably impossible to make the idea of 'justification' precise enough to support a claim that every program ought to be justified. And yet it is always satisfying when beside a program we can point to a clean, independent theory of why it works. In vision research, for instance, it was a major step just to move from 'heterarchical' models, with their air of mystery, to models justified by physics and psychophysics. In the domain of qualitative envisioning (de Kleer and Brown 1985; Forbus 1984), there is nothing wrong with the programs that have been written, but it is clarifying to have Kuipers's (1985) analysis of their meaning and limits.

But there are large classes of programs that lack any kind of theoretical underpinnings, especially those concerned with inference to the best explanation, or *abduction*. It would be nice if we could go back to the philosophers and mine their wisdom again. Surely if they could come up with such a great theory of deductive inference they must have done just as well on other kinds, too. Unfortunately, the philosophers have let us down. A theory of abduction might start with answers to questions like these:

What sorts of things need to be explained?
What counts as an explanation?
What counts as evidence for an explanation?
How do you measure the strength of evidential support?
When is evidence strong enough to justify belief in a hypothesis?

So far these questions have received only vague, unmechanizable, piecemeal, or ridiculous answers. We have Bayesian theories, Dempster–Shafer theories, deductive-nomological theories, local induction theories, and a lot of arguments about which is best, but none of them answers more than one or two of the questions above, and none seems entirely correct.

This state of affairs does not stop us from writing medical-diagnosis programs. But it does keep us from understanding them. There is no independent theory to appeal to that can justify the inferences a program makes. One medical-diagnosis program is better than another if fewer of its patients die in clinical trials, I suppose. Actually, what's really bothering me is that these programs embody *tacit* theories of abduction; these theories would be the first non-trivial formal theories of abduction, if only we could make them explicit.

There is an optimistic way and a pessimistic way to view this situation. The pessimistic view is that AI researchers are merely being·naïve about their chances, buoyed by simple ignorance of the past failures of philosophers. The reason why we cannot extract theories from our programs is that there are no theories to extract. Fodor (1983) puts this conclusion rather grandiloquently at the end of his book *The Modularity of Mind*:

Localness . . . is a leading characteristic of the sorts of computations that we know to think about. Consider . . . [the] contrast . . . between deductive logic—the history of which is, surely, one of the great success stories of human inquiry—and confirmation theory [i.e. what I was calling abduction theory above] which, by fairly general consensus, is a field that mostly does not exist. My point is that this asymmetry . . . is likely no accident. Deductive logic is the logic of validity, and validity is a *local* property of sentences. . . . The validity of a sentence contrasts starkly with its level of confirmation, since the latter . . . is highly sensitive to global properties of belief systems. . . . We have, to put it bluntly, no computational formalisms that show us how to do this, and we have no idea how such formalisms might be developed. . . . In this respect, cognitive science hasn't even started; we are literally no farther advanced than we were in the darkest days of behaviorism. . . . If someone—a Dreyfus, for example—were to ask us why we should even suppose that the digital computer is a plausible mechanism for the simulation of global processes, the answering silence would be deafening.

The optimistic view, of course, is that AI researchers can make much faster progress than all those philosophers because we are equipped with 'powerful ideas' they didn't have, especially the idea of sophisticated autonomous computation. I hope this is right. But if all we do is go on writing programs, without any general theories emerging, then I am going to get increasingly uncomfortable.

6. CONCLUSIONS

To summarize: the logicist project of expressing 'naïve physics' in first-order logic has not been very successful. One reason may be that the basic argument was flawed. You cannot write down axioms independent of a program for manipulating them if the inferences you are interested in are not deductions. Unfortunately, very few interesting inferences are deductions, and the attempts by logicists to extend logic to cover more territory have been disappointing. Hence we must resign ourselves to writing programs, and viewing knowledge representations as entities to be manipulated by the programs.

In many respects this is not a critique of logic *per se*. When you sit down to express a body of knowledge, the notation you use recedes quickly into the background. If you are trying to develop a theory of shape, the constraints imposed by the notational conventions of logic soon dwindle beside the task of trying to express what you know at all. Hence, as I mentioned before, I consider Lenat *et al.*'s (1986) CYC project to be under much the same shadow as the logicists' project.

However, there is one respect in which logic is peculiarly vulnerable, and that is in its resting on denotational semantics. One can accept my conclusions about the futility of formalizing knowledge without a program, and yet still, as I do, have a strong intuition that it is better for a notation to have a denotational semantics than not to. One reason for this might be that at least a sound semantics helps ensure that the deductive inferences done by a program will be right; they may be trivial, but at least they will not be wrong.

Another way of justifying formal semantics has recently been pointed out by Shoham (1986). Suppose a program manipulates a notation, and you can show that the program's conclusions are just those that are true in all A-models of its premises, where what an A-model is depends on the class of inferences you are trying to capture. If the characterization of A-models is intuitively appealing, then you will have provided an independent justification for the operation of the program. If we plug 'minimal model' into the schema, we get a program justified, in a way, by circumscription, *except that we dispense with the circumscription axiom*, and just use the semantic notion directly. In the case of temporal inference, the notion of model we need is different; see Shoham (1986) for one proposal. Does this idea apply to a wide variety of types of inference? If so, it provides a way of justifying the ontological and semantic parts of the logicist project, while, alas, dispensing with the idea of programless knowledge representation.

As a tool for studying issues in the semantics and mechanics of knowledge representation, logic still seems unsurpassed. I have in mind examples like Moore's (1980, 1985) work on a computational version of Hintikka's logic of knowledge, which explained how a thinker can refer to unidentified entities whose identities are known by someone else; and Charniak's (1986) work explaining 'script variables' as Skolem terms. The insights these papers provide apply to a variety of reasoning programs. Anyone who ignores them just because they are expressed in terms of logic is risking writing an inelegant, irrelevant program.

Finally, I should admit that I am still doing work in the paradigm that I criticize here. In the domain of shape representation, so little is known that focusing on an idealization cannot but help teach us something. The problem I would like to tackle is representing the knowledge required to answer questions like, 'Could a paper clip be used as a key ring?' The idealization I have been forced to fall back on is to prove that a paper clip of a certain size and shape could fit through the hole of a typical key. It should be obvious how much of the original problem this leaves out. Still, the territory is so unexplored that a tour through the idealized fragment could turn up something interesting. What one cannot hope for is to express as logical axioms everything there is to know about using shapes in unusual ways, before designing programs for this task. This will probably come as a shock to no one but me and a few friends.[4]

[4] I thank Larry Birnbaum, Steve Hanks, Pat Hayes, Yoav Shoham, and many others for help, some of it against their will. I should also point out that Carl Hewitt, Marvin Minsky, and Bill Woods have been saying similar things for a long time. This work was supported by the National Science Foundation under grant number DCR-8407077.

REFERENCES

Allen, J. (1984). 'Towards a General Theory of Action and Time.' *Artificial Intelligence* 23(2): 123–54.
Birnbaum, L. (1986). 'Integrated Processing in Planning and Understanding.' Yale Computer Science Technical Report 489. New Haven, Conn.: Yale University.
Brownston, L., Farrell, R., Kant, E., and Martin, N. (1985). *Programming Expert Systems in OPS5: An Introduction to Rule-based Programming*. Reading, Mass.: Addison-Wesley.

Charniak, E. (1986). 'Motivation Analysis, Abductive Unification, and Nonmonotonic Equality.' *Artificial Intelligence* [34(1988): 275–96].

—— and McDermott, D. (1985). *Introduction to Artificial Intelligence*. Reading, Mass.: Addison-Wesley.

Clark, K. L. (1978). 'Negation as Failure.' In H. Gallaire and J. Minker (eds.), *Logic and Databases*, pp. 293–322. New York: Plenum Press.

Clocksin, W., and Mellish, C. (1981). *Programming in Prolog*. Berlin: Springer-Verlag.

Davis, R., and Lenat, D. B. (1982). *Knowledge-based Systems in Artificial Intelligence*. New York: McGraw-Hill.

de Kleer, J., and Brown, J. S. (1985). 'A Qualitative Physics Based on Confluences.' In J. Hobbs and R. C. Moore (eds.), *Formal Theories of the Commonsense World*, pp. 231–72. Norwood, NJ: Ablex.

Doyle, J. (1979). 'A Truth Maintenance System.' *Artificial Intelligence* 12(3): 231–72.

Duda, R. O., Gaschnig, J. G., and Hart, P. E. (1980). 'Model Design in the Prospector Consultant System for Mineral Exploration.' In D. Michie (ed.), *Expert Systems in the Micro-Electronic Age*, pp. 153–67. Edinburgh: Edinburgh University Press.

Fodor, J. (1983). *The Modularity of Mind*. Cambridge, Mass.: MIT Press/Bradford Books.

Forbus, K. (1984). 'Qualitative Process Theory.' *Artificial Intelligence* 24: 85–168.

Genesereth, M. R. (1983). 'An Overview of Meta-level Architecture.' *Proc. Nat. Conf. AI*, pp. 119–24. Washington, DC.

Hanks, S., and McDermott, D. (1985). 'Temporal Reasoning and Default Logics.' Computer Science Department Technical Report 430. New Haven, Conn.: Yale University.

—— (1986). 'Default Reasoning and Temporal Logics.' *Proc. Nat. Conf. AI*, pp. 328–33. Philadelphia.

Hayes, P. J. (1985a). 'The Second Naïve Physics Manifesto.' In J. Hobbs and R. C. Moore (eds.), *Formal Theories of the Commonsense World*, pp. 1–20. Norwood, NJ: Ablex.

—— (1985b). 'The Ontology of Liquids.' In J. Hobbs and R. C. Moore (eds.), *Formal Theories of the Commonsense World*, pp. 71–107. Norwood, NJ: Ablex.

Hempel, C. G. (1965). *Aspects of Scientific Explanation*. New York: Free Press.

—— and Oppenheim, P. (1948). 'Studies in the Logic of Explanation.' *Philosophy of Science* 15: 135–75.

Hewitt, C. (1969). 'PLANNER: A Language for Proving Theorems in Robots.' *Proc. 1st IJCAI Conference*, pp. 295–301. Washington, DC.

Hobbs, J. (1986). 'Commonsense Summer: Final Report.' AI Center, SRI International, Technical Note. Menlo Park, Calif.

—— and Moore, R. C. (eds.) (1985). *Formal Theories of the Commonsense World*. Norwood, NJ: Ablex.

Israel, D. (1980). 'What's Wrong with Non-Monotonic Logic?' *Proc. Nat. Conf. AI*, pp. 99–101. Stanford, Calif.

Kuipers, B. (1985). 'The Limits of Qualitative Simulation.' *Proc. IJCAI Conference*, pp. 128–36. Los Angeles, Calif.

Lenat, D. B. (1982). 'AM: Discovery in Mathematics as Heuristic Search.' In R. Davis and D. B. Lenat (eds.), *Knowledge-based Systems in Artificial Intelligence*, pp. 1–225. New York: McGraw-Hill.

Lenat, D. B., Prakash, M., and Shepherd, M. (1986). 'CYC: Using Commonsense Knowledge to Overcome Brittleness and Knowledge Acquisition Bottlenecks.' *AI Magazine* 6(4): 65–85.

Lifschitz, V. (1985). 'Computing Circumscription.' *Proc. IJCAI Conference*, pp. 121–27. Los Angeles, Calif.

Luckham, D. C., and Nilsson, N. J. (1971). 'Extracting Information from Resolution Proof Trees.' *Artificial Intelligence* 2(1): 27–54.

Marcus, M. P. (1980). *A Theory of Syntactic Recognition for Natural Language.* Cambridge, Mass.: MIT Press.

Marr, D. (1977). 'Artificial Intelligence—A Personal View.' *Artificial Intelligence* 9: 37–48.

McCarthy, J. (1980). 'Circumscription: A Nonmonotonic Inference Rule.' *Artificial Intelligence* 13: 27–40.

McDermott, D. (1978). 'Tarskian Semantics or, No Notation Without Denotation!' *Cognitive Science* 2(3): 277–82.

—— (1982). 'A Temporal Logic for Reasoning about Processes and Plans.' *Cognitive Science* 6: 101–55.

—— (1985). 'Reasoning about Plans.' In J. Hobbs and R. C. Moore (eds.), *Formal Theories of the Commonsense World*, pp. 269–317. Norwood, NJ: Ablex.

—— and Doyle, J. (1980). 'Non-Monotonic Logic I.' *Artificial Intelligence* 13: 41–72.

Moore, R. C. (1980). 'Reasoning about Knowledge and Action.' AI Center Technical Report 191, SRI International. Menlo Park, Calif.

—— (1985). 'A Formal Theory of Knowledge and Action.' In J. Hobbs and R. C. Moore (eds.), *Formal Theories of the Commonsense World*, pp. 319–58. Norwood, NJ: Ablex.

Newell, A. (1981). 'The Knowledge Level.' *AI Magazine* 1(3): 1–20.

Reiter, R. (1980). 'A Logic for Default Reasoning.' *Artificial Intelligence* 13: 81–132.

Rosenschein, S. J. (1981). 'Plan Synthesis: A Logical Perspective.' *Proc. IJCAI Conference*, pp. 331–7. Vancouver, BC.

Salmon, W. C. (1967). *The Foundation of Scientific Inference.* Pittsburgh: University of Pittsburgh Press.

—— (1975). 'Theoretical Explanation.' In S. Koerner (ed.), *Explanation*, pp. 118–45. New Haven, Conn.: Yale University Press.

Shoham, Y. (1985). 'Naïve Kinematics: One Aspect of Shape.' *Proc. IJCAI Conference*, pp. 436–42. Los Angeles, Calif.

—— (1986). 'Time and Causality from the Standpoint of Artificial Intelligence.' Ph.D. diss. Yale University.

—— and McDermott, D. (1984). 'Knowledge Inversion.' *Proc. Nat. Conf. AI*, pp. 295–9. Austin, Tex.

Shortliffe, E. (1976). *Computer-based Medical Consultations: MYCIN.* New York: Elsevier.

10

MOTIVES, MECHANISMS, AND EMOTIONS

AARON SLOMAN

1. INTRODUCTION

Ordinary language makes rich and subtle distinctions between different sorts of mental states and processes such as mood, emotion, attitude, motive, character, personality, and so on. Our words and concepts have been honed for centuries against the intricacies of real life under pressure of real needs and therefore give deep hints about the human mind.

Yet actual usage is inconsistent, and our ability to articulate the distinctions we grasp and use intuitively is as limited as our ability to recite rules of English syntax. Words like 'motive' and 'emotion' are used in ambiguous and inconsistent ways. The same person will tell you that love is an emotion, that she loves her children deeply, and that she is not in an emotional state. Many inconsistencies can be explained away if we rephrase the claims using carefully defined terms. As scientists we need to extend colloquial language with theoretically grounded terminology that can be used to mark distinctions and describe possibilities not normally discerned by the populace. For instance, it will be seen that love is an attitude, not an emotion, although deep love can easily trigger emotional states. In the jargon of philosophers (Ryle 1949), attitudes are dispositions, emotions are episodes, although with dispositional elements.

For a full account of these episodes and dispositions we require a theory about how mental states are generated and controlled, and how they lead to action—a theory about the mechanisms of mind. The theory should explain how internal representations are built up, stored, compared, and used to make inferences, formulate plans, or control actions. Outlines of a theory are given later. Design constraints for intelligent animals or machines are sketched, then design solutions are related to the structure

Aaron Sloman, 'Motives, Mechanisms, and Emotions' from *Cognition and Emotion* 1 (1987): 217–33. Reprinted by permission of the author, and Lawrence Erlbaum Associates Ltd.

of human motivation and to computational mechanisms underlying familiar emotional states.

Emotions are analysed as states in which powerful motives respond to
relevant beliefs by triggering mechanisms required by resource-limited
intelligent systems. New thoughts and motives get through various filters
and tend to disturb other ongoing activities. The effects may interfere
with, or modify the operation of other mental and physical processes,
sometimes fruitfully sometimes not. These are states of being 'moved'.
Physiological changes need not be involved. Emotions contrast subtly
with related states and processes such as feeling, impulse, mood, attitude,
temperament; but there is no space for a full discussion here.

On this view we need posit no special subsystem to account for
emotions since mechanisms underlying intelligence suffice (cf. Oatley and
Johnson-Laird 1985). If emotional states arise from mechanisms required
for coping intelligently in a complex and rapidly changing world, this
challenges the common separation of emotion and cognition. This applies
equally to human beings, other animals, or intelligent machines to come.

2. DESIGN CONSTRAINTS FOR A MIND

The enormous variety of animal behaviours indicates that there are different ways of designing agents that take in information from the environment and are able to act on it individually or co-operatively. Human
beings merely occupy one corner of this 'space of possible minds'.
Elsewhere I have sketched constraints determining the design solutions
embodied in the human mind. I have space only to summarize the key
results relevant to emotions.

Constraints include: a multiplicity of internal and external sources of
motivation (often inconsistent), speed limitations, inevitable gaps and
errors in beliefs about the environment, and varying degrees of urgency
associated with motives. Resource limits and urgency render inevitable
the use of potentially unreliable 'rule-of-thumb' strategies. Unpredictabil-
ity of new information and new goals implies a need to be able to interrupt, modify, suspend, or abort ongoing activities, whether external or
internal. This includes such things as hardware and software 'reflex'
actions, some of which should be modified in the light of experience.

Reflexes are inherently fast-acting but stupid. They may be partly
controlled by context-sensitive filters using rules-of-thumb to assess
priorities rapidly and allow extremely important, urgent, or dangerous
ongoing activities to proceed without disturbance while allowing new,

specially important, or urgent, motives to interrupt them. (The 'insistence' of a motive is defined later in terms of its ability to get past such filters.) A major conclusion is that intelligent systems will have fast but stupid subsystems, including filters for new motives. Fast, dumb filters will sometimes let in undesirables.

Incomplete information and the need to cope with long-term change in the social or physical environment require higher-order sources of action that provide learning: not only generators and comparators of motives, but generators and comparators for the generators and comparators themselves.

Although several independent subsystems can execute plans in parallel, like eating and walking, conflicts among requirements can generate incompatible goals, necessitating a decision-making mechanism. The two main options are a 'democratic' voting scheme, and a centralized decision-maker. If subsystems do not all have access to the full store of available information or not all have equal reasoning powers, a 'democratic' organization may be dangerous. Instead, a specialized central mechanism is required for major decisions (Sloman 1978: chs. 6 and 10). This seems to be how normal human minds are organized.

Similar constraints determine the design of intelligent artefacts. Physical limitations of biological or artificial computing equipment necessitate major divisions of functions, including the allocation of the highest level control to a part with access to most information and the most powerful inference mechanisms. However, an occasional urgent need for drastic action requires overriding hardware or software reflexes that operate independently of higher-level control—a mechanism enabling emotional processes described in the following sections.

Goal Generators

Many different sorts of motivators are pointed to by ordinary words and phrases such as:

aims, attitudes, desires, dislikes, goals, hates, hopes, ideals, impulses, likes, loves, preferences, principles, alluring, amusing, bitter, boring, charming, cheering, depressing, distressing.

There are many more. They mark subtle distinctions between different springs of action and the various ways things affect us. Conceptual analysis (Sloman 1978: ch. 4) brings out their presuppositions. A key concept is having a goal.

To a first approximation, to have a goal is to use a symbolic structure, represented in some formalism, to describe a state of affairs to be produced, preserved, or prevented. The symbols need not be physical structures: virtual formalisms will do as well (Sloman 1984). Goals can use exactly the same descriptive formalism as beliefs and hypotheses. The difference is solely in the roles they play.

A representation of a state of affairs functions as a goal if it tends (subject to many qualifications) to produce behaviour that changes reality to conform to the representation.

A representation functions as a belief if it is produced or modified by perceptual and reasoning processes which tend (subject also to many qualifications) to alter the representations to conform to reality.

(What 'conform' means here cannot be explained without a lengthy digression.) The same representations may also be used in other roles, as instructions, hypothesized situations, rules, etc.

Some new goals subserve a prior goal and are generated by planning processes. Some are responses to new information, such as wanting to know what caused the loud noise around the corner. Goals are not triggered only by external events: a thought, inference, or recollection may have the same effect.

How can a goal be produced by a belief or thought? If goals involve symbolic structures, a computational explanation might be that a 'goal-generating' condition-action rule is used. For example, a benevolence rule might be: 'If X is in distress generate the goal [X is not distressed]'. A retribution goal generator underlying anger might be 'If X harms me generate the goal [X suffers].' A full analysis would describe various 'goal generators', 'goal generator generators', and so on. A learning system would produce new goal generators in the light of experience, using generator generators.

Goal Comparators

Generators do not always produce consistent goals. Different design constraints lead to different co-existing goal generators. Social animals or machines need goal generators that produce goals for the benefit of others, and these can conflict with the individual's own goals and needs. Goal comparators are therefore needed for selection between different ends.

Some comparators apply constraint goals in planning, for instance using a 'minimize cost' rule to select the cheaper of two subgoals. Others directly order ends, like a rule that saving life is always more important than any other goal, but not because of some common measure applicable to both. As there are different incommensurable sources of motivation and different bases for comparison, there need not be any *optimal* resolution of a conflict.

Higher-Order Motivators

Despite possibly confusing colloquial connotations, the general term 'motivator' is used to refer to mechanisms and representations that tend to produce, or modify, or select between actions, in the light of beliefs. Motivators recursively include generators and comparators of motivators. Some motivators are transient, like the goal of picking up a particular piece of cake, while others are long term, like an ambition to be slim.

Motivators should not be static—motivator generators are required for flexible production of new goals. Still higher intelligence involves the ability to learn from experience and modify the generators. Thus the requirement for generators is recursive. The same applies to comparators: If two generators regularly come into conflict by generating conflicting goals, then it may be necessary to suppress or modify one of them. This requires a generator comparator. Comparator generators and comparator comparators are also needed. Higher-order generators and comparators account for some personality differences. Their effects account for some of the subtleties of emotional states.

Theoretical research is needed to design generally useful higher-level generators and comparators. Empirical research is needed to establish what the mechanisms are in people. Do we have a limit to levels of generators and comparators or can new levels be recursively generated indefinitely?

Varieties of Motivators

'Derivative' and 'non-derivative' motivators can be distinguished. Roughly, a motivator is derivative if it is explicitly derived from another motivator by means–ends analysis and this origin is recorded and plays a role in subsequent processing. A desire to drink when thirsty would be non-derivative, whereas a desire for money to buy the drink would be derivative. A motive can be partly derivative, partly non-derivative, like a desire to quench one's thirst with whisky to impress others. An attempt will be made to show how non-derivative motivators are central to emotional states.

The distinction has behavioural implications. Derivative goals are more readily abandoned and their abandonment has fewer side-effects, e.g. if they appear to be unattainable or if the goals from which they are derived are satisfied or abandoned. An unpromising derivative goal can easily be replaced by another if it serves the supergoal as well. If a non-derivative goal is abandoned because it conflicts with something regarded as more important it may continue to demand attention—one source of emotions. Abandonment will produce regret and a disposition to revive the goal if the inconsistency can be removed.

Human non-derivative goals include bodily needs, desire for approval, curiosity, aesthetic wishes, and the desire to succeed in tasks undertaken. Because these goals serve more general biological purposes some theorists regard them as derivative. However, the mechanisms that create a goal need not explicitly associate it with higher-level goals, but simply give it the causal power to produce planning and action, for instance, by simply inserting a representation of the goal in a data base whose contents constantly drive the system. Despite its implicit function such a goal is non-derivative for the individual.

Quantitative Dimensions of Variation

Motives can be compared on different dimensions, definable in terms of the mechanism sketched above. *Insistence* of a motive is its interrupt priority level. Insistent desires, pains, fears, etc., are those that more easily get through interrupt filters, depending on the threshold set in relation to current activities.

Goals that get through filters need to be compared to assess their relative *importance*. This (sometimes partial) ordering is determined by beliefs and comparators, and may change if they do. Complex inferences may be required. Importance of a derivative goal is linked to beliefs about effects of achieving or not achieving it. Insistence concerns how likely a goal is to get through the interrupt filter in order to be considered, whereas importance concerns how likely it is to be adopted as something to be achieved if considered. Insistence has to be assessed very quickly, and should correlate with importance but sometimes will not. A bad filter will assign low priorities to important goals, and vice versa. A desire to sneeze does not go away just because silence is essential for survival. (Not all animals have such complex motivational systems.)

Urgency is a measure of how much time is left before it is too late. This is not the same as insistence or importance: something not wanted very much may be urgent, and vice versa.

Intensity of a goal determines how actively or vigorously it is pursued if adopted. It is partly related to urgency and importance, and partly independent of them. Obstacles to an intense goal tend to be treated as a challenge rather than a reason to abandon the goal. Often a long-term important goal will lose out to something much less important but more intense—the age-old conflict between desire and duty. Ideally insistence, intensity, and importance should be correlated, but the relationship can be upset by interactions with urgency and the way reflex assignments of insistence or intensity derive from prior experience or evolutionary origins.

Another measure of a motive is how distressing or disruptive failure to achieve it is. Different again is how much pleasure is derived from fulfilment. This can be assessed by how much effort tends to go into preserving the state of fulfilment, or achieving a repetition at a later time. Both are normally expressed as how much someone 'cares', and relate to the potential to generate emotional states as described later.

These different kinds of 'strength' of motives all play a role in cognitive functioning and may be needed in sophisticated robots. They can have subjective correlates in a system with self-monitoring, although self-monitoring is not always totally reliable. Objectively they are defined in terms of dispositions to produce effects or resist changes of various sorts, internal or external. Different combinations of strengths will affect what happens at various stages in the evolution of a goal, from initial conception to achievement, abandonment, or failure.

3. SUMMARY OF PROCESSES INVOLVING MOTIVES

So far the following intermediate processes through which motives may go have been sketched:

1. Initiation—by a body monitor, motive generator, or planner creating a new subgoal.

2. Reflex prioritization of a new goal—assigning insistence.

3. Suppression or transmission by the interrupt filter.

4. Triggering a reflex action (internal or external, hardware or software).

5. Evaluation of relative importance, using comparators.

6. Adoption, rejection, or deferred consideration—adopted motives are generally called 'intentions'. Desires may persist as desires, although not adopted for action.

7. Planning—'intrinsic' planning is concerned with how to achieve the goal, 'extrinsic' planning with when, and how to relate it to other activities, e.g. should it be postponed?

8. Activation—starting to achieve the motive, or re-activating temporarily suspended motives.
9. Plan execution.
10. Interruptions—abandonment or suspension.
11. Comparison with new goals.
12. Plan or action modification in the light of new information or goals, including changes of speed, style, or subgoals.
13. Satisfaction (complete or partial).
14. Frustration or violation.
15. Internal monitoring (self-awareness).
16. Learning—modification of generators and comparators in the light of experience.

These are all computational processes, capable of being expressed in terms of rule-governed manipulation of representations of various sorts, although filling out the details is not a trivial task. I will now try to indicate how they relate to emotions. The full story is very complicated.

An Example: Anger. What is meant by 'X is angry with Y'. This implies that X believes that there is something Y did or failed to do and as a result one of X's motives has been violated. This combination of belief and motive does not suffice for anger, since X might merely regret what happened or be disappointed in Y, without being angry. Anger also requires a new motive in X: a desire to hurt or harm Y. Most people and many animals seem to have retributive motive generators that react like this, alas. The new motive is not necessarily selected for action however intense it may be: fear of consequences and appropriate comparators may keep it inoperative.

Production of the new desire is still not sufficient for anger. X may have the desire, yet put it out of mind and calmly get on with something else: In that case he is not angry. Alternatively, X's desire to do something unpleasant to Y may be entirely derivative: purely a practical measure to reduce the likelihood of future occurrence, without any ill-will felt towards Y. Then if X can be assured somehow that there will be no recurrence, the motive will be dropped. That is not anger.

Anger involves an *insistent* and *intense* non-derivative desire to do something to make Y suffer. High insistence means the desire frequently gets through X's filters to 'request attention' from X's decision-making processes. So even after rejection by comparators, the desire frequently comes back into X's thoughts, making it hard for him to concentrate on other activities. Filters designed for speed can be too stupid to reject motives already ruled out by higher levels. Moreover, the desire must not be derivative, that is a subgoal that will disappear if a supergoal is removed. In socially sophisticated agents, anger may include a belief that Y's action had no social or ethical justification.

So emotions are states produced by motivators, and involve production of new motivators.

The violation of the original motive, and the insistence of the new motive, may be associated with additional secondary effects. For example, if X becomes aware of his anger this can make him annoyed with himself. If other people perceive his state, this can also affect the nature of the emotion. The episode can revive memories of other situations which enhance the anger.

Sometimes, in human beings, emotional states produce physiological disturbances too, probably as a result of the operation of physical and chemical reflexes driven by 'rule-of-thumb' strategies, as suggested earlier. However, if X satisfied enough of the other conditions he could rightly be described as very angry, even without any physical symptoms. Strong anger can exist without any physical side-effects in so far as it constantly intrudes into X's thoughts and decisions, and in so far as he strongly desires to make Y suffer, and suffer a great deal. Although non-physical anger might be called 'cold', it would still have all the socially significant aspects of anger.

Anger is partly dispositional in that it need not *actually* interfere with other motives: for instance, if the new motive to punish Y is acted on, there need be no further disturbance. However, the anger has the *potential* to disturb other activities if the new motive has high insistence.

Anger is sometimes *felt*, as a result of self-monitoring. However, it is possible to be angry, or be in other emotional states, without being aware of the fact. For example, I suspect that dogs and very young children are unaware of their anger (although very much aware of whatever provoked it).

Emotions like anger can vary along different quantitative and qualitative dimensions, such as: how certain X is about what Y has done, how much X cares about it (i.e. how important and intense the violated motive is); how much harm X wishes to do to Y; how important this new desire is, how intense it is, how insistent it is, how long lasting it is; how much mental disturbance is produced in X; how much physiological disturbance there is; which aspects of the state X is aware of; how many secondary motives and actions are generated. Different dimensions will be appropriate to different emotions.

Variations at different stages of the scenario correspond to different states, some not emotional. When there is no desire to cause harm to Y, the emotion is more like exasperation than anger. If there is no attribution of responsibility, then the emotion is simply some form of annoyance, and if the motive that is violated is very important, and cannot readily be satisfied by some alternative, then the emotion involves dismay. Because arbitrarily many motives, beliefs, and motive generators can be involved, with new reactions triggered by the effects of old ones, the range of variation covered by this theory is bound to be richer than the set of labels in ordinary language. It will also be richer than the range of physiological responses.

4. TOWARDS A GENERATIVE GRAMMAR FOR EMOTIONS

Analysing anger and other emotions in the light of the mechanisms sketched above, suggests the following components of emotional states:

1. There is at least one initiating motive, M1, with a high level of importance and intensity.

2. A belief, B1, about real or imagined or expected satisfaction or violation of M1 triggers generators of various kinds, often producing new motives.

3. Different sorts of cases depend on; (a) whether M1 is concerned with something desired or disliked; (b) whether B1 is a belief about M1 being satisfied or violated; (c) whether B1 concerns past, present, or future; (d) whether B1 involves uncertainty or not; (e) whether the agent is aware of his emotion or not; (f) whether other agents are thought be to involved or not; (g) whether M1 is concerned with how other agents view one (cf. Roseman 1979).

4. In more complex situations several motives simultaneously interact with beliefs, e.g. a situation where B1 implies that important motives M1(a) and M1(b) are inconsistent, e.g. in dilemmas.

5. Sometimes M1 and B1 trigger a generator that produces a secondary motive, M2, for instance a desire to put things right, preserve a delight, punish a perpetrator, or inform others. This in turn can interact with other beliefs, to disturb, interrupt, or otherwise affect cognitive processes. This would be a 'two-level' emotional state. Several levels are possible.

6. Sometimes M1 and B1 trigger several motive generators simultaneously. The resulting interactions can be very complex especially when new motives are in conflict, e.g. a desire to undo the damage and to catch the culprit.

7. Sometimes the newly generated motives conflict with previously existing motives.

8. New motives with high insistence get through interrupt filters and tend to produce (although they need not actually produce) a *disturbance*, i.e. continually interrupting thinking and deciding, and influencing decision-making criteria and perceptions.

9. Thoughts as well as motives can interrupt. Even with no new motive there may simply be a constant dwelling on M1 and B1. This is especially true of emotions like grief, involving what cannot be undone. Such compulsive dwelling might derive from triggering of automatic learning mechanisms concerned with re-programming generators.

10. New motives need not be selected for action. M2 may be considered and rejected as unimportant, yet continue to get through interrupt filters if its insistence is high.

11. In some emotional states, like fright, M2 triggers reflex action, bypassing deliberation and planning, and interrupting other actions (Sloman 1978: ch. 6). 'Software reflexes' are called 'impulsive' actions. Reflexes make it possible to take very rapid remedial action or grasp sudden opportunities. Sometimes they are disastrous, however. Some reflexes are purely mental: a whole barrage of thoughts and feelings may be triggered.

12. Some emotional states arise out of the individual's own thoughts or actions, for instance, fear generated by contemplating possible errors. Secondary motives may be generated to take extra care, etc. These secondary motives may generate so much disturbance that they lead to disaster.

13. Some emotions involve interrupting and re-directing many ongoing processes, for instance, processes controlling different parts of the body in restoring balance. If sensory detectors record local changes, the system's perception of its own state will be changed.

14. Self-monitoring processes may or may not detect the new internal state. If not, X will not be conscious of, or feel, the emotion. Internal monitoring need not produce recognition, e.g. relevant schemata might not have been learnt (Sloman 1978: ch. 10). People have to learn to discriminate and recognize complex internal states, using perceptual processes no less complex than recognizing a face or a typewriter.

15. Recognition of an emotion can produce further effects, e.g. if the internal state fulfils or violates some motive. It may activate dormant motives or motive generators and possibly lead to successively higher-order emotions (recursive escalation).

The interruptions, disturbances, and departures from rationality that characterize some emotions are a natural consequence of the sorts of mechanisms arising from constraints on the design of intelligent systems, especially the inevitable stupidity of resource-limited interrupt filters that have to act quickly. A robot with an infinitely fast computer and perfect knowledge and predictive power would not need such mechanisms. However, not all emotions are dysfunctional: when walking on a narrow ledge it is important that you do not forget the risks.

These mechanisms allow so many different subprocesses in different situations that no simple table of types of emotions can do justice to the variety. The same rich variation could characterize the detailed phenomenology of emotions in clever robots with self-monitoring abilities.

A full account of how people typically *feel* anger, elation, fear, etc. would have to include bodily awareness. Yet what makes many emotions important in our lives is not this sort of detail, but the more global cognitive structure. Fury matters because it can produce actions causing harm to the hater and hated, not because there is physical tension and sweating. Grief matters because the beloved child is lost, not because there is a new feeling in the belly. So it would be reasonable for us to use terms like 'afraid', 'disappointed', 'ecstatic', 'furious', or 'grief-stricken' to describe the state of mind of an alien being, or even a sufficiently sophisticated robot, without the physiological responses (contrast Lyons 1980).

5. MOODS, ATTITUDES, AND PERSONALITY

A *mood* is partly like an emotion: it involves some kind of global disturbance of, or disposition to disturb, mental processes. However, it need not include any specific beliefs, desires, inclinations to act, etc. In humans, moods can be induced by chemical or by cognitive factors, for instance, drinking or hearing good or bad news. A mood can colour the way one perceives things, interprets the actions of others, predicts the outcome of actions, makes plans, etc. As with an emotional state, a mood may or may not be perceived and classified by the individual concerned. A more detailed theory would have to distinguish different mechanisms, for instance, global 'hardware-induced' speed changes of certain subprocesses and global 'software-induced' changes in relative priorities of motives or inference strategies.

An *attitude,* such as love or admiration, is a collection of beliefs, motives, motive generators, and comparators focused on some individual, object, or idea. People who love their children will acquire new goals when they detect dangers or opportunities that might affect their wellbeing. The strength of the love determines the importance and interrupt priority levels assigned to such goals. Selfishness is a similar attitude to oneself. In communities of intelligent systems, able to think and care about the mental states of others, the richness and variety of attitudes makes them an inexhaustible topic for study by poets, novelists, and social scientists. Attitudes are often confused with emotions. It is possible to love, pity, admire, or hate someone without being at all emotional about it. Attitudes are expressed in tendencies to make certain choices *when the opportunity arises*, but need not include continual disturbance of thoughts and decisions. One can love one's children without having them constantly in mind, although news of danger to loved ones may trigger emotions.

Character and personality include long-term attitudes. Generosity, for example, is not a goal but a cluster of goal generators that produce new goals in response to information about another's needs and comparators that select them over more self-centred goals. Hypocrites produce similar goals but never adopt them for action. A personality or character is a vast collection of unfocused general dispositions to produce certain goals in specific situations. The set of such collections is too rich for ordinary adjectives. A whole novel may be required to portray a complex personality. More generally, the space of possible mental states and processes is too rich and complex for colloquial labels like 'attitude', 'emotion', 'mood' to survive in an adequate scientific theory.

There are many kinds of deep and moving experiences that we describe as emotions, for lack of a richer, more fine-grained vocabulary: for instance, delight in a landscape, reading poetry, hearing music, being absorbed in a film or a problem. These involve powerful interactions between perception and a large number of additional processes, some physical as well as mental. Listening to music can produce a tendency to move physically and also a great deal of mental 'movement': memories, perceptions, ripples of association, all controlled by the music. Such processes might be accounted for in terms of aspects of the design of intelligent systems not discussed here, such as the need for associative memories and subtle forms of integration and synchronization in controlling physical movements. The synchronization is needed both within an individual and between individuals engaged in co-operative tasks. Music seems to take control of some such processes.

I conjecture that the mechanisms sketched here are capable of generating states we ordinarily describe as emotional—fear, anger, frustration, excitement, dismay, grief, joy, etc. The mechanisms are generative in the sense that the relevant motives, beliefs, plans, and social contexts can be indefinitely complex and the emotional processes they generate can be correspondingly complex and varied (Abelson 1973; Dyer 1981; Lehnert,

Black, and Reiser 1981). This means that no simple bounded taxonomy of emotional states can begin to capture the variety, any more than a taxonomy can capture the variety of sentences of English (cf. Roseman 1979).

Does a Scientific Theory of Mind Need Such Concepts?

It is sometimes suggested that although concepts like 'belief', 'desire', 'emotion' play an important role in individual thoughts about other people, they are not required for a fully developed scientific theory of the mind. In its extreme form this is materialist reductionism, but that is as implausible for psychology as the suggestion that concepts of software design can be replaced by concepts relating only to computing hardware.

A more subtle suggestion (S. Rosenschein, SRI, pers. commun.) is that an entirely new collection of 'intermediate level' concepts, unrelated to beliefs, desires, intentions, etc., will suffice for a predictively and explanatorily successful scientific theory of how people and other intelligent organisms work. Because it is unlikely that ordinary concepts can be dispensed with entirely in expressing significant generalizations about human behaviour (Pylyshyn 1986) I have taken a weaker stand: instead of totally replacing ordinary concepts we need to extend and refine them, showing how they relate to a working design specification.

Even if this sort of theory is wrong, it may be deeply implicated in semantics of natural-language concepts concerning human mental states and actions. If so, a machine able to understand ordinary language and simulate human communication will require at least an implicit grasp of the theory.

Implications

Not all these mechanisms can be found in all animals. In some less intelligent creatures, selection of a motive might be inseparable from the process of initiating action: operative motives could not be dormant. In such animals or machines lacking the mechanisms required for flexibility in a complex environment, emotions in the sense described here would be impossible.

It is also unlikely that all of this richness exists in young children. By investigating the development of the cognitive and computational mechanisms in children, including the motivational mechanisms, we can hope to understand more about their emotional states. In particular, it seems that many higher-order generators and comparators are not available to infants, and that interrupt filters are far less selective than in most adults, which is not surprising if software filters are the result of learning.

The very complexity of the mechanisms described reveals enormous scope for 'bugs'. Motive generators and comparators could produce unfortunate desires and preferences. Interrupt priorities may be assigned in a way that doesn't correlate well with reflective judgements of importance. Thresholds for interrupts may be set too high or too low. Learning processes that modify generators and comparators may be too quick to change things on flimsy evidence. Given the inevitable stupidity of some of the faster reflexes and filters, we can expect some kinds of malfunctions of generators and comparators to lead to intense emotions that interfere with normal cognitive or social functioning. Reactions to unfulfilled motives may be too strong, or too weak for the long-term good of the individual or his associates. The relative importance assigned to different sorts of motives by the goal assessment procedures may produce a tendency to select goals that are unachievable or achievable only at enormous cost. Dormant, temporarily suspended motives may too often go unattended because the monitoring process fails to detect opportunities, perhaps because of inadequate indexing. The pervasiveness of 'rules-of-thumb' for coping with inadequate information, limited resources, and the need for speed, provides enormous scope for systematic malfunction. Recursive escalation of emotions might account for some catatonic states.

The inevitability of familiar types of fallibility should be a matter of concern to those who hope that important decisions can be taken very rapidly by machines in the not too distant future.

In fact, if people are as complex and intricate as we have suggested, it is amazing that so many are stable and civilized. Perhaps this theory will reveal types of disturbance we previously could not recognize.

The theory implies that processes of learning and cognitive development, occur in a framework of a complex and frequently changing collection of motivators. These and the processes they generate must have a profound influence on what is learnt when, and it is to be expected that there will be enormous variation between individuals. The implications for educators have yet to be explored.

6. CONCLUSION

A theory of this general sort is a *computational* theory of mind. The computations may occur in a *virtual* machine implemented in lower-level machines, brainlike or computerlike: they need not be implemented directly in physical processes. Thus, the theory is neutral between physically explicit representations as found at low levels in conventional computers and implicit or distributed representations studied in neural-net models.

The test of this approach will be the explanatory power of the theories based on it. We need both a systematic explanation of the whole range of possibilities we find in human behaviour and an account of how people differ from one another and from other actual and possible behaving systems. (Concerning explanations of possibilities, see Sloman 1978: ch. 2.)

Understanding computational mechanisms behind familiar mental processes may enable us to reduce suffering from emotional disturbances, learning disabilities, and a range of social inadequacies. Some problems may be due to brain damage or neural malfunction. Other problems seem more like software faults in a computer. I conjecture that many emotionally disturbed people are experiencing such software 'bugs'.

The analysis still has many gaps. In particular, an account of pleasure and pain is missing, and I am not yet able to give an acceptable analysis of what it is to find something funny. There are states like being thrilled by rapid motion, spellbound by a sunset, moved to tears by reading a book or watching a play, that require more detailed analysis. I have not discussed the many aspects of human emotional life that arise contingently from our evolutionary history and would not necessarily be found in well-designed robots. So there is much yet to be done. Nevertheless, the theory provides a framework for thinking about a range of possible types of intelligent systems, natural and artificial—part of our general study of the space of possible minds. Attempting to test the ideas in working computer simulations will surely reveal gaps and weaknesses.[1]

[1] This work was supported by a fellowship from the GEC Research Laboratories and a grant from the Renaissance Trust. Some of the ideas derive from Simon's seminal paper. My ideas have benefited from discussions over several years with Monica Croucher and her thesis (Croucher 1985) developed the ideas of this paper in greater depth. The view of ordinary language as a source of information about the human mind, sketched here and expanded in Chapter 4 of Sloman (1978) owes much to Austin (1961). Keith Oatley's editorial comments have been most helpful.

REFERENCES

Abelson, R. A. (1973). 'The Structure of Belief Systems.' In R. C. Schank and K. M. Colby (eds.), *Computer Models of Thought and Language*, pp. 287–340. San Francisco: Freeman.
Austin, J. L. (1961). 'A Plea for Excuses.' In *Philosophical Papers*. Repr. in A. R. White (ed.), *Philosophy of Action*, pp. 19–42. Oxford: Oxford University Press.

Boden, M. (1972). *Purposive Explanation in Psychology*. Cambridge, Mass.: Harvard University Press.

—— (1977). *Artificial Intelligence and Natural Man*. Hassocks, Sussex: Harvester Press.

Croucher, M. (1985). 'A Computational Approach to Emotions.' Unpublished thesis. University of Sussex.

Dennett, D. C. (1979). *Brainstorms*. Hassocks, Sussex: Harvester Press.

Dyer, M. G. (1981). 'The Role of TAUs in Narratives.' *Proceedings Cognitive Science Conference*, pp. 225–7. Berkeley, Calif.

Edelson, T. (1986). 'Can a System Be Intelligent If It Never Gives a Damn?' *Proc. 5th Nat. Conf. AI* (AAAI-86), pp. 298–302. Philadelphia.

Heider, F. (1958). *The Psychology of Interpersonal Relations*. New York: Wiley.

Lehnert, W. G., Black, J. B., and Reiser, B. J. (1981). 'Summarising Narratives.' *Proc. 7th IJCAI Conference*, pp. 184–9. Vancouver, BC.

Lyons, W. (1980). *Emotion*. Cambridge: Cambridge University Press.

Oatley, K., and Johnson-Laird, P. N. (1985). 'Sketch for a Cognitive Theory of the Emotions.' *Cognitive Science Research Paper No. CSRP.045*. University of Sussex, Cognitive Studies.

Pylyshyn, Z. W. (1986). *Computation and Cognition: Toward a Foundation For Cognitive Science*. Cambridge, Mass.: MIT Press.

Roseman, I. (1979). 'Cognitive Aspects of Emotions and Emotional Behaviour.' Paper presented to the *87th Annual Convention of the American Psychological Association*.

Ryle, G. (1949). *The Concept of Mind*. London: Hutchinson.

Simon, H. A. (1979). 'Motivational and Emotional Controls of Cognition.' In H. A. Simon, *Models of Thought*, pp. 23–38. New Haven, Conn.: Yale University Press.

Sloman, A. (1969). 'How to Derive "Better" from "Is".' *American Philosophical Quarterly* 6: 43–52.

—— (1978). *The Computer Revolution in Philosophy: Philosophy, Science and Models of Mind*. Hassocks, Sussex: Harvester Press.

—— (1981). 'Skills Learning and Parallelism.' *Proceedings Cognitive Science Conference*, pp. 284–5. Berkeley, Calif.

—— (1984). 'Why We Need Many Knowledge Representation Formalisms.' In M. Bramer (ed.), *Research and Development in Expert Systems*, pp. 163–83. Cambridge: Cambridge University Press.

—— (1985). 'Real-Time Multiple-Motive Expert Systems.' In M. Merry (ed.), *Expert Systems 85*, pp. 213–24. Cambridge: Cambridge University Press.

—— and Croucher, M. (1981). 'Why Robots Will Have Emotions.' *Proc. 7th IJCAI Conference*, pp. 197–202. Vancouver, BC.

11

DISTRIBUTED REPRESENTATIONS

GEOFFREY E. HINTON, JAMES L. McCLELLAND,
and DAVID E. RUMELHART

Given a network of simple computing elements and some entities to be represented, the most straightforward scheme is to use one computing element for each entity. This is called a *local* representation. It is easy to understand and easy to implement because the structure of the physical network mirrors the structure of the knowledge it contains. The naturalness and simplicity of this relationship between the knowledge and the hardware that implements it have led many people to simply assume that local representations are the best way to use parallel hardware. There are, of course, a wide variety of more complicated implementations in which there is no one-to-one correspondence between concepts and hardware units, but these implementations are only worth considering if they lead to increased efficiency or to interesting emergent properties that cannot be conveniently achieved using local representations.

This chapter describes one type of representation that is less familiar and harder to think about than local representations. Each entity is represented by a pattern of activity distributed over many computing elements, and each computing element is involved in representing many different entities. The strength of this more complicated kind of representation does not lie in its notational convenience or its ease of implementation in a conventional computer, but rather in the efficiency with which it makes use of the processing abilities of networks of simple, neurone-like computing elements.

Every representational scheme has its good and bad points. Distributed representations are no exception. Some desirable properties arise very naturally from the use of patterns of activity as representations. Other properties, like the ability to temporarily store a large set of arbitrary

G. E. Hinton *et al.*, 'Distributed Representations', in D. E. Rumelhart and J. E. McClelland (eds.), *Parallel Distributed Processing: Explorations in the Microstructure of Cognition*, Vol. 1, *Foundations*, pp. 77–109. Reprinted by permission of the MIT Press, Cambridge, Mass. Chapter references given in this article are to other chapters in *Parallel Distributed Processing*.

associations, are much harder to achieve. As we shall see, the best psychological evidence for distributed representations is the degree to which their strengths and weaknesses match those of the human mind.

The first section of this chapter stresses some of the virtues of distributed representations. The second section considers the efficiency of distributed representations, and shows clearly why distributed representations can be better than local ones for certain classes of problems. A final section discusses some difficult issues which are often avoided by advocates of distributed representations, such as the representation of constituent structure and the sequential focusing of processing effort on different aspects of a structure object.

Disclaimers.   Before examining the detailed arguments in favour of distributed representations, it is important to be clear about their status within an overall theory of human information-processing. It would be wrong to view distributed representations as an *alternative* to representational schemes like semantic networks or production systems that have been found useful in cognitive psychology and artificial intelligence. It is more fruitful to view them as one way of implementing these more abstract schemes in parallel networks, but with one proviso: distributed representations give rise to some powerful and unexpected emergent properties. These properties can therefore be taken as primitives when working in a more abstract formalism. For example, distributed representations are good for content-addressable memory, automatic generalization, and the selection of the rule that best fits the current situation. So if one assumes that more abstract models are implemented in the brain using distributed representations, it is not unreasonable to treat abilities like content-addressable memory, automatic generalization, or the selection of an appropriate rule as primitive operations, even though there is no easy way to implement these operations in conventional computers. Some of the emergent properties of distributed representations are not easily captured in higher-level formalisms. For example, distributed representations are consistent with the simultaneous application of a large number of partially fitting rules to the current situation, each rule being applied to the degree that it is relevant. We shall examine these properties of distributed representations in the chapter on schemata (ch. 14). There we will see clearly that schemata and other higher-level constructs provide only approximate characterizations of mechanisms which rely on distributed representations. Thus, the contribution that an analysis of distributed representations can make to these higher-level formalisms is to legitimize certain powerful, primitive operations which would otherwise appear to be an appeal to magic; to enrich our repertoire of primitive operations beyond those which can conveniently be captured in many higher-level formalisms; and to suggest that these higher-level formalisms may only capture the coarse features of the computational capabilities of the underlying processing mechanisms.

Another common source of confusion is the idea that distributed representations are somehow in conflict with the extensive evidence for localization of function in the brain (Luria 1973). A system that uses distributed representations still requires many different modules for representing completely different kinds of thing at the same time. The distributed representations occur *within* these localized modules. For example, different modules would be devoted to things as different as mental images and sentence structures, but two different mental images would correspond to *alternative* patterns of activity in the same module. The representations advocated here are local at a global scale but global at a local scale.

1. VIRTUES OF DISTRIBUTED REPRESENTATIONS

This section considers three important features of distributed representations: (*a*) their essentially constructive character; (*b*) their ability to generalize automatically to novel situations; and (*c*) their tunability to changing environments. Several of these virtues are shared by certain local models, such as the interactive activation model of word perception, or McClelland's (1981) model of generalization and retrieval described in Chapter 1.

Memory as Inference

People have a very flexible way of accessing their memories: they can recall items from partial descriptions of their contents (Norman and Bobrow 1979). Moreover, they can do this even if some parts of the partial description are wrong. Many people, for example, can rapidly retrieve the item that satisfies the following partial description: It is an actor, it is intelligent, it is a politician. This kind of *content-addressable* memory is very useful and it is very hard to implement on a conventional computer because computers store each item at a particular address, and to retrieve an item they must know its address. If all the combinations of descriptors that will be used for access are free of errors and are known in advance, it is possible to use a method called *hash coding* that quickly yields the address of an item when given part of its content. In general, however, content-addressable memory requires a massive search for the item that best fits the partial description. The central computational problem in memory is how to make this search efficient. When the cues can contain errors, this is very difficult because the failure to fit one of the cues cannot be used as a filter for quickly eliminating inappropriate answers.

Distributed representations provide an efficient way of using parallel hardware to implement best-fit searches. The basic idea is fairly simple, though it is quite unlike a conventional computer memory. Different items correspond to different patterns of activity over the very same group of hardware units. A partial description is presented in the form of a partial activity pattern, activating some of the hardware units.[1] Interactions between the units then allow the set of active units to influence others of the units, thereby completing the pattern, and generating the item that best fits the description. A new item is 'stored' by modifying the interactions between the hardware units so as to create a new stable pattern of activity. The main difference from a conventional computer memory is that patterns which are not active do not exist anywhere. They can be re-created because the connection strengths between units have been changed appropriately, but each connection strength is involved in storing many patterns, so it is impossible to point to a particular place where the memory for a particular item is stored.

Many people are surprised when they understand that the connections between a set of simple processing units are capable of supporting a large number of different patterns. Illustrations of this aspect of distributed models are provided in a number of papers in the literature (e.g. Anderson 1977; Hinton 1981); this property is illustrated in the model of memory and amnesia described in chs. 17 and 25.

One way of thinking about distributed memories is in terms of a very large set of plausible inference rules. Each active unit represents a 'microfeature' of an item, and the connection strengths stand for plausible 'microinferences' between microfeatures. Any particular pattern of activity of the units will satisfy some of the microinferences and violate others. A stable pattern of activity is one that violates the plausible microinferences

---

[1] This is easy if the partial description is simply a set of features, but it is much more difficult if the partial description mentions relationships to other objects. If, for example, the system is asked to retrieve John's father, it must represent John, but if John and his father are represented by mutually exclusive patterns of activity in the very same group of units, it is hard to see how this can be done without preventing the representation of John's father. A distributed solution to this problem is described in the text.

less than any of the neighbouring patterns. A new stable pattern can be created by changing the inference rules so that the new pattern violates them less than its neighbours. This view of memory makes it clear that there is no sharp distinction between genuine memory and plausible reconstruction. A genuine memory is a pattern that is stable because the inference rules were modified when it occurred before. A 'confabulation' is a pattern that is stable because of the way the inference rules have been modified to store several different previous patterns. So far as the subject is concerned, this may be indistinguishable from the real thing.

The blurring of the distinction between veridical recall and confabulation or plausible reconstruction seems to be characteristic of human memory (Bartlett 1932; Neisser 1981). The reconstructive nature of human memory is surprising only because it conflicts with the standard metaphors we use. We tend to think that a memory system should work by storing literal copies of items and then retrieving the stored copy, as in a filing cabinet or a typical computer data base. Such systems are not naturally reconstructive.

If we view memory as a process that constructs a pattern of activity which represents the most plausible item that is consistent with the given cues, we need some guarantee that it will converge on the representation of the item that best fits the description, though it might be tolerable to sometimes get a good but not optimal fit. It is easy to *imagine* this happening, but it is harder to make it actually work. One recent approach to this problem is to use statistical mechanics to analyse the behaviour of groups of interacting *stochastic* units. The analysis guarantees that the better an item fits the description, the more likely it is to be produced as the solution. This approach is described in Chapter 7, and a related approach is described in Chapter 6. An alternative approach, using units with continuous activations (Hopfield 1984) is described in Chapter 14.

Similarity and Generalization

When a new item is stored, the modifications in the connection strengths must not wipe out existing items. This can be achieved by modifying a very large number of weights very slightly. If the modifications are all in the direction that helps the pattern that is being stored, there will be a conspiracy effect: the total help for the intended pattern will be the sum of all the small separate modifications. For unrelated patterns, however, there will be very little transfer of effect because some of the modifications will help and some will hinder. Instead of all the small modifications conspiring together, they will mainly cancel out. This kind of statistical reasoning underpins most distributed memory models, but there are many variations of the basic idea (See Hinton and Anderson 1981, for several examples).

It is possible to prevent interference altogether by using orthogonal patterns of activity for the various items to be stored (a rudimentary example of such a case is given in ch. 1). However, this eliminates one of the most interesting properties of distributed representations: they automatically give rise to generalizations. If the task is simply to remember accurately a set of unrelated items, the generalization effects are harmful and are called interference. But generalization is normally a helpful phenomenon. It allows us to deal effectively with situations that are similar but not identical to previously experienced situations.

People are good at generalizing newly acquired knowledge. If you learn a new fact about an object, your expectations about other similar objects tend to change. If, for example, you learn that chimpanzees like onions you will probably raise your estimate of the probability that gorillas like onions. In a network that uses distributed representations, this kind of generalization is automatic. The new knowledge about chimpanzees is incorporated by modifying some of the connection strengths so as to alter the causal effects of the distributed pattern of activity that represents chimpanzees.[2] The modifications automatically change the causal effects of all similar activity patterns. So if the representation of gorillas is a similar activity pattern over the same set of units, its causal effects will be changed in a similar way.

The very simplest distributed scheme would represent the concept of onion and the concept of chimpanzee by *alternative* activity patterns over the very same set of units. It would then be hard to represent chimps and onions at the same time. This problem can be solved by using separate modules for each possible role of an item within a larger structure. Chimps, for example, are the 'agent' of the liking and so a pattern representing chimps occupies the 'agent' module and the pattern representing onions occupies the 'patient' module (see Fig. 1). Each module can have alternative patterns for all the various items, so this scheme does not involve local representations of items. What is localized is the role.

If you subsequently learn that gibbons and orangutans do not like

---

The internal structure of this pattern may also change. There is always a choice between changing the weights on the outgoing connections and changing the pattern itself so that different outgoing connections become relevant. Changes in the pattern itself alter its similarity to other patterns and thereby alter how generalization will occur in the future. It is generally much harder to figure out how to change the pattern that represents an item than it is to figure out how to change the outgoing connections so that a particular pattern will have the desired effects on another part of the network.

RELATIONSHIP

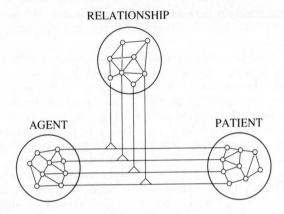

AGENT   PATIENT

FIG. 1. In this simplified scheme there are two different modules, one of which represents the agent and the other the patient. To incorporate the fact that chimpanzees like onions, the pattern for chimpanzees in one module must be associated with the pattern for onions in the other module. Relationships other than 'liking' can be implemented by having a third group of units whose pattern of activity represents the relationship. This pattern must then 'gate' the interactions between the agent and patient groups. Hinton (1981) describes one way of doing this gating by using a fourth group of units.

onions your estimate of the probability that gorillas like onions will fall, though it may still remain higher than it was initially. Obviously, the combination of facts suggests that liking onions is a peculiar quirk of chimpanzees. A system that uses distributed representations will automatically arrive at this conclusion, provided that the alternative patterns that represent the various apes are related to one another in a particular way that is somewhat more specific than just being similar to one another: there needs to be a part of each complete pattern that is identical for all the various apes. In other words, the group of units used for the distributed representations must be divided into two subgroups, and all the various apes must be represented by the same pattern in the first subgroup, but by different patterns in the second subgroup. The pattern of activity over the first subgroup represents the *type* of the item, and the pattern over the second subgroup represents additional microfeatures that discriminate each instance of the type from the other instances. Note that any subset of the microfeatures can be considered to define a type. One subset might be common to all apes, and a different (but overlapping) subset might be common to all pets. This allows an item to be an instance of many different types simultaneously.

When the system learns a new fact about chimpanzees, it usually has no way of knowing whether the fact is true of all apes or is just a property of chimpanzees. The obvious strategy is therefore to modify the strengths of the connections emanating from *all* the active units, so that the new knowledge will be partly a property of apes in general and partly a property of whatever features distinguish chimps from other apes. If it is subsequently learned that other apes do not like onions, correcting modifications will be made so that the information about onions is no longer associated with the subpattern that is common to all apes. The knowledge about onions will then be restricted to the subpattern that distinguishes chimps from other apes. If it had turned out that gibbons and orangutans also like onions, the modifications in the weights emanating from the subpattern representing apes would have reinforced one another, and the knowledge would have become associated with the subpattern shared by all apes rather than with the patterns that distinguish one ape from another.

A very simple version of this theory of generalization has been implemented in a computer simulation (Hinton 1981). Several applications that make use of this property can be found in Part IV of this book.

There is an obvious generalization of the idea that the representation of an item is composed of two parts, one that represents the type and another that represents the way in which this particular instance differs from others of the same type. Almost all types are themselves instances of more general types, and this can be implemented by dividing the pattern that represents the type into two subpatterns, one for the more general type of which this type is an instance, and the other for the features that discriminate this particular type from other instances of the same general type. Thus the relation between a type and an instance can be implemented by the relationship between a set of units and a larger set that includes it. Notice that the more general the type, the *smaller* the set of units used to encode it. As the number of terms in an *intensional* description gets smaller, the corresponding *extensional* set gets larger.

In traditional semantic networks that use local representations, generalization is not a direct consequence of the representation. Given that chimpanzees like onions, the obvious way of incorporating the new knowledge is by changing the strengths of connections belonging to the chimpanzee unit. But this does not automatically change connections that belong to the gorilla unit. So extra processes must be invoked to implement generalization in a localist scheme. One commonly used method is to allow activation to spread from a local unit to other units that represent similar concepts (Collins and Loftus 1975; Quillian 1968). Then when one concept unit is activated, it will partially activate its neighbours and so any knowledge stored in the connections emanating from these neighbours will be partially effective. There are many variations of this basic idea (Fahlman 1979; Levin 1976; McClelland 1981).

It is hard to make a clean distinction between systems that use local representations plus spreading activation and systems that use distributed representations. In both cases the result of activating a concept is that many different hardware units are active. The distinction almost completely disappears in some models such as McClelland's (1981) generalization model, where the properties of a concept are represented by a pattern of activation over feature units and where this pattern of activation is determined by the interactions of a potentially very large number of units for instances of the concept. The main difference is that in one case there is a particular individual hardware unit that acts as a 'handle' which makes it easy to attach purely conventional properties like the name of the concept and easier for the theorist who constructed the network to know what the individual parts of the network stand for.

If we construct our networks by hand-specifying the connections between the units in the network, a local representation scheme has some apparent advantages. First, it is easier to think one understands the behaviour of a network if one has put in all the 'knowledge'—all the connections—oneself. But if it is the entire, distributed pattern of interacting influences among the units in the network that is doing the work, this understanding can often be illusory. Second, it seems intuitively obvious that it is harder to attach an arbitrary name to a distributed pattern than it is to attach it to a single unit. What is intuitively harder, however, may not be more efficient. We will see that one can actually implement arbitrary associations with fewer units using distributed representations. Before we turn to such considerations, however, we examine a different advantage of distributed representations: they make it possible to create new concepts without allocating new hardware.

Creating New Concepts

Any plausible scheme for representing knowledge must be capable of learning novel concepts that could not be anticipated at the time the network was initially wired up. A scheme that uses local representations must first make a discrete decision about *when* to form a new concept, and then it must find a spare hardware unit that has suitable connections for implementing the concept involved. Finding such a unit may be difficult if we assume that, after a period of early development, new knowledge is incorporated by changing the strengths of the existing connections rather than by growing new ones. If each unit only has connections to a small fraction of the others, there will probably not be any units that are connected to just the right other ones to implement a new concept. For example, in a collection of a million units each connected at random to 10,000 others, the chance of there being *any* unit that is connected to a particular set of six others is only one in a million.

In an attempt to rescue local representations from this problem, several clever schemes have been proposed that use two classes of units. The units that correspond to concepts are not directly connected to one another. Instead, the connections are implemented by indirect pathways through several layers of intermediate units (Fahlman 1980; Feldman 1982). This scheme works because the number of *potential* pathways through the intermediate layers far exceeds the total number of physical connections. If there are k layers of units, each of which has a fan-out of n connections to randomly selected units in the following layer, there are n^k potential pathways. There is almost certain to be a pathway connecting any two concept-units, and so the intermediate units along this pathway can be dedicated to connecting those two concept-units. However, these schemes end up having to dedicate several intermediate units to each effective connection, and once the dedication has occurred, all but one of the actual connections emanating from each intermediate unit are wasted. The use of several intermediate units to create a single effective connection may be appropriate in switching networks containing elements that have units with relatively small fan-out, but it seems to be an inefficient way of using the hardware of the brain.

The problems of finding a unit to stand for a new concept and wiring it up appropriately do not arise if we use distributed representations. All we need to do is modify the interactions between units so as to create a new stable pattern of activity. If this is done by modifying a large number of connections very slightly, the creation of a new pattern need not disrupt the existing representations. The difficult problem is to choose an appropriate pattern for the new concept. The effects of the new representation on representations in other parts of the system will be determined by the units that are active, and so it is important to use a collection of active units that have roughly the correct effects. Fine-tuning of the effects of the new pattern can be achieved by slightly altering the effects of the active units it contains, but it would be unwise to choose a *random* pattern for a new concept because major changes would then be needed in the weights, and this would disrupt other knowledge. Ideally, the distributed representation that is chosen for a new concept should be the one that requires the least modification of weights to make the new pattern stable and to make it have the required effects on other representations.

Naturally, it is not necessary to create a new stable pattern all in one step. It is possible for the pattern to emerge as a result of modifications on many separate occasions. This alleviates an awkward problem that arises with local representations: the system must make a discrete all-or-none decision about when to create a new concept. If we view concepts as stable patterns, they are much less discrete in character. It is possible, for example, to differentiate one stable pattern into two closely related but different variants by modifying some of the weights slightly. Unless we are allowed to clone the hardware units (and all their connections), this kind of gradual, conceptual differentiation is much harder to achieve with local representations.

One of the central problems in the development of the theory of distributed representation is the problem of specifying the exact procedures by which distributed representations are to be learned. All such procedures involve connection strength modulation, following 'learning rules' of the type outlined in Chapter 2. Not all the problems have been solved, but significant progress is being made on these problems. (See the chapters in Part II.)

2. DISTRIBUTED REPRESENTATIONS THAT WORK EFFICIENTLY

In this section, we consider some of the technical details about the implementation of distributed representations. First, we point out that certain distributed representation schemes can fail to provide a sufficient basis for differentiating different concepts, and we point out what is required to avoid this limitation. Then, we describe a way of using distributed representations to get the most information possible out of a simple network of connected units. The central result is a surprising one: if you want to encode features accurately using as few units as possible, it pays to use units that are very coarsely tuned, so that each feature activates many different units and each unit is activated by many different features. A specific feature is then encoded by a pattern of activity in many units rather than by a single active unit, so coarse coding is a form of distributed representation.

To keep the analysis simple, we shall assume that the units have only two values, on and off.[3] We shall also ignore the dynamics of the system because the question of interest, for the time being, is how many units it takes to encode features with a given accuracy. We start by considering the kind of feature that can be completely specified by giving a type (e.g. line-segment, corner, dot) and the values of some continuous parameters that distinguish it from other features of the same type (e.g. position, orientation, size). For each type of feature there is a space of possible instances. Each continuous parameter defines a dimension of the feature space, and each particular feature corresponds to a point in the space. For features like dots in a plane, the space of possible features is 2-dimensional. For features like stopped, oriented edge-segments in 3-dimensional space, the feature space is 6-dimensional. We shall start by considering 2-dimensional feature spaces and then generalize to higher dimensionalities.

Suppose that we wish to represent the position of a single dot in a plane, and we wish to achieve high accuracy without using too many units. We define the accuracy of an encoding scheme to be the number of different encodings that are generated as the dot is moved a standard distance through the space. One encoding scheme would be to divide the units into an X group and a Y group, and dedicate each unit to encoding a particular X or Y interval as shown in Figure 2. A given dot would then be encoded by activity in two units, one from each group, and the accuracy would be proportional to the number of units used. Unfortunately, there are two problems with this. First, if two dots have to be encoded at the same time, the method breaks down. The two dots will activate two units in each group, and there will be no way of telling, from the active units, whether the dots were at $(x1, y1)$ and $(x2, y2)$ or at $(x1, y2)$ and $(x2, y1)$. This is called the *binding problem*. It arises because the representation does not specify what goes with what.

The second problem arises even if we allow only one point to be represented at a time. Suppose we want certain representations to be associated with an overt response, but not others: We want $(x1, y1)$ and $(x2, y2)$ to be associated with a response, but not $(x1, y2)$ or $(x2, y1)$. We cannot implement this association using standard weighted connections to response units from units standing for the values on the two dimensions separately. For the unit for $x1$ and the unit for $x2$ would both have to activate the response, and the unit for $y1$ and the unit for $y2$ would both

[3] Similar arguments apply with multivalued activity levels, but it is important not to allow activity levels to have arbitrary precision because this makes it possible to represent an infinite amount of information in a single activity level. Units that transmit a discrete impulse with a probability that varies as a function of their activation seem to approximate the kind of precision that is possible in neural circuitry (see chs. 20 and 21).

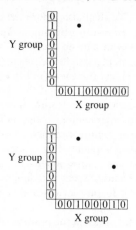

FIG. 2. *A*: A simple way of using two groups of binary units to encode the position of a point in a 2-dimension space. The active units in the X and Y groups represent the *x*- and *y*-coordinates. *B*: When two points must be encoded at the same time, it is impossible to tell which *x*-coordinate goes with which *y*-coordinate.

have to activate the response. There would be no way of preventing the response from being activated when the unit for $x1$ and the unit for $y2$ were both activated. This is another aspect of the binding problem since, again, the representation fails to specify what must go with what.

In a conventional computer it is easy to solve the binding problem. We simply create two records in the computer memory. Each record contains a pair of coordinates that go together as coordinates of one dot, and the binding information is encoded by the fact that the two coordinate values are sitting in the same record (which usually means they are sitting in neighbouring memory locations). In parallel networks it is much harder to solve the binding problem.

Conjunctive Encoding

One approach is to set aside, in advance, one unit for each possible *combination* of X and Y values. This amounts to covering the plane with a large number of small, non-overlapping zones and dedicating a unit to each zone. A dot is then represented by activity in a single unit so this is a *local* representation. The use of one unit for each discriminable feature solves the binding problem by having units which stand for the conjunction of values on each of two dimensions. In general, to permit an arbitrary association between particular combinations of features and some output or other pattern of activation, some conjunctive representation may be required.

However, this kind of local encoding is very expensive. It is much less efficient than the previous scheme because the accuracy of pinpointing a point in the plane is only proportional to the square root of the number of units. In general, for a k-dimensional feature space, the local encoding yields an accuracy proportional to the kth root of the number of units. Achieving high accuracy without running into the binding problem is thus very expensive.

The use of one unit for each discriminable feature may be a reasonable encoding if a very large number of features are presented on each occasion, so that a large fraction of the units are active. However, it is a very inefficient encoding if only a very small fraction of the possible features are presented at once. The average amount of information conveyed by the state of a binary unit is 1 bit if the unit is active half the time, and it is much less if the unit is only rarely active.[4] It would therefore be more efficient to use an encoding in which a larger fraction of the units were active at any moment. This can be done if we abandon the idea that each discriminable feature is represented by activity in a single unit.

Coarse Coding

Suppose we divide the space into larger, overlapping zones and assign a unit to each zone. For simplicity, we will assume that the zones are circular, that their centres have a uniform random distribution throughout the space, and that all the zones used by a given encoding scheme have the same radius. The question of interest is how accurately a feature is encoded as a function of the radius of the zones. If we have a given number of units at our disposal is it better to use large zones so that each feature point falls in many zones, or is it better to use small zones so that each feature is represented by activity in fewer but more finely tuned units?

The accuracy is proportional to the number of different encodings that are generated as we move a feature point along a straight line from one side of the space to the other. Every time the line crosses the boundary of a zone, the encoding of the feature point changes because the activity of the unit corresponding to that zone changes. So the number of discriminable features along the line is just twice the number of zones that the line

---

[4] The amount of information conveyed by a unit that has a probability of p of being on is $-p \log p - (1 - p) \log (1 - p)$.

penetrates.[5] The line penetrates every zone whose centre lies within one radius of the line (see Fig. 3). This number is proportional to the radius of

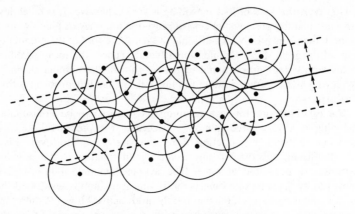

FIG. 3. The number of zone boundaries that are cut by the line is proportional to the number of zone centres within one-zone radius of the line.

the zones, r, and it is also proportional to their number, n. Hence the accuracy, a, is related to the number of zones and to their radius as follows:

$$a \propto nr$$

In general, for a k-dimensional space, the number of zones whose centres lie within one radius of a line through the space is proportional to the volume of a k-dimensional hypercylinder of radius r. This volume is equal to the length of the cylinder (which is fixed) times its $(k - 1)$-dimensional cross-sectional area which is proportional to r^{k-1}. Hence, the accuracy is given by

$$a \propto nr^{k-1}$$

So, for example, doubling the radius of the zones increases by a factor of 32, the *linear* accuracy with which a 6-dimensional feature like a stopped oriented 3-dimensional edge is represented. The intuitive idea

[5] Problems arise if you enter and leave a zone without crossing other zone borders in between because you revert to the same encoding as before, but this effect is negligible if the zones are dense enough for there to be many zones containing each point in the space.

that larger zones lead to sloppier representations is entirely wrong because distributed representations hold information much more efficiently than local ones. Even though each active unit is less specific in its meaning, the combination of active units is far more specific. Notice also that with coarse coding the accuracy is proportional to the number of units, which is much better than being proportional to the kth root of the number.

Units that respond to complex features in retinotopic maps in visual cortex often have fairly large receptive fields. This is often interpreted as the first step on the way to a translation invariant representation. However, it may be that the function of the large fields is not to achieve translation invariance but to pinpoint accurately where the feature is!

Limitations on coarse coding. So far, only the advantages of coarse coding have been mentioned, and its problematic aspects have been ignored. There are a number of limitations that cause the coarse coding strategy to break down when the 'receptive fields' become too large. One obvious limitation occurs when the fields become comparable in size to the whole space. This limitation is generally of little interest because other, more severe, problems arise before the receptive fields become this large.

Coarse coding is only effective when the features that must be represented are relatively sparse. If many feature points are crowded together, each receptive field will contain many features and the activity pattern in the coarse-coded units will not discriminate between many alternative combinations of feature points. (If the units are allowed to have integer activity levels that reflect the number of feature points falling within their fields, a few nearby points can be tolerated, but not many.) Thus there is a resolution/accuracy trade-off. Coarse coding can give high accuracy for the parameters of features provided that features are widely spaced so that high resolution is not also required. As a rough rule of thumb, the diameter of the receptive fields should be of the same order as the spacing between simultaneously present feature points.[6]

The fact that coarse coding only works if the features are sparse should be unsurprising given that its advantage over a local encoding is that it uses the information capacity of the units more efficiently by making each unit active more often. If the features are so dense that the units would be

---

[6] It is interesting that many of the geometric visual illusions illustrate interactions between features at a distance much greater than the uncertainty in the subjects' knowledge of the position of a feature. This is just what would be expected if coarse coding is being used to represent complex features accurately.

active for about half the time using a local encoding, coarse coding can only make things worse.

A second major limitation on the use of coarse coding stems from the fact that the representation of a feature must be used to affect other representations. There is no point using coarse coding if the features have to be recoded as activity in finely tuned units before they can have the appropriate effects on other representations. If we assume that the effect of a distributed representation is the *sum* of the effects of the individual active units that constitute the representation, there is a strong limitation on the circumstances under which coarse coding can be used effectively. Nearby features will be encoded by similar sets of active units, and so they will inevitably tend to have similar effects. Broadly speaking, coarse coding is only useful if the required effect of a feature is the average of the required effects of its neighbours. At a fine enough scale this is nearly always true for spatial tasks. The scale at which it breaks down determines an upper limit on the size of the receptive fields.

Another limitation is that whenever coarse-coded representations interact, there is a tendency for the coarseness to increase. To counteract this tendency, it is probably necessary to have lateral inhibition operating within each representation. This issue requires further research.

Extension to Non-Continuous Spaces.   The principle underlying coarse coding can be generalized to non-continuous spaces by thinking of a set of items as the equivalent of a receptive field. A local representation uses one unit for each possible item. A distributed representation uses a unit for a set of items, and it implicitly encodes a particular item as the intersection of the sets that correspond to the active units.

In the domain of spatial features there is generally a very strong regularity: sets of features with similar parameter values need to have similar effects on other representations. Coarse coding is efficient because it allows this regularity to be expressed in the connection strengths. In other domains, the regularities are different, but the efficiency arguments are the same: it is better to devote a unit to a set of items than to a single item, provided that the set is chosen in such a way that membership in the set implies something about membership in other sets. This implication can then be captured as a connection strength. Ideally, a set should be chosen so that membership of this set has strong implications for memberships of other sets that are also encoded by individual units.

We illustrate these points with a very simple example. Consider a microlanguage consisting of the three-letter words of English made up of *w* or *l*, followed by *i* or *e*, followed by *g* or *r*. The strings *wig* and *leg* are words, but *weg*, *lig*, and all strings ending in *r* are not. Suppose we wanted to use a distributed representation scheme as a basis for representing the words, and we wanted to be able to use the distributed pattern as a basis for deciding whether the string is a word or a non-word. For simplicity we will have a single 'decision' unit. The problem is to find connections from the units representing the word to the decision unit such that it fires whenever a word is present but does not fire when no word is present.[7]

Figure 4 shows three representation schemes: a distributed scheme that does not work, a distributed scheme that does work, and a local scheme. In the first scheme, each letter/position combination is represented by a

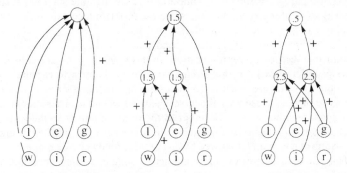

FIG. 4. Three networks applied to the problem of determining which of the strings that can be made from *w* or *l*, followed by *i* or *e*, followed by *g* or *r* form words. Numbers on the connections represent connection strengths; numbers on the units represent the units' thresholds. A unit will take on an activation equal to 1 if its input exceeds its threshold; otherwise, its activation is 0.

different unit. Since there are only five letter/position possibilities, only five units have connections to the output unit. Each word and non-word produces a different and unique pattern over these five units, but the connections from the five units to the decision unit cannot be set in such a way as to make the decision unit fire whenever one of the words is present and fail to fire whenever one of the non-words is present.

The reason for the problem is simply that the connections between the letter/position units and the decision units can only capture the degree to

---

[7] Note that the problem remains the same if the decision unit is replaced by a set of units and the task of the network is to produce a different pattern for the word and non-word decisions. For when we examine each unit, it either takes the same or a different value in the two patterns: in the cases where the value is the same, there is no problem, but neither do such units differentiate the two patterns. When the values are different, the unit behaves just like the single decision unit discussed in the text.

which each letter indicates whether the string is a word or not. The *g* tends to indicate that a word is present, whereas the *r* indicates that the item is not a word; but each of the other letters, taken individually, has absolutely no predictive ability in this case.

Whether a letter string is a word or not cannot be determined conclusively from the individual letters it contains; it is necessary to consider also what *combinations* of letters it contains. Thus, we need a representation that captures what combinations of letters are present in a way that is sufficient for the purposes of the network. One could capture this by using local representations and assigning one node to each word, as in the third panel of Figure 4. However, it is important to see that one need not go all the way to local representations to solve the problem facing our network. Conjunctive distributed representations will suffice.

The scheme illustrated in the second panel of the figure provides a conjunctive distributed representation. In this scheme, there are units for pairs of letters which, in this limited vocabulary, happen to capture the combinations that are essential for determining whether a string of letters is a word or not. These are, of course, the pairs *wi* and *le*. These conjunctive units, together with direct input to the decision unit from the *g* unit, are sufficient to construct a network which correctly classifies all strings consisting of a *w* or an *l*, followed by an *i* or an *e*, followed by a *g* or *r*.

This example illustrates that conjunctive coding is often necessary if distributed representations are to be used to solve problems that might easily be posed to networks. This same point could be illustrated with many other examples—the *exclusive or* problem is the classic example (Minsky and Papert 1969). Other examples of problems requiring some sort of conjunctive encoding can be found in Hinton (1981) and in Chapters 7 and 8. An application of conjunctive coding to a psychological model is found in Chapter 18.

Some problems (mostly very simple ones) can be solved without any conjunctive encoding at all, and others will require conjuncts of more than two units at a time. In general, it is hard to specify in advance just what 'order' of conjunctions will be required. Instead, it is better to search for a learning scheme that can find representations that are adequate. The mechanisms proposed in Chapters 7 and 8 represent two steps towards this goal.

Implementing an Arbitrary Mapping Between Two Domains

The attentive reader will have noticed that a *local* representation can always be made to work in the example we have just considered.

However, we have already discussed several reasons why distributed representations are preferable. One reason is that they can make more efficient use of parallel hardware than local representations.

This section shows how a distributed representation in one group of units can cause an appropriate distributed representation in another group of units. We consider the problem of implementing an *arbitrary* pairing between representations in the two groups, and we take as an example an extension of the previous one: the association between the visual form of a word and its meaning. The reason for considering an arbitrary mapping is that this is the case in which local representations seem most helpful. If distributed representations are better in this case, then they are certainly better in cases where there are underlying regularities that can be captured by regularities in the patterns of activation on the units in one group and the units in another. A discussion of the benefit distributed representations can provide in such cases can be found in Chapter 18.

If we restrict ourselves to monomorphemic words, the mapping from strings of graphemes onto meanings appears to be arbitrary in the sense that knowing what some strings of graphemes mean does not help one predict what a new string means.[8] This arbitrariness in the mapping from graphemes to meanings is what gives plausibility to models that have explicit word units. It is obvious that arbitrary mappings can be implemented if there are such units. A grapheme string activates exactly one word unit, and this activates whatever meaning we wish to associate with it (see Fig. 5A). The semantics of similar grapheme strings can then be completely independent because they are mediated by separate word units. There is none of the automatic generalization that is characteristic of distributed representations.

Intuitively, it is not at all obvious that arbitrary mappings can be implemented in a system where the intermediate layer of units encodes the word as a distributed pattern of activity instead of as activity in a single local unit. The distributed alternative appears to have a serious drawback. The effect of a pattern of activity on other representations is the combined result of the individual effects of the active units in the pattern. So similar patterns tend to have similar effects. It appears that we are not free to make a given pattern have whatever effect we wish on the meaning representations without thereby altering the effects that other patterns have. This kind of interaction appears to make it difficult to implement

---

[*] Even for monomorphemic words there may be particular fragments that have associated meaning. For example, words starting with *sn* usually mean something unpleasant to do with the lips or nose (*sneer*, *snarl*, *snigger*), and words with long vowels are more likely to stand for large, slow things than words with short vowels (George Lakoff, personal communication). Much of Lewis Carroll's poetry relies on such effects.

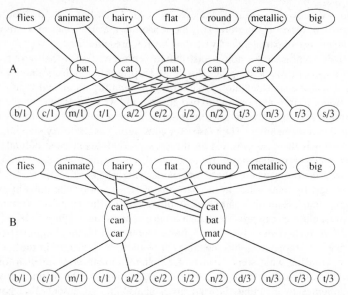

FIG. 5. *A*: A three-layer network. The bottom layer contains units that represent particular graphemes in particular positions within the word. The middle layer contains units that recognize complete words, and the top layer contains units that represent semantic features of the meaning of the word. This network uses local representations of words in the middle layer. *B*: The top and bottom layers are the same as in (*A*), but the middle layer uses a more distributed representation. Each unit in this layer can be activated by the graphemic representation of any one of a whole set of words. The unit then provides input to every semantic feature that occurs in the meaning of *any* of the words that activate it. Only those word sets containing the word *cat* are shown in this example. Notice that the only semantic features which receive input from *all* these word sets are the semantic features of cat.

arbitrary mappings from distributed representations of words onto meaning representations. We shall now show that these intuitions are wrong and that distributed representations of words can work perfectly well and may even be more efficient than single word units.

Figure 5B shows a three-layered system in which grapheme/position units feed into *word-set* units which, in turn, feed into *semantic* or *sememe* units. Models of this type, and closely related variants, have been analysed by Willshaw (1981), V. Dobson (personal communication 1984), and by David Zipser (personal communication 1981); some further relevant analyses are discussed in Chapter 12. For simplicity, we shall assume that each unit is either active or inactive and that there is no feedback or cross-connections. These assumptions can be relaxed without substantially affecting the argument. A word-set unit is activated whenever the pattern of the grapheme/position units codes a word in a particular set. The set could be all the four-letter words starting with *HE*, for example, or all the words containing at least two Ts. All that is required is that it is possible to decide whether a word is in the set by applying a simple test to the activated grapheme/position units. So, for example, the set of all words meaning 'nice' is *not* allowed as a word set. There is an implicit assumption that word meanings can be represented as sets of sememes. This is a contentious issue. There appears to be a gulf between the componential view in which a meaning is a set of features and the structuralist view in which the meaning of a word can only be defined in terms of its *relationships* to other meanings. Later in this chapter we consider one way of integrating these two views by allowing articulated representations to be built out of a number of different sets of active features.

Returning to Figure 5B, the question is whether it is possible to implement an arbitrary set of associations between grapheme/position vectors and sememe vectors when the word-set units are each activated by more than one word. It will be sufficient to consider just one of the many possible specific models. Let us assume that an active word-set unit provides positive input to all the sememe units that occur in the meaning of any word in the word set. Let us also assume that each sememe unit has a variable threshold that is dynamically adjusted to be just slightly less than the number of active word-set units. Only sememe units that are receiving input from every active word-set unit will then become active.

All the sememes of the correct word will be activated because each of these sememes will occur in the meaning of one of the words in the active word sets. However, additional sememes may also be activated because, just by chance, they may receive input from every active word-set unit. For a sememe to receive less input than its threshold, there must be at least one active word set that does not contain any word which has the sememe as part of its meaning. For each active word set the probability, i, of this happening is

$$i = (1 - p)^{(w - 1)}$$

where p is the proportion of words that contain the sememe and w is the number of words in the word set of the word-set unit. The reason for the term $w - 1$ is that the sememe is already assumed not to be part of the meaning of the correct word, so there are only $w - 1$ remaining words that could have it in their meaning.

Assume that when a word is coded at the graphemic level it activates u units at the word-set level. Each sememe that is not part of the word's meaning has a probability i of failing to receive input from each word-set unit. The probability, f, that all of these word-set units will provide input to it is therefore

$$f = (1 - i)^u$$
$$= [1 - (1 - p)^{(w-1)}]^u.$$

By inspection, this probability of a 'false-positive' sememe reduces to zero when w is 1. Table 1 shows the value of f for various combinations of values of p, u, and w. Notice that if p is very small, f can remain negligible even if w is quite large. This means that distributed representations in which each word-set unit participates in the representation of many words do not lead to errors if the semantic features are relatively sparse in the sense that each word meaning contains only a small fraction of the total set of sememes. So the word-set units can be fairly non-specific provided the sememe units are fairly specific (not shared by too many different word meanings). Some of the entries in the table make it clear that for some values of p, there can be a negligible chance of error even though the number of word-set units is considerably less than the number of words (the ratio of words to word-set units is w/u).

The example described above makes many simplifying assumptions. For example, each word-set unit is assumed to be connected to *every* relevant sememe unit. If any of these connections were missing, we could not afford to give the sememe units a threshold equal to the number of active word-set units. To allow for missing connections we could lower the threshold. This would increase the false-positive error rate, but the effect may be quite small and can be compensated by adding word-set units to increase the specificity of the word-level representations (Willshaw 1981). Alternatively, we could make each word-set unit veto the sememes that do not occur in any of its words. This scheme is robust against missing connections because the absence of one veto can be tolerated if there are other vetos (V. Dobson, personal communication, 1984).

There are two more simplifying assumptions both of which lead to an underestimate of the effectiveness of distributed representations for the arbitrary mapping task. First, the calculations assume that there is no fine-tuning procedure for incrementing some weights and decrementing others to improve performance in the cases where the most frequent errors occur. Second, the calculations ignore cross-connections among the sememes. If each word meaning is a familiar stable pattern of sememes, there will be a strong 'clean-up' effect which tends to suppress erroneous

TABLE 1.

u	w	p	f	u	w	p	f	u	w	p	f
5	5	.2	0.071	5	5	.1	0.0048	5	5	.01	9.5×10^{-8}
5	10	.2	0.49	5	10	.1	0.086	5	10	.01	4.8×10^{-6}
5	20	.2	0.93	5	20	.1	0.48	5	20	.01	0.00016
5	40	.2	1.0	5	40	.1	0.92	5	40	.01	0.0036
5	80	.2	1.0	5	80	.1	1.0	5	80	.01	0.049
10	10	.2	0.24	10	10	.1	0.0074	10	10	.01	2.3×10^{-11}
10	20	.2	0.86	10	20	.1	0.23	10	20	.01	2.5×10^{-8}
10	40	.2	1.0	10	40	.1	0.85	10	40	.01	1.3×10^{-5}
10	80	.2	1.0	10	80	.1	1.0	10	80	.01	0.0024
10	160	.2	1.0	10	160	.1	1.0	10	160	.01	0.10
40	40	.2	0.99	40	40	.1	0.52	40	40	.01	2.7×10^{-20}
40	80	.2	1.0	40	80	.1	0.99	40	80	.01	3.5×10^{-11}
40	160	.2	1.0	40	160	.1	1.0	40	160	.01	0.00012
40	320	.2	1.0	40	320	.1	1.0	40	320	.01	0.19
40	640	.2	1.0	40	640	.1	1.0	40	640	.01	0.94
100	100	.2	1.0	10	100	.1	0.99	100	100	.01	9.0×10^{-21}
100	200	.2	1.0	10	200	.1	1.0	100	200	.01	4.8×10^{-7}
100	400	.2	1.0	100	400	.1	1.0	100	400	.01	0.16
100	800	.2	1.0	100	800	.1	1.0	100	800	.01	0.97

The probability, f, of a false-positive sememe as a function of the number of active word-set units per word, u, the number of words in each word-set, w, and the probability, p, of a sememe being part of a word meaning.

sememes as soon as the pattern of activation at the sememe level is sufficiently close to the familiar pattern for a particular word meaning. Interactions among the sememes also provide an explanation for the ability of a single grapheme string (e.g. *bank*) to elicit two quite different meanings. The *bottom-up* effect of the activated word-set units helps both sets of sememes, but as soon as *top-down* factors give an advantage to one meaning, the sememes in the other meaning will be suppressed by competitive interactions at the sememe level (Kawamoto and Anderson 1984).

A Simulation. As soon as there are cross-connections among the sememe units and fine-tuning of individual weights to avoid frequent errors, the relatively straightforward probabilistic analysis given above breaks down. To give the cross-connections time to clean up the output, it is necessary to use an iterative procedure instead of the simple 'straight-through' processing in which each layer completely determines the states of all the units in the subsequent layer in a single, synchronous step.

Systems containing cross-connections, feedback, and asynchronous processing elements are probably more realistic, but they are generally very hard to analyse. However, we are now beginning to discover that there are subclasses of these more complex systems that behave in tractable ways. One example of this subclass is described in more detail in Chapter 7. It uses processing elements that are inherently stochastic. Surprisingly, the use of stochastic elements makes these networks *better* at performing searches, *better* at learning, and *easier* to analyse.

A simple network of this kind can be used to illustrate some of the claims about the ability to 'clean up' the output by using interactions among sememe units and the ability to avoid errors by fine-tuning the appropriate weights. The network contains 30 grapheme units, 20 word-set units, and 30 sememe units. There are no direct connections between grapheme and sememe units, but each word-set unit is connected to all the grapheme and sememe units. The grapheme units are divided into three sets of ten, and each three-letter word has one active unit in each group of ten (units can only have activity levels of 1 or 0). The 'meaning' of a word is chosen at random by selecting each sememe unit to be active with a probability of 0.2. The network shown in Figure 6 has learned to associate 20 different grapheme strings with their chosen meanings. Each word-set unit is involved in the representation of many words, and each word involves many word-set units.

The details of the learning procedure used to create this network and the search procedure which is used to settle on a set of active sememes when given the graphemic input are described in Chapter 7. Here we simply summarize the main results of the simulation.

After a long period of learning, the network was able to produce the correct pattern of sememes 99.9 per cent of the time when given a graphemic input. Removal of any one of the word-set units after the learning typically caused a slight rise in the error rate for several different words rather than the complete loss of one word. Similar effects have been observed in other distributed models (Wood 1978). In our simulations, some of the erroneous responses were quite interesting. In 10,000 tests with a missing word-set unit there were 140 cases in which the model failed to recover the right sememe pattern. Some of these consisted of one or two missing or extra sememes, but 83 of the errors were exactly the pattern of sememes of some other word. This is a result of the co-operative interactions among the sememe units. If the input coming from the word-set units is noisy or underspecified as it may be when units are knocked out, the clean-up effect may settle on a similar but incorrect meaning.

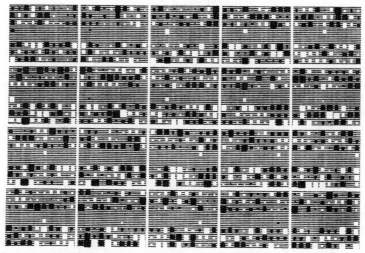

FIG. 6. A compact display that shows all the connection strengths of the 20 units in the middle layer of a three-layer network. The network can map from a pattern of activity over the 30 units in the bottom layer (representing graphemes) to an associated pattern of activity over the 30 units of the top layer (representing sememes). Within each of the large rectangles that are used to depict middle-layer units, the 30 black and white rectangles at the top depict the weights of the connections to the top layer, and the 30 rectangles at the bottom depict the weights from the bottom layer. White rectangles are positive weights, black are negative, and the area of a rectangle depicts the magnitude of the weight. The single weight that occurs somewhere in the middle of a unit is its threshold (black means a positive threshold). The weights between the 30 units in the top layer are not shown in this display.

This effect is reminiscent of a phenomenon called *deep dyslexia* which occurs with certain kinds of brain damage in adults. When shown a word and asked to read it, the subject will sometimes say a different word with a very similar meaning. The incorrect word sometimes has a very different sound and spelling. For example, when shown the word PEACH, the subject might say APRICOT. (See Coltheart, Patterson, and Marshall 1980, for more information about acquired dyslexia.) Semantic errors of this kind seem bizarre because it seems as if the subject must have accessed the lexical item PEACH in order to make the semantically related error, and if he can get to the lexical item why can't he say it? (These subjects may know and be able to say the words that they misread.) Distributed representations allow us to dispense with the rigid distinction between accessing a word and not accessing it. In a network that has learned the word PEACH, the graphemic representation of PEACH will cause approximately the right input to the sememe units, and interactions at the sememe level can then cause exactly the pattern of sememes for APRICOT. Another psychologically interesting effect occurs when the network relearns after it has been damaged. The network was damaged by adding noise to every connection that involved a word-set unit. This reduced the performance from 99.3 per cent correct to 64.3 per cent.[9] The network was then retrained and it exhibited very rapid relearning, much faster than its original rate of learning when its performance was 64.3 per cent. This rapid recovery was predicted by a geometrical argument which shows that there is something special about a set of connection strengths that is generated by adding noise to a near-perfect set. The resulting set is very different from other sets of connection strengths that exhibit the same performance. (See Chapter 7 for further discussion.)

An even more surprising effect occurs if a few of the words are omitted from the retraining. The error rate for these words is substantially reduced as the retraining proceeds, even though the other grapheme–sememe pairings have no intrinsic relation to them because all the pairings were selected randomly. The 'spontaneous' recovery of words that the network is not shown again is a result of the use of distributed representations. *All* the weights are involved in encoding the subset of the words that are shown during retraining, and so the added noise tends to be removed from *every* weight. A scheme that used a separate unit for each word would not behave in this way, so one can view spontaneous recovery of unrehearsed items as a qualitative signature of distributed representations.

3. STRUCTURED REPRESENTATIONS AND PROCESSES

In this section we consider two extensions of distributed representations. These extensions illustrate that the idea of distributed representations is consistent with some of the major insights from the field of artificial intelligence concerning the importance of structure in representations and processes. Perhaps because some proponents of distributed representations have not been particularly attuned to these issues, it is often unclear how structure is to be captured in a distributed representational scheme. The two parts of this section give some indication of the directions that can be taken in extending distributed representations to deal with these important considerations.

---

[9] The error rate was 99.3% rather than 99.9% in this example because the network was forced to respond faster, so the co-operative effects had less time to settle on the optimal output.

Representing Constituent Structure

Any system that attempts to implement the kinds of conceptual structures that people use has to be capable of representing two rather different kinds of hierarchy. The first is the 'IS-A' hierarchy that relates types to instances of those types. The second is the part/whole hierarchy that relates items to the constitutent items that they are composed of. The most important characteristics of the IS-A hierarchy are that known properties of the types must be 'inherited' by the instances, and properties that are found to apply to all instances of a type must normally be attributed to the type. Earlier in this chapter we saw how the IS-A hierarchy can be implemented by making the distributed representation of an instance *include*, as a subpart, the distributed representation for the type. This representational trick automatically yields the most important characteristics of the IS-A hierarchy, but the trick can only be used for one kind of hierarchy. If we use the part/whole relationship between patterns of activity to represent the type/instance relationship between items, it appears that we cannot also use it to represent the part/whole relationship between items. We cannot make the representation of the whole be the sum of the representations of its parts.

The question of how to represent the relationship between an item and the constituent items of which it is composed has been a major stumbling block for theories that postulate distributed representations. In the rival, localist scheme, a whole is a node that is linked by labelled arcs to the nodes for its parts. But the central tenet of the distributed scheme is that different items correspond to *alternative* patterns of activity in the same set of units, so it seems as if a whole and its parts cannot both be represented at the same time.

Hinton (1981) described one way out of this dilemma. It relies on the fact that wholes are not simply the sums of their parts. They are composed of parts that play particular roles within the whole structure. A shape for example, is composed of smaller shapes that have a particular size, orientation, and position relative to the whole. Each constituent shape has its own spatial role, and the whole shape is composed of a set of shape/role pairs.[10] Similarly, a proposition is composed of objects that occupy particular semantic roles in the whole propositional structure. This suggests a way of implementing the relationship between wholes and

[10] Relationships between parts are important as well. One advantage of explicitly representing shape/role pairs is that it allows different pairs to support each other. One can view the various different locations within an object as slots and the shapes or parts of an object as the fillers of these slots. Knowledge of a whole shape can then be implemented by positive interactions between the various slot-fillers.

parts: the identity of each part should first be combined with its role to produce a single pattern that represents the *combination* of the identity and the role, and then the distributed representation for the whole should consist of the sum of the distributed representations for these identity/role combinations (plus some additional 'emergent' features). This proposal differs from the simple idea that the representation of the whole is the sum of the representations of its parts because the subpatterns used to represent identity/role combinations are quite different from the patterns used to represent the identities alone. They do not, for example, contain these patterns as parts.

Naturally, there must be an access path between the representation of an item as a whole in its own right and the representation of that same item playing a particular role within a larger structure. It must be possible, for example, to generate the identity/role representation from two separate, explicit, distributed patterns one of which represents the identity and the other of which represents the role. It must also be possible to go the other way and generate the explicit representations of the identity and role from the single combined representation of the identity/role combination (see Fig. 7).

The use of patterns that represent identity/role combinations allows the part/whole hierarchy to be represented in the same way as the type/instance hierarchy. We may view the whole as simply a particular instance of a number of more general types, each of which can be defined as the type that has a particular kind of part playing a particular role (e.g. men with wooden legs).

Sequential Symbol-Processing

If constituent structure is implemented in the way described above, there is a serious issue about how many structures can be active at any one time. The obvious way to allocate the hardware is to use a group of units for each possible role within a structure and to make the pattern of activity in this group represent the identity of the constituent that is currently playing that role. This implies that only one structure can be represented at a time, unless we are willing to postulate multiple copies of the entire arrangement. One way of doing this, using units with programmable rather than fixed connections, is described in Chapter 16. However, even this technique runs into difficulties if more than a few modules must be 'programmed' at once. However, people do seem to suffer from strong constraints on the number of structures of the same general type that they can process at once. The sequentiality that they exhibit at this high level of

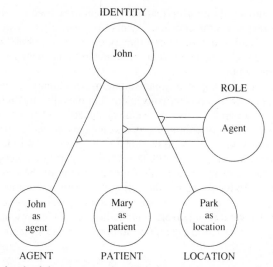

FIG. 7. A sketch of the apparatus that might be necessary for combining separate representations of an identity and a role into a single pattern. Only one identity and only one role can be explicitly represented at a time because the identity and role groups can each have only one pattern of activity at a time. However, the various role groups allow many identity/role combinations to be encoded simultaneously. The small triangular symbols represent the ability of the pattern of activity in the group that explicitly represents a role to determine which one of the many role groups is currently interacting with the identity group. This allows the identity occupying a particular role to be 'read out' as well as allowing the reverse operation of combining an identity and a role.

description is initially surprising given the massively parallel architecture of the brain, but it becomes much easier to understand if we abandon our localist predelictions in favour of the distributed alternative which uses the parallelism to give each active representation a very rich internal structure that allows the right kinds of generalization and content-addressability. There may be some truth to the notion that people are sequential *symbol* processors if each 'symbolic representation' is identified with a successive state of a large interactive network. See Chapter 14 for further discussion of these issues.

One central tenet of the sequential symbol-processing approach (Newell 1980) is the ability to focus on any part of a structure and to expand that into a whole that is just as rich in content as the original whole of which it was a part. The recursive ability to expand parts of a structure for indefinitely many levels and the inverse ability to package up whole structures into a reduced form that allows them to be used as constituents of larger structures is the essence of symbol-processing. It allows a system to build structures out of things that refer to other whole structures without requiring that these other structures be represented in all their cumbersome detail.

In conventional computer implementations, this ability is achieved by using pointers. These are very convenient, but they depend on the use of addresses. In a parallel network, we need something that is functionally equivalent to arbitrary points in order to implement symbol-processing. This is exactly what is provided by subpatterns that stand for identity/role combinations. They allow the full identity of the part to be accessed from a representation of the whole and a representation of the role that the system wishes to focus on, and they also allow explicit representations of an identity and a role to be combined into a less cumbersome representation, so that several identity/role combinations can be represented simultaneously in order to form the representation of a larger structure.

SUMMARY

Given a parallel network, items can be represented by activity in a single, local unit or by a pattern of activity in a large set of units with each unit encoding a microfeature of the item. Distributed representations are efficient whenever there are underlying regularities which can be captured by interactions among microfeatures. By encoding each piece of knowledge as a large set of interactions, it is possible to achieve useful properties like content-addressable memory and automatic generalization, and new items can be created without having to create new connections at the hardware level. In the domain of continuously varying spatial features it is relatively easy to provide a mathematical analysis of the advantages and drawbacks of using distributed representations.

Distributed representations *seem* to be unsuitable for implementing purely arbitrary mappings because there is no underlying structure and so generalization only causes unwanted interference. However, even for this task, distributed representations can be made fairly efficient and they exhibit some psychologically interesting effects when damaged.

There are several difficult problems that must be solved before distributed representations can be used effectively. One is to decide on the pattern of activity that is to be used for representing an item. The similarities between the chosen pattern and other existing patterns will determine the kinds of generalization and interference that occur. The search for good patterns to use is equivalent to the search for the underlying regularities of the domain. This learning problem is addressed in the chapters of Part II.

Another hard problem is to clarify the relationship between distributed representations and techniques used in artificial intelligence like schemas, or hierarchical structural descriptions. Existing artificial intelligence programs have great difficulty in rapidly finding the schema that best fits the current situation. Parallel networks offer the potential of rapidly applying a lot of knowledge to this best-fit search, but this potential will only be realized when there is a good way of implementing schemas in parallel networks. A discussion of how this might be done can be found in Chapter 14.[11]

[11] This chapter is based on a technical report by the first author, whose work is supported by a grant from the System Development Foundation. We thank Jim Anderson, Dave Ackley, Dana Ballard, Francis Crick, Scott Fahlman, Jerry Feldman, Christopher Longuet-Higgins, Don Norman, Terry Sejnowski, and Tim Shallice for helpful discussions.

REFERENCES

Anderson, J. A. (1977). 'Neural Models with Cognitive Implications.' In D. LaBerge and S. J. Samuels (eds.), *Basic Processes in Reading Perception and Comprehension*, pp. 27–90. Hillsdale, NJ: Erlbaum.

Bartlett, F. C. (1932). *Remembering*. Cambridge: Cambridge University Press.

Collins, A. M., and Loftus, E. A. (1975). 'A Spreading-Activation Theory of Semantic Processing.' *Psychological Review* 82: 407–25.

Coltheart, M., Patterson, K., and Marshall, J. C. (1980). *Deep Dyslexia*. London: Routledge & Kegan.

Fahlman, S. E. (1979). *NETL: A System for Representing and Using Real-World Knowledge*. Cambridge, Mass.: MIT Press.

—— (1980). *The Hashnet Interconnection Scheme*. Tech. Rep. CMU-CS-80-125. Pittsburgh: Carnegie-Mellon University, Department of Computer Science.

Feldman, J. A. (1982). 'Dynamic Connections in Neural Networks.' *Biol. Cybernetics* 46: 27–39.

Hinton, G. E. (1981). 'Implementing Semantic Networks in Parallel Hardware.' In G. E. Hinton and J. A. Anderson (eds.), *Parallel Models of Associative Memory*, pp. 161–88. Hillsdale, NJ: Erlbaum.

—— and Anderson, J. A. (eds.) (1981). *Parallel Models of Associative Memory*. Hillsdale, NJ: Erlbaum.

Hopfield, J. J. (1984). 'Neurons with Graded Response Have Collective Computational Properties Like Those of Two-State Neurons.' *Proc. Nat. Acad. Sci. (USA)* 81: 3088–92.

Kawamoto, A. H., and Anderson J. A. (1984). 'Lexical Access Using a Neural Network.' *Proc. Sixth Annual Conference of the Cognitive Science Society*, 204–13.

Levin, J. A. (1976). *Proteus: An Activation Framework for Cognitive Process Models*. Tech. Rep. No. ISI/WP-2. Marina del Rey, Calif. University of Southern California, Information Sciences Institute.

Luria, A. R. (1973). *The Working Brain.* London: Penguin.

McClelland, J. L. (1981). 'Retrieving General and Specific Information from Stored Knowledge of Specifics.' *Proc. Third Annual Meeting of the Cognitive Science Society,* 170–2.

Minsky, M., and Papert, S. (1969). *Perceptrons.* Cambridge, Mass.: MIT Press.

Neisser, U. (1981). 'John Dean's Memory: A Case-Study.' *Cognition* 9: 1–22.

Newell, A. (1980). 'Physical Symbol Systems.' *Cognitive Science* 4: 135–83.

Norman, D. A., and Bobrow, D. G. (1979). 'Descriptions: An Intermediate Stage in Memory Retrieval.' *Cognitive Psychology* 11: 107–23.

Quillian, M. R. (1968). 'Semantic Memory.' In M. Minsky (ed.), *Semantic Information Processing,* pp. 227–70. Cambridge, Mass.: MIT Press.

Willshaw, D. J. (1981). 'Holography, Associative Memory, and Inductive Generalization.' In G. E. Hinton and J. A. Anderson (eds.), *Parallel Models of Associative Memory,* pp. 83–104. Hillsdale, NJ: Erlbaum.

Wood, C. C. (1978). 'Variations on a Theme by Lashley: Lesion Experiments on the Neural Model of Anderson, Silverstein, Ritz, & Jones.' *Psychological Review* 85: 582–91.

12

CONNECTIONISM, COMPETENCE, AND EXPLANATION

ANDY CLARK

1. LEVELS OF EXPLANATION AND THE IDEA OF AN EQUIVALANCE CLASS

Explanation, it seems, is a many-levelled thing. A single phenomenon may be subsumed under a panoply of increasingly general explanatory schemas. On the swings and roundabouts of explanation, we trade the detailed descriptive/explanatory power of lower levels for a satisfying width of application at higher ones. And at each such level there are virtues and vices; some explanations may be available only at a certain level; but individual cases thus subsumed may vary in ways explicable only by descending the ladder of explanatory generality.

For example, the Darwinian, or neo-Darwinian theory of natural selection is pitched at a very high level of generality. It pictures some very general circumstances under which 'blind' selection can yield apparently teleological (or purposeful) evolutionary change. What is required for this miracle to occur is differential reproduction according to fitness and some mechanism of transmission of characteristics to progeny. This is an extremely general and potent idea. The virtue of this top-level explanation, then, lies in its covering an open-ended set of cases in which very different actual *mechanisms* (e.g. of transmission) may be involved. In this way it defines an *equivalence set* of mechanisms, that is, a set of mechanisms which may be disparate in many ways but which are united by their ability to satisfy the Darwinian demands.

The natural accompaniment to virtue is, of course, vice, and the vice of the general Darwinian account is readily apparent. We don't yet know, in any given case, *how* the Darwinian demands are satisfied. That is to say, we don't yet have the foggiest idea of the actual mechanisms of heritability and transmission in any given case. Moreover, there may well be facts

Andrew Clark, 'Connectionism, Competence, and Explanation', to appear in a forthcoming issue of *The British Journal for the Philosophy of Science*. Reprinted by permission of the author.

about some specific class of cases (e.g. recessive characteristics in Mendel's peas), which are not predicted by the general Darwinian theory, which give us still further reason to seek a more specific and detailed account.

Mendelian genetics offers just such an account. It posits a class of theoretical entities (genes, as they are now called) controlling each trait, and describes the way such entities must combine to explain various observed facts concerning evolution in successive generations of pea plants. The specification included, for example, the idea of pairs of genes (genotypes) in which one gene may be dominant, thus explaining the facts about recessive characteristics. (For an accessible account of evolutionary theory and Mendelian genetics, see Ridley 1985.)

We may note in passing that between any two levels (e.g. Darwinism and Mendelian genetics) there will almost certainly be other, theoretically significant levels. Thus Mendelian inheritance is in fact an *instance* of a more general mechanism called Weismannist inheritance (see Ridley 1985: 23). But Weismannist inheritance is still *less* general than Darwinian inheritance. Weismannism carves off a theoretically unified subset of general Darwinian cases. And Mendelism carves off a theoretically unified subset of Weismannism. At each stage the equivalence class is strategically redefined to exclude a number of previous members. We can visualize this as a gradual shrinking of the size of the equivalence class, although this may not be *strictly* true, since each new class has a possible infinity of members and so they are, I suppose, identical in size!

Mendelian genetics provides an interesting case for one further reason. It was originally conceived as *neatly specifying* the details of lower level DNA-based inheritance (i.e. of the hardware realization of an inheritance mechanism). As Dennett puts it, Mendelian genes were seen as specifying 'the language of inheritance, straightforwardly realised in hunks of DNA' (Dennett 1988b: 385). This corresponds to what we shall be terming the classicist vision of the relation between a certain level of abstract theorizing in cognitive science (competence theorizing) and actual processing strategies. But in fact, according to Dennett:

there are theoretically important mismatches between the language of 'bean-bag genetics' and the molecular and developmental details—mismatches serious enough to suggest that all things considered, there don't turn out to be genes (classically understood) at all (Dennett 1988b: 385).

This looks like (and is regarded by Dennett as) an analogue of the connectionist's view of the fate of the constructs of a classical competence theory.

Be that as it may, the point for now is simply this, that beneath the level of Mendelian genetics there is some further level of physical implementation (with God knows what in between, as remarked earlier), and this completes our descent down the ladder of explanatory generality. We start at the top level (level-1) with the general Darwinian theory defining a large and varied equivalence class of instantiating mechanisms. We descend to a more detailed specification of a subclass of mechanisms (Mendelian theory) and thence, one way or another, to the details of the implementation of those mechanisms in DNA. The effect is a kind of triangulation upon the actual details of earth-animal inheritance from much broader explanatory principles governing whole sets of possible worlds (see Fig. 1).

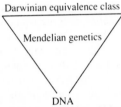

FIG. 1. The swings and roundabouts of explanatory generality. What DNA-based stories gain in detailed power they lose in cross-world scope.

Explanation in cognitive science, as conceptualized by, among others, Marr, Chomsky, and Newell and Simon, has a similar multilayered structure. For a given class, or class of tasks (e.g. vision, parsing, etc.) there will be a top-level story which comprises 'an abstract formulation of *what* is being computed and *why*', a lower level which specifies a particular algorithm for carrying out the computation, and (still lower) an account of how that algorithm is to be realized by physical hardware. To illustrate this, Marr gives the example of Fourier analysis. At the top level we have the general idea of a Fourier analysis. This can be realized by several different algorithms. And each algorithm in turn can be implemented in many different kinds of hardware organization (see Marr 1977: 129).

There is an important gap between the 'official' account of the top level (level-1 as Marr calls it) and the actual practice of giving 'level-1' theories. For although the official line is that a level-1 account specifies only the *what* and the *why* of a computation, this specification can be progressively refined so as to define a more informative (i.e. more restrictive) equivalence-class. This more refined version of level-1 theorizing (which yet falls short of a full algorithmic account) has been persuasively defended by Christopher Peacocke under the title of 'Level-1.5' (see Peacocke 1986).

The contrast Peacocke highlights is between an equivalence class generated by defining a function *in extension* (i.e. by its results—the *what*, in

Marr's terms) and a more restrictive (and informative) equivalence class generated by specifying the *body of information* upon which an algorithm draws. Thus, to adapt one of Peacocke's own examples, suppose the goal of a computation is to compute depth D and physical size P from retinal size R. And suppose, in addition, that this computation is to occur inside a restricted universe of values of D, P, and R. Specifying the function in extension merely tells us that whenever the system is given some D and P as input, it should yield some specified R as output. One way of doing this (and I here adapt a strategy used by Martin Davies (see Davies forthcoming, and §2 following)), is to store the set of legal values for R for every combination of values of D and P—a simple look-up table. A second way is to process data in accordance with the equation $P = D \times R$. In saying that the system draws on the information that $P = D \times R$, we are, as Peacocke insists, doing *more* than specifying a function in extension. For the look-up table *does not* draw on that information, yet it falls within the equivalence class generated by the function in extension specification. But we are doing *less* than specifying a particular algorithm, since there will be many ways of computing the equation in question (e.g. using different algorithms for multiplication).

It is this grain of analysis (i.e. what Peacocke calls level-1.5) that I will have in mind when speaking, in the remainder of this paper, of a *competence theory*. This seems to accord at least with the practice of Chomsky, who coined the term competence theory to describe the pitch of his own distinctive investigations into the structure of linguistic knowledge. And it may well accord with Marr's actual practice at 'level-1' though not with the official dogma.

A Chomskian competence theory does far more than specify a function in extension. Instead it seeks to answer (at a level of abstraction from the physical mechanisms of the brain and from specific algorithms) the question 'what constitutes knowledge of language?'. In so doing it seeks a 'framework of principles and elements common to attainable human languages' (Chomsky 1986: 3). And (in its most recent incarnation) it characterizes that framework as a quite specific 'system of principles associated with certain parameters of variation and a markedness system with several components of its own' (Chomsky 1986: 221). It does not matter, for the purposes of this paper, just what principles and parameters Chomsky actually suggests. Rather, we should merely note that if a competence theory is as definite and structured as a Chomskian model (and it's his word, after all!) then it is more like a level-1.5 analysis than a simple level-1 account. For it describes, at a certain level of abstraction, the structure of a form of processing (by specifying the information drawn on by the processes) and hence helps 'guide the search for mechanisms' (Chomsky 1986: 221). In short, it is more like Mendelian genetics than General Darwinism. Rather than being merely *descriptive* of a class of results, it is meant also to be *suggestive* of the processing structure of a class of mechanisms of which we are a member. Whether competence theories (at least as we currently know them) actually *are* suggestive of the form of human processing is the topic of this paper.

2. THE CLASSICAL CASCADE

A competence theory, then, leads a double life. It both specifies the function to be computed *and* it specifies the body of knowledge or information which is used by some class of algorithms. In classical cognitive science, these two roles can easily be simultaneously discharged. For the competence theory is just an articulated set of rules and principles defined over symbolic data-structures. Since classical cognitive science relies on symbol-processing architecture, it is natural (at level-2) to represent directly the data-structures (e.g. structural descriptions of sentences) and then carry out the processing by the explicit *or* tacit representation of the rules and principles defined (in the competence theory) to operate on those structures. Thus, given a structural description of an inflected verb as comprising a stem plus an ending, the classicist can go on to define a level-2 computational process to take the stem and add -ed to form the past tense (or whatever). The classicist, then, is (by virtue of using a symbol-processing architecture to implement level-2 algorithms) uniquely well placed to preserve a very close relation between a competence theory and its level-2 implementations. Indeed, it begins to seem as if that close relation is what is *constitutive* of a classical approach. Thus Dennett visualizes the classicist dream as involving 'a triumphant cascade through Marr's three levels' (Dennett 1987: 227). Such a characterization of the essential classicist vision seems to me to fit very nicely with Fodor and Pylyshyn's recent account of the classical/connectionist divide.

Fodor and Pylyshyn argue that there are two fundamental differences between truly connectionist and classical approaches to cognitive modelling. ('Truly connectionist' here rules out those cases where a units and connections substructure is used to implement a classical theory.) The differences are:

1. 'Classical theories—but not connectionist theories—posit a "language of thought".'

This means they posit mental representations (data-structures) with a certain form. Such representations are *syntactically structured,* i.e. they are systematically built by combining atomic constituents into molecular assemblies which in turn (in complex cases) make up whole data-structures. In short, they posit *symbol systems* with a combinatorial syntax and semantics.

2. 'In classical models, the principles by which mental states are transformed, or by which an input selects the corresponding output, are defined over structural properties of mental representations. Because classical mental *representations* have combinatorial structure, it is possible for classical mental *operations* to apply to them by reference to their form' (quotes are from Fodor and Pylyshyn 1988: 12–13).

This means that *given* that you have a certain (languagelike) kind of structured representation available (as demanded by point 1), it is possible to define computational operations on those representations so as the operations are sensitive to that very structure. If the structure wasn't there (i.e. if there was no symbolic representation) you couldn't do it! (Though you might make it *look* as if you had by fixing a suitable function in extension.)

In short, a classical system is one which posits syntactically structured symbolic *representations* and which defines its computational *operations* to apply to such representations in virtue of their structure.

The computational operations, in any such case, can be described by transition or derivation rules defined over syntactically structured representations. For example:

> If $(A$ and $B)$ then (A)
> If $(A$ and $B)$ then (B)
> If (stem + ending) then (stem + -ed)

The bracketed items are structural descriptions which will pick out open-ended classes of classical representations. The If–then specifies the operation. But note that the classicist, under the terms of the act, is *not* committed to the systems *explicitly* representing the if–then clause. All that needs to be explicit is the structured description upon which it operates. Thus a machine could be hard-wired so as to take expressions of the form $(A$ and $B)$ and transform them into the expressions (A) and (B). The derivation rules may thus be implicit, or *tacit*; but the data-structures must be explicit. On this matter, Fodor and Pylyshyn are rightly insistent:

Classical machines can be *rule implicit* with respect to their programs . . . What *does* need to be explicit in a classical machine is not its program but the symbols that it writes on its tapes (or stores in its registers). These, however, correspond not to the machine's rules of state transition but to its data structures (Fodor and Pylyshyn 1988: 61).

As an example they point out that the grammar posited by a linguistic theory need not be explicitly represented in a classical machine. But the *structural descriptions of sentences* over which the grammar is defined (e.g. in terms of verb stems, subclauses etc.) must be. A successful 'classical cascade' from a linguistic competence theory to a level-2 processing story can thus tolerate having the rules of the grammar built into the machine. Those attempts to characterize the classicist/connectionist contrast solely by reference to the explicitness or otherwise of rules are thus shown to be in error.

Now, however, there is a danger of losing sight of the way in which (for a classicist) the competence theory (or set of derivation rules and data-structures) is meant to bear a close relation to the level-2 implementation. For we said (§1 above) that given, say, a simple competence theory like '$P = D \times R$', it *wouldn't* do to have a system which simply stored, for some finite universe of discourse, all legal values of P, D, and R. Yet such a system certainly has explicit representations of P, D, and R. So *if* it won't do, it must be because it lacks *even tacit* knowledge of the derivation rule '$P = D \times R$'. The question then is, how do we motivate this difference? What are the constraints on tacit knowledge ascription such that the rule '$P = D \times R$' *needn't* be explicitly represented, but which rules out the look-up table as an instance of tacit knowledge of *that very rule*? The answer will be significant when we come to ask (§3 and §4) whether connectionist systems have tacit knowledge of classical rules.

Martin Davies (drawing on information provided by Gareth Evans) offers the following suggestion: 'For a Speaker to have tacit knowledge of a particular articulated theory, there must be a causal-explanatory structure in the Speaker which mirrors the derivational structure in the theory' (Davies forthcoming: 4). By 'the derivational structure in the theory' Davies means the transition rules (e.g. $P = D \times R$). What is it, then, to embody a 'causal-explanatory structure' which 'mirrors' such a derivational structure? Simply, according to Davies, for there to be a *causal common factor* (Davies's phrase) in the processing story told for each instance which, at the higher level, is seen as involving the rule of derivation. Thus, in the case of the look-up table, there need be no causal common factor in the processing of *all* the instances of various values of P, D, and R. Conversely, if there *is* a causal common factor through which all processing is routed, then (subject to some niggling provisos, see Davies forthcoming, and Davies 1987) the system is rightly said to have tacit knowledge of the rule. A result which, as Davies notes, sits nicely with our cognitive neuropsychological intuitions. For systems which meet the tacit knowledge constraint so construed are prima-facie candidates for a type of breakdown in which damage to the causal common factor causes total loss of capacity to solve a whole class of problems (e.g. P, D, and R specifying). Whereas systems which fail to meet the constraint are prima-facie candidates for less systematic deficits (e.g. the look-up system may lose its knowledge of some legal combinations of P, D, and R but preserve its knowledge of others). Similar comments apply to the past-tense generation case. Systems with tacit knowledge of the rule 'take stem and add -ed' could lose all capacity to form regular pasts. Systems which do it by look-up wouldn't.

Davies's account (modulo a quibble about *virtual* 'causal' common factors, see §3 following) seems convincing. If we take it on board, we end up with the following characterization of any properly *classical* cognitive model:

> (Classicism defined by an attitude to competence theories.)
> A cognitive model is classical if it has a processing-level description which bears a certain rather close relation to the structure of a standard competence theory. A standard competence theory posits a set of rules or principles of derivation defined to apply to a class of structured, symbolic representations according to their form. The close relation required involves (1) the explicit representation, in the processing-level description, of the structured representations over which the rules are defined, and it involves (2) the explicit *OR* tacit representation of those rules and principles themselves. A rule or principle is judged to be tacitly represented just in case there is a causal common factor in the processing-level description which is in play whenever the rule or principle is invoked in a competence-level specification of a transition.

Such, in tortuous detail (and apologies for that) is the substance of the 'classical cascade' through Marr's levels of explanation. Connectionism dams the cascade. How it does so, and what watercourses result, will occupy us for the remainder of this paper.

3. NEWTONIAN COMPETENCE

The connectionist vision of the relation between a structured competence theory and a level-2 processing story is radically unlike the neat 'cascade' imagined in §2. Instead of the level-2 story mirroring the derivational form of the competence theory, it is seen as relating to it in rather the way Newtonian mechanics relates to quantum physics. The physical universe is not, in fact, Newtonian. But under certain specifiable conditions, it behaves very much *as if* it were. Newtonian principles thus describe and predict the behaviour of physical systems in a range of cases. But, in some intuitive but slightly elusive sense, those principles do not describe the actual forces which determine physical behaviour. The analogy is much favoured by Rumelhart and McClelland who write:

It might be argued that conventional symbol processing models are macroscopic accounts, analogous to Newtonian mechanics, whereas our models offer more microscopic accounts, analogous to quantum theory—Through a thorough understanding of the relationship between the Newtonian mechanics and quantum theory we can understand that the macroscopic level of description may be *only an approximation* to the more microscopic theory (Rumelhart and McClelland 1986a: i. 125).

To illustrate this point, consider a simple example due to Paul Smolensky. Imagine that the cognitive task to be modelled involves answering qualitative questions concerning the behaviour of a particular electrical circuit. (The restriction to a single circuit may appal classicists, although it is defended by Smolensky on the grounds that a small number of such representations may act as the 'chunks' utilized in general-purpose expertise, see Smolensky 1986: ii, 241.) Given a description of the circuit, an expert can answer questions such as 'If we increase the resistance at point A what effect will that have on the voltage? (i.e. will the voltage increase, decrease, or remain the same).

Suppose, as seems likely, that a high-level competence-theoretic specification of the information to be drawn on by an algorithm tailored to this task cites various laws of circuitry in its derivations (what Smolensky refers to as the 'hard laws' of circuitry; Ohm's law and Kirchoff's law). For example, derivations involving Ohm's law would invoke the equation

Voltage (V) = Current (C) × Resistance (R)

We recognized, in §2 above, just two ways in which a level-2 processing story might bear an appropriately close relation to such a competence theory. In the simplest case, the processing might involve a symbolic representation of Ohm's law which is read and followed by the system. In the more complex case, it might involve tacit knowledge of Ohm's law unpacked in terms of a causal common factor in a set of state transitions. (Note in passing: Smolensky's own treatment here seems to place uncalled for emphasis on the simple option, see Fodor and Pylyshyn 1988; Pinker and Prince 1988; Davies forthcoming; Clark 1989.)

Neither cascade is operative in the case of Smolensky's connectionist model of simple circuit problem-solving. To see why, we need to look at the form of the model in question. The model represents the state of the circuit by a pattern of activity over a set of feature units. These encode the qualitative changes found in the circuit variables, i.e. in training instances, they encode whether when the resistance at R_1 goes up, the overall voltage falls or rises and so forth. These feature units are connected to a set of what Smolensky calls 'knowledge atoms' which represent patterns of activity across subsets of the feature units. These in fact encode the legal combinations of feature unit states allowed by the actual laws of circuitry. Thus, for example: 'The system's knowledge of Ohm's law . . . is distributed over the many knowledge atoms whose subpatterns encode the legal feature combinations for current, voltage and resistance' (Smolensky 1988: 19). In short, there is a subpattern for every legal combination of qualitative changes (GS subpatterns, or 'knowledge atoms' for the circuit in question).

It might seem, at first sight, that the system is merely a units and connections implementation of a look-up table. But this is not so. In fact, connectionist networks act as look-up tables only when they are provided with an over-abundance of hidden units and hence simply memorize input–output pairings. By contrast, the system in question encodes what Smolensky terms 'soft constraints', i.e. patterns of relations which usually obtain between the various feature units (microfeatures). It thus has 'general knowledge' of qualitative relations among circuit microfeatures. But it does *not* have the general knowledge encapsulated in *hard* constraints like Ohm's law. The soft constraints are two-way connections between feature units and knowledge atoms which *incline* the network one way or another, but do not *compel* it; that is, they can be overwhelmed by the activity of other units—that's why they are 'soft'. And as in all connectionist networks, the system computes by trying simultaneously to satisfy as many of these soft constraints as it can. To see that it is not a mere look-up tree of legal combinations we need only note that it is capable of giving sensible answers to (inconsistent or incomplete) questions which *have* no answer in a simple look-up table of legal combinations.

The soft constraints are numerically encoded as weighted inter-unit connection strengths. Thus problem-solving is achieved by 'a series of many node (i.e. unit) updates, each of which is a *microdecision* based on formal *numerical* rules and numerical computations' (Smolensky 1986: ii. 246).

The network has two properties of special interest to us. First, it can be shown that *if* it is given a well-posed problem *and* unlimited processing time it will *always* give the correct answer as predicted by the hard laws of circuitry. But, as already remarked, it is by no means bound by such laws. Give it an ill-posed or inconsistent problem and it will satisfy as many of the soft constraints (which are all it really knows about) as it can. Thus, 'Outside of the idealised domain of well-posed problems and unlimited processing time, the system gives sensible performance' (Smolensky 1988: 20). The hard rules (Ohm's law etc.) can thus be viewed as an external theorist's characterization of an idealized subset of its actual performance (it is no accident if this puts us in mind of Dennett's claims about the 'intentional stance', see Dennett 1987).

Second, the network exhibits interesting *serial* behaviour as it repeatedly tries to satisfy all the soft constraints. This serial behaviour is characterized by Smolensky as a set of *macrodecisions* each of which amounts to a 'commitment of part of the network to a portion of the solution'. These macrodecisions, Smolensky notes, are 'approximately like the firing of production rules. In fact, these "productions" "fire" in essentially the same order as in a symbolic forward-chaining inference system' (Smolensky 1988: 19). Thus the network will look as if it is sensitive to hard, symbolic rules at quite a fine grain of description. It will not *simply* be that it solves the problem 'in extension' as if it knew hard rules. Even the *stages* of problem-solving may look as if they are caused by the system's running a processing analogue of the steps in the symbolic derivations available in the competence theory.

But the appearance is, on the terms set out in §2 above, an illusion. The system has neither explicit nor tacit knowledge of the hard rules. It is not hard to see why. Quite clearly, it does not explicitly represent Ohm's law to itself. There is, for example, no neat subpattern of units which can be seen to stand for the general ideal of Resistance which figures in Ohm's law. Instead, sets of units stand for Resistance-at-R_1, and other sets for Resistance-at-R_2. In more complex networks, the coalition of units which, when active, stand in for a top (or conceptual) level concept like resistance are highly *context-sensitive*. That is, they vary according to context of occurrence. Thus, to use Smolensky's own example, the representation of *coffee* in such a network would not comprise a single recurrent syntactic item but a coalition of smaller items (microfeatures) which shift according to context. Coffee in the context of cup may be represented by a coalition which includes (liquid) (contacting-porcelain). Coffee in the context of jar may include (granule) (contacting-glass). There is thus only an 'approximate equivalence of the "coffee vectors" across contexts' unlike the 'exact equivalence of the coffee tokens across different contexts in a symbolic processing system' (Smolensky 1988: 16). By thus replacing the conceptual-level symbol 'coffee' with a shifting coalition of microfeatures, the so-called 'dimension-shift', such systems deprive themselves of the structured mental representations which are deployed both in a classical *competence* theory *and* in a classical symbol-processing (*level-2*) account. Likewise, there is no stable representational entity in the simple network described which stands for Resistance (just as in the infamous past-tense network there is no stable, recurrent entity which stands for 'verb-stem' (see Rumelhart and McClelland 1986*b*; Pinker and Prince 1988; Clark 1989). The immediate result is that there can be no explicit representation of rules which involve reference to the conceptual-level constructs. The lack of *tacit* representation is almost immediate, since the processing can hardly be sensitive to structures which aren't there.

To put the point in our favoured terms, the system cannot be said tacitly to represent the rules since there is no causal common factor in its problem-solving such that whenever, for example, Ohm's law would be cited in the competence theory, that single factor is pivotal in the processing which yields the actual result. To see this we need only reflect that different feature units and knowledge atoms will be pivotal in solving problems which relate to the fate of R_1 and ones which relate to the fate of R_2. In this (restricted) sense it *does* have something in common with the look-up tree. For the network fails to embody strict tacit knowlege of the rule because it fails to route all its actual processing through a causal bottleneck corresponding to the derivational bottleneck marked by the repeated citing of Ohm's law. By having multiple causal routes where the competence theory has a single derivational equation, the network loses its claim to strict tacit knowledge of the rule. In that respect, it fails to embody tacit knowledge of the rule for the same reason as does the look-up tree.

Now for the quibble promised earlier. In adopting, as far as I understand it, Davies's characterization of tacit knowledge, I am uneasy about the use of the phrase 'causal common factor'. It has the advantage of making neuropsychological implications seem very immediate. But it may paper over some of the complexities of stacked virtual machines. For my guess is that what would need to be common for the classical cascade to be realized, is *not* a simple physical state so much as a state of *the virtual machine over which the processing story is defined*. After all, even a classical system, courtesy of various niceties of operating systems, may not use the same *physical* state every time it goes through a processing transition marked (in the competence theory) by Ohm's law. However, the level-2 processing description need not (and ought not) signal the difference, since it has no implications as far as the actual algorithm is concerned. It is merely an implementation detail. Contrariwise, the variety of states which, in a connectionist story, may correspond to a single symbolic transition, *must* be signalled in the processing/algorithmic description. After all, the system's real knowledge *is* the knowledge so encoded—a fact which is directly responsible for the much-vaunted fluidity and context-sensitivity of connectionist processing. I am not sure how much of a difference this makes since virtual machines, as much as real ones, can exhibit distinctive breakdown patterns and hence tie in with the cognitive neuropsychology.

Quibbling aside, we are now in a position to sum up the Newtonian attitude to competence theorizing. A Newtonian connectionist will regard a competence theory as *descriptive* (perhaps at a quite fine grain—recall the discussion of 'macrodecisions') of the course of processing. But she will not regard it as *suggestive* of the actual processing involved. It is not suggestive because the behaviour is not dependent on the system's having explicit or tacit knowledge of the symbolic derivation rules; a fact evidenced in its behaviour outside the idealized, 'Newtonian' domain of well-posed problems and unlimited processing time. This behaviour shows that 'it's really been a "quantum" system all along' (Smolensky 1988: 20).

In a revealing footnote (Smolensky 1986: ii. 246) the point is cast in terms highly appropriate to our discussion. The characterization of competence as a set of derivation rules applied to a symbol system can be viewed, Smolensky suggests, as providing a *grammar* for generating the high-harmony (= maximal soft constraint satisfaction) states of a system. Thus a competence theory emerges as a body of laws which serve to pick out the states into which the system will settle in certain ideal conditions. This, then, is the full Newtonian attitude to a competence theory: a competence theory is a kind of grammar which fixes on certain stable states of the system. As such it is, in a central range of cases, descriptively adequate. But it does not reveal what Smolensky calls the *dynamics*, or actual processing strategies, of the system. It is not a properly suggestive guide to the level-2 processing story. For the Newtonian then, competence theorizing just ain't what it used to be.

4. ROGUE COMPETENCE

On the Newtonian connectionist model, then, the competence theory functions as a descriptively adequate guide to the output in a somewhat idealized range of cases. This, however, is not the only understanding of competence theories available to a connectionist. And indeed, it is not the understanding implicit in some *other* connectionist treatments of high-level problem-solving. In this section I look at a class of alternative treatments which I shall call *rogue* models of competence.

The basic difference between Newtonian and rogue models is simply this. In a Newtonian model, the connectionist network is *itself* capable, under idealized conditions, of behaving in all the ways specified by the competence theory. In a rogue model, by contrast, the basic connectionist network does not *itself* have the capacity (even under idealizations of processing time and well-posed problems) to produce the full range of results required by (i.e. derivable in) the competence theory. Instead, it will be claimed that in so far as human beings actually exhibit the full-scale classical competence they do so only by deploying *other resources* (for example, a linked symbol-processing *or* real-world structures—like pen and paper—for manipulating symbols). The view of competence models which emerges from a rogue approach is thus that they involve pressing into service extra resources which are not on-line in fast daily problem-solving in the domain.

An example of a rogue model can be found in Rumelhart, Smolensky, McClelland, and Hinton (1986). The example concerns our capacity to multiply numbers. We might imagine a symbolic competence theory here appealing to the laws of arithmetic. But a basic connectionist model will not resemble such a symbolic store. Rather, it will amount to a well trained pattern matcher which can 'see' the results of some multiplication right away. For example, most of us can 'see' the answer to 7×7, but not to 7984×5431. How then, do we solve the latter kind of problem?

The conjecture is that: 'The answer comes from our ability to create artifacts—that is, our ability to create physical representations that we can manipulate in simple ways to get answers to very difficult and abstract problems' (Rumelhart, Smolensky, McClelland, and Hinton 1986: ii. 44). Thus, to solve 7984×5431 we might write down the question and *then* solve it by the careful deployment of a series of the simple pattern-matching steps we are good at, e.g. beginning by multiplying 4×1 and so on;

```
    7  9  8  4
    5  4  3  1
   ───────────
    .  .  .  . 4

       .  .  .  .
```

We may, they go on to say, even learn to do this *in our head* by representing the external symbols to ourselves in some manner. But it is still an essentially 'external' symbolic medium which we are manipulating, and it still constitutes a resource built *on top of* the basic connectionist pattern-matching capacity which we deploy. (Daniel Dennett has recently been saying very similar things about the cases where *sentences* seem to run through our heads. In these cases, we do indeed do classical symbol-processing. But such processing may constitute an extra resource, not implicated in all our daily, non-linguistic reasoning, see Dennett 1987: 233, 114–15; also Clark 1988.)

The account of complex multiplication is of course highly problematic since the whole thing *seems* to involve knowing symbolic rules governing the serial deployment of the pattern-matching capacities! But we have seen already that much *apparently* symbol-reliant behaviour may be subsymbolically produced. (But see Clark 1989 for a detailed discussion.) And at any rate, I use the example merely as a gesture at the *kind* of account which would constitute a rogue model.

To give one final example (which I owe to Martin Davies), consider our capacity to parse garden-path sentences like, 'the horse raced past the barn fell.' A rogue model of parsing might go something like this. We have on-line a quick and dirty connectionist network which can parse most of the sentences we encounter in daily speech. But it does not have the capacity (even in principle, subject to idealization) to parse a garden-path sentence. However, we *also* have (not on-line, but in the background) a classical symbolic parser (something like an ATN?) which can parse such cases. And when the quick and dirty network fails, this back-up comes on-line to save the day. This fits the phenomenology, in which the sentence at first looks like nonsense, then falls into place. In such a case the classical competence theory correctly describes the structure of the *back-up system*. But it does not describe the on-line network. If, in addition, we imagine that the classical back-up system was active in training up the network, the partial confluence of the two systems over a range of simple cases is rendered unsurprising.

An obvious and related advantage of the rogue approach concerns the psychological plausibility of so-called supervised learning algorithms. There are procedures for training connectionist networks which rely on the back-propagation of error messages, and hence rely on a *teacher* (usually a conventional computer) which looks at the system's output and tells it what the output *should* have been like. (For a little more detail, see the discussion of *NETtalk* in §5.) Such set-ups have often appeared deeply psychologically unrealistic. For example, when we learn a language, we can do so by being given *positive* examples only (as Chomskians are fond of pointing out). Whence, then, the teacher and the error messages?

The possibility which rogue models open up is that a separate system stores a set of input–output pairings (e.g. a set of observed print–phoneme pairings) and uses these to train a connectionist network. The negative instances are thus generated and spotted by the brain itself, rather than by other agents. Terence Sejnowski has recently endorsed such a picture and illustrates it by citing the case of the white crown sparrow which hears its father's song one year but does not sing it until the next. The hypothesis is that the bird somehow *stores* the song, but must train up a network to reproduce it—a process which explains the long gap between exposure and reproduction. White crown sparrows aside, rogue approaches clearly offer the best hope for the psychological respectability of the back-propagation method of connectionist learning.

At its most extreme, a rogue model may divorce human on-line processing from the strict competence model, but *reinstate* the classical competence as a full and proper description of a back-up system. Note that the status of the classical competence theory on a rogue model, is quite different from its status on a Newtonian one. For the rogue modeller, the classical competence theory properly describes an important, though not constantly on-line, class of processing systems. In fact, the importance of these classical resources is, I suspect, not yet fully appreciated even where lip service is paid to their presence. Thus Smolensky (1988) introduces the idea of language as a special medium of knowledge transmission involving processing by a classical virtual machine called the Conscious Rule Interpreter. But the role of linguistic instruction is still presented as somewhat second-grade. Language allows us to formulate rules which, for example, help the novice in the early stages of training (see also Smolensky 1986: ii. 251–2 where essentially the same picture is applied to the previously discussed case of electric circuit problem-solving). The *expert,* however, is pictured as using a powerful connectionist network, and seems to need language only to transmit potted elements of her insights to others. This may severely underestimate the contribution of symbol-processing. Such processing may also help the expert to understand and extend her own skills by providing a kind of meta-reflection on her own on-line reasoning. (For some related hypotheses see Karmiloff-Smith 1987 and Dennett 1988*a*.)

The most potent effect of the adoption of a rogue approach is vastly to complicate the currently fashionable debate concerning the 'correct' cognitive architecture of mind (see Fodor and Pylyshyn 1988). For if a rogue model is adopted, there is no unique answer to such questions. Any good account of human cognitive skills will need to employ *both* kinds of model, and the classical version will not be just a convenient approximation. It is as if the physical world turned out to be Newtonian in some areas and quantum in others, rather than being uniformly quantum-describable but in some circumstances *looking* Newtonian.

To sum up, rogue models deny even the *descriptive* adequacy of classical competence models to on-line processing. But they allow that the classical theory is both descriptive *and* suggestive of the processing of an *additional resource system*. This additional resource system guarantees what might be called our *canonical* reasoning abilities in a given domain. In rogue cases, the competence model *is* what it used to be (an accurate description of *some* processing strategy), but it isn't *where* it used to be—for it does not describe the computational form of daily on-line processing.

5. THE METHODOLOGY OF CONNECTIONIST EXPLANATION

Connectionist explanatory strategies, it seems, *cannot* fit into the mould suggested by Newell and Simon. A connectionist cannot begin with a Newell and Simon style competence theory and then simply implement it in a level-2 algorithmic model. The reason, we saw, is straightforward. Such a competence theory consists of a set of transition rules defined to apply to standard symbolic representations on data-structures. In a classical model, these data-structures are explicitly represented in the machine (classical functional architectures are *precisely* those architectures which make this possible). And the machine then manipulates them in accordance with the rules (which need not *themselves* be explicitly tokened in any such data-structures). In a distinctively connectionist model, by contrast, there will be nothing which neatly corresponds to the classical symbolic data-structures. Instead, context-sensitive, shifting coalitions of units will correspond to single classical representations. This is the dimension-shift described earlier. Since there are thus no neat analogues to the classical symbolic structures, the system *cannot* (not even tacitly) embody knowledge of transition rules defined over *those very structures*. So a classical competence theory can't be richly suggestive of a connectionist level-2 processing story. If it were, then the 'connectionist' system would amount merely to a fast, robust, implementation of a *classical* cognitive model (see Fodor and Pylyshyn 1988).

Given all this, we saw that the devout connectionist could adopt one of two positions regarding the classical competence model. These were the Newtonian and Rogue positions discussed above. But a deeper, foundational issue remains unresolved. For the Newtonian and Rogue positions are united in denying that any top-level classical competence theory can be richly suggestive of the level-2 processing strategies of the central on-line connectionist network which carries out a given cognitive task. But this (recall §1 above) now looks to be a doubly embarrassing loss. For the classical competence theory performed two tasks. First, it figured in a picture of the proper form of investigations in cognitive science (i.e. delineate the task at the level of competence theorizing and then write algorithms to carry it out). And second, it figured in a picture of what *explanation* in cognitive science involved. Just having a working program was not, in itself, to be regarded as having an explanation of how we perform a given cognitive task. Rather, we wanted some high-level understanding of what constraints the program was meeting and why they had to be met—an understanding naturally provided by giving the top-level competence theory which a given *class* of programs could be seen to implement. The unavailability of the classical competence theory thus threatens to render connectionist models *non-explanatory* in a very deep sense. And it leaves the actual *methodology* of connectionist investigations obscure.

As a brief illustration of the problem, consider an example of Good Old Fashioned Explanation In Cognitive Science (GOFEICS, apologies to John Haugeland). Take Naïve Physics. Naïve Physics, as everyone knows, is the attempt to discover the knowledge which enables a mobile, embodied being to negotiate its way around a complex physical universe. A well-known instance of this general project is Hayes's work on the naïve physics of liquids (Hayes 1985*b*). This involved trying to compile a 'taxonomy of the possible states liquid can be in' and formulating a set of rules concerning movement, change, and liquid geometry. The final theory included specifications of fifteen states of liquid and seventy-four numbered rules or axioms written out in predicate calculus. This amounts to a detailed competence specification which might eventually be given full level-2 algorithmic form. Indeed, Hayes is quite explicit about the high level of the investigative project, insisting that it is a mistake to seek a working program too soon (see Hayes 1985*a*: 3). The explanatory strategy of naïve physics is thus a paradigm example of the official classical methodology recommended by Newell and Simon. First, seek a high-level competence theory involving symbolic representations and a set of state transition rules. Then write level-2 algorithms implementing the competence theory, secure in the knowledge that we have a precise higher-level understanding of the requirements which the algorithms meet and hence a real grasp of why they are capable of carrying out the task in question. It is this security which the connectionist lacks, since she does not *(cannot)* proceed by formulating a detailed classical competence theory and then neatly implementing it on a classical symbol-processing architecture.

Hence the problem: how *should* the connectionist proceed, and what constitutes the higher-level understanding of the processing which we need in order to claim to have really *explained* how a task is performed. What is needed, it seems, is some kind of connectionist analogue to the classical competence theoretic level of explanation.

I believe that such an analogue exists. But it remains invisible until we perform a kind of Copernican revolution in our picture of explanation in Cognitive Science. For the connectionist effectively inverts the usual temporal and methodological order of explanation, much as Copernicus inverted the usual astronomical model of the day by having the earth revolve around the sun instead of the other way round. Likewise, in connectionist theorizing, the high-level understanding will be made to revolve around a working program which has learnt how to negotiate some cognitive terrain. This inverts the official Marr-style ordering in which the high-level understanding (i.e. competence theory) comes first and closely guides the search for algorithms. To make this clear, and to see how the connectionist's high-level theory will depart from the form of a classical competence theory, I propose to take a look at Sejnowski's NETtalk project.

NETtalk is a large, distributed connectionist model which aims to investigate part of the process of turning written input (i.e. words) into phonemic output (i.e. sounds or speech). The network architecture comprises a set of input units which are stimulated by seven letters of text at a time, a set of hidden units, and a set of output units which code for phonemes. The output is fed into a voice synthesizer which produces the actual speech sounds.

The network began with a random distribution of hidden unit weights and connections (within chosen parameters), i.e. it had no 'idea' of any rules of text-to-phoneme conversion. Its task was to learn, by repeated exposure to training instances, to negotiate its way around this particularly tricky cognitive domain (tricky because of irregularities, sub-regularities, and context-sensitivity of text $\longrightarrow$ phoneme conversion). And learning proceeded in the standard way, i.e. by a back-propagation learning rule. This works by giving the system an input, checking (this is done automatically by a computerized 'supervisor') its output, and telling it what output (i.e. what phonemic code) it *should* have produced. The learning rule then causes the system to minutely adjust the weights on the hidden units in a way which would tend towards the correct output. This procedure is repeated many thousands of times. Uncannily, the system slowly and audibly learns to pronounce English text, moving from babble to half-recognizable words and on to a highly creditable final performance. For a full account, see Rosenberg and Sejnowski (1987) and Sejnowski and Rosenberg (1986).

Consider now the methodology of the NETtalk project. It begins, to be sure, by invoking the results of some fairly rich prior analysis of the domain. This is reflected in the author's choice of input representation (e.g. the choice of a seven letter window, and a certain coding for letters and punctuation), in the choice of output representation (the coding for phonemes), and in the choice of hidden unit architecture (e.g. the number of hidden units) and learning rule. These choices highlight the continued importance of some degree of prior task-analysis in connectionist modelling. But they are a far cry from any fully articulated competence theory of text-to-phoneme conversion. For what is noticeably lacking is any set of special-purpose state transition rules defined over the input and output representations. Instead, the system will be set the task of learning a set of weights over its hidden units such that the weights perform the task of mediating the desired state transitions. For this reason I shall characterize the connectionist as beginning her investigations with a level-0.5 'task-analysis', as opposed to a level-1 (or 1.5) competence theory. It is worth remarking, however, that the level-0.5 specification, though *less* than a full-blown symbolic competence theory, may *still* embody a psychologically unrealistic amount of prior information. For when a human learns to perform a task she does not know, in advance, how many hidden units to allocate (too many and you form an uninformative 'look-up tree', too few and you fail to deal with the data) or the best way to represent the solution. In this sense, the level-0.5 specification may be doing more of the problem-solving work than some connectionists would like to admit. For present purposes, however, the point is just that the level-0.5 model forms the basis upon which, courtesy of the powerful connectionist learning rules, the system comes to be able (after much training) to negotiate the targetted cognitive terrain. At this point, the connectionist has in her hand a working system—a full-scale level-3 implementation.

Suppose we were to stop there. We would have a useful toy, but very little in the way of increased understanding of the phenomenon of text–phoneme conversion. But, of course, the connectionist *doesn't* stop there. From the up and running level-3 implementation she must now work *backwards* to a higher-level understanding of the task. This is Marr-through-the-looking-glass. How is this higher-level understanding to be obtained?

There are a variety of strategies in use and many more to be discovered. I shall mention just three. First, there is simple *watching*, but at a microscopic level. Given a particular input, the connectionist can see the patterns of unit activity (in the hidden units) which result. (This at any rate, will be the case if the network is simulated on a conventional machine which can keep a record of such activity.) This, as Sejnowski points out, provides a kind of data which neuroscientists are hard pressed to gather. For neuroscience has excellent techniques for recording single cell activity. But it is not well placed to record patterns of simultaneous activity across large numbers of cells. (See also Churchland forthcoming: 1989.) Second, there is *network pathology*. While it is obviously unethical deliberately to damage human brains to help us see what role sub-assemblies of cells play in various tasks, it seems far more acceptable to damage artificial neural networks. Lastly, and perhaps most significantly, the connectionist can generate a picture of the way in which the system has learnt to divide up the cognitive space it is trying to negotiate. It is this picture, given by so-called 'hierarchical cluster-analysis', which seems to me to offer the closest connectionist analogue to a high-level, competence-theoretic understanding.

Cluster-analysis is an attempt to answer the question, 'What kinds of representation have become encoded in the network's hidden units?'. This is a hard question since the representations, as noted earlier, will in general be of somewhat complex, unobvious, dimension-shifted features. To see how cluster-analysis works, consider the task of the network to be that of setting hidden unit weights in a way which will enable it to perform a kind of set partitioning. The goal is for the hidden units to respond in distinctive ways when, and only when, the input is such as to deserve a distinctive output. Thus in text-to-phoneme conversion, we want the hidden units to perform very differently when given 'the' as input than they would if given 'sail' as input. But we want them to perform *identically* if given 'sail' and 'sale' as inputs. So the hidden units' task is to partition a space (defined by the number of such units and their possible levels of activation) in a way which is geared to the job in hand. A very simple system, such as the rock/mine network described in Churchland (forth-coming: 1989) may need only to partition the space defined by its hidden units into two major subvolumes—one distinctive pattern for inputs signifying mines and one for those signifying rocks. The complexities of text–phoneme conversion being what they are, NETtalk must partition its hidden unit space more subtly (in fact, into a distinctive pattern for each of 79 possible letter-to-phoneme pairings). Cluster-analysis, as carried out by Rosenberg and Sejnowski (1987) in effect constructs a hierarchy of partitions on top of this base level of 79 distinctive stable patterns of hidden unit activation. The hierarchy is constructed by taking each of the 79 patterns and pairing it with its closest neighbour, i.e. with the pattern which has most in common with it. These pairings act as the building blocks for the next stage of analysis, in which an average activation profile (between the members of the original pair) is calculated and paired with *its* nearest neighbour drawn from the pool of secondary figures generated by averaging each of the original pairs. The process is repeated until the final pair is generated. This represents the grossest division of the hidden unit space which the network learnt—a division which in the case of NETtalk turned out to correspond to the division between vowels and consonants.

Cluster-analysis thus provides a kind of picture of the shape of the space of the possible hidden unit activations which power the network's performance. By reflecting on the various aspects of this space (i.e. the various clusterings) the theorist can hope to obtain some insight into what the system is doing. It may, for example, turn out to be highly sensitive to some subregularity which had hitherto been unnoticed or considered unimportant. It is as if we are provided with a tracing of the shape of the cognitive space we are attempting to understand. The tracing must be interpreted and that is a real and at times difficult task. But it is not shooting in the dark, for we can see what inputs are associated with that configuration (even if it is a higher-level configuration revealed by cluster-analysis).

We are thus given members of each class in question—the task is then to find perspicuous, conceptual-level terms in which to describe the conditions of class membership.

A fully interpreted cluster-analysis, I would like to suggest, constitutes the nearest connectionist analogue to a classical competence theory. Like a competence theory, it provides a level of understanding which is higher than (i.e. more general than) the algorithmic level. For the 'algorithmic' specification, for a connectionist, must be a specification of (*a*) the network configuration and (*b*) the unit rules and connection strengths. But there is a many–one mapping between such algorithmic specifications and a particular cluster-analysis. For example, a network which started out with a different random set of weights would, after training, exhibit the *same* partitioning profile (hence have an identical cluster-analysis) but do so using a very different set of individual weights. Unlike a classical competence theory, however, the cluster-analysis will typically *not* look like a set of state transition rules defined over conceptual-level entities. Instead, it will be more like a kind of geometric picture of the shape of a piece of cognitive terrain. Those theorists who think that a high-level explanation must be like a set of sentences and rules may find this hard to adjust to.

On the other hand some radically anti-sentential theorists (e.g. Churchland forthcoming: 1989) may consider that an interpreted cluster-analysis gives away *too much* to ordinary propositional discourse. Churchland argues that the correct level of understanding lies at the level of the connection weights. For, he insists, those are all the system 'really' knows about; it has no representation of its own partitionings. Moreover, the way two systems learn given new inputs can vary even if they have identical cluster analyses at time t_1. For the connection weights (which, we saw, stand in many–one relations to cluster-analyses) are the pivotal unit of cognitive evolution.

This, however, looks like the ordinary swings and roundabouts of high-level explanation. In opting at times for a level of analysis which groups particular connection weight specifications into equivalence classes governed by common cluster analyses we naturally trade specificity for generality. Just as pure Darwinism leaves recessive characteristics unexplained, but highlights general principles covering a class of evolutionary mechanisms, so cluster-analysis leaves some details of cognitive evolution unexplained but highlights the gross sensitivity which enables a class of networks to negotiate successfully a given cognitive terrain. Some such high-level understanding seems essential if connectionism is to be deeply explanatory of cognitive performance. A mere specification of a set of connection weights is surely not an *explanation*, even for the anti-sententialist.

The main point I want to stress is, however, independent of any view about the merits or demerits of cluster-analysis. It concerns the methodological inversion of traditional cognitive science. The connectionist, by whatever means, achieves her high-level understanding of a cognitive task by reflecting on, and tinkering with, a network which has *learnt* to perform the task in question. Unlike the classical, Marr-inspired theorist, she does not begin with a well worked out (sentential, symbolic) competence theory and then give it algorithmic flesh. Instead, she begins at level-0.5, trains a network, and then seeks to grasp the high-level principles it has come to embody. This is an almost miraculous boon for cognitive science. For the discipline has been dogged by the (related) evils of *ad-hoc*ery and sententialism. Forced to formulate competence theories as sets of rules defined over classical, symbolic data-structures, theorists have plucked principles out of thin air to help organize their work. Connectionist methodology, by contrast, allows the task demands to trace themselves and thus suggest the shape of the space in a way uncontaminated by the demands of standard symbolic formulation. We thus avoid imposing the form of our conscious, sentential thought on our models of unconscious processing—an imposition which was generally as practically unsuccessful as it was evolutionarily bizarre.

In sum, the connectionist, in being compelled to make do without the comfort of a classical competence theory is deprived neither of high-level explanatory power nor of methodological soundness. On the contrary, the methodology of connectionist explanation is perfectly geared to the avoidance of *ad-hoc* organizing principles and sentential, linguistic bias. There remain important and unresolved questions concerning the best ways to extract and couch such high-level explanations as connectionism may provide. But techniques such as cluster-analysis, network pathology, and activation recording are already being developed and will no doubt become well-understood. Once they do, the Copernican revolution in cognitive explanation will be well under way.

6. CONCLUSIONS: THE CASCADE, THE DAM, AND THE DIVIDED STREAM

Classicists and connectionists, it seems, must differ fundamentally in the way they expect actual processing (level-2) models to relate to traditional competence theories. The paper began by displaying the classicist vision of this relation and two connectionist alternatives. These may conveniently be pictured as follows.

Relation One: The Cascade

Dennett describes the classicists' vision as one of a 'triumphant cascade through Marr's three levels' (Dennett 1987: 227). The cascade flows easily given the presence of a classical symbol-processing architecture. The axioms or rules of the competence theory are linguistically expressed formulas for deriving one symbol from another. Various algorithms (level-2) may then implement that derivational structure. They may do so explicitly (by tokening the rule) or tacitly (by processing explicit symbol strings in accordance with the rule). In this classical vision, level-2 is a neat echo of level-1.

Relation Two: The Dam

Newtonian connectionism dams the classical cascade by introducing a dimension-shift between the items (symbol strings) operated on by the level-1 derivational rules and the items (subsymbols) 'operated on' by a connectionist network. The level-1 theory may describe some (idealized) aspects of the network's behaviour. But the network embodies neither explicit nor tacit knowlege of the derivational rules nor the conceptual-level structures over which they are defined.

Relation Three: The Divided Stream

Rogue models represent a more complex state of affairs in which actual performance is dependent on *two* systems. One, the daily, on-line system relates to the competence theory in the way described by the Newtonian connectionist, i.e. it matches some of the implied behaviour, but without embodying the classical knowledge. The other is an additional resource, perhaps created by the exploitation of external symbols, which simulates a classical machine. As such, it is capable of embodying the derivational rules and conceptual-level structures specified in the competence theory. Rogue models complicate the debate over the 'correct' architecture of cognition by suggesting a multiplicity of interactive (virtual) architectures.

One way or another, then, the connectionist must distance herself from the details of the classical competence model. Such models are not properly suggestive of the form of on-line connectionist processing, though they may be either descriptive of (a subset of) the results of such processing, or descriptive of some other cognitive resource. But this dis-location of connectionism and competence theorizing raised a serious problem. For the classicist had a methodology which guaranteed a useful and accurate higher-level understanding of the cognitive phenomenon modelled. The connectionist, by contrast, may seem to have working systems but no higher-level understanding of them—hence, in a certain sense, no *explanations* of cognitive phenomena.

This worry loses some of its force once we manage to perform a kind of Copernican revolution in our thinking about explanation in cognitive science. Under Marr's influence, cognitive scientists are likely to expect some high-level understanding of a task to *precede* and *inform* the writing of algorithms. Classical competence theoretic specifications aim to do just that job. The connectionist, however, effectively inverts this strategy. She begins with a minimal understanding of the task, trains a network to perform it, and *then* seeks, in various principled ways, to achieve a higher-level understanding of what it's doing and why. This may involve careful recording of network activity, the examination of the network's behaviour after various forms of damage, and plotting the way the network's hidden units divide up the cognitive space they are negotiating. This last activity (cluster-analysis, as described in the text), clearly provides a kind of higher-level understanding since there is a many–one relation between a given cluster-analysis and the set of connection weights which could implement it. The connectionist starts with a level-0.5 model, moves rapidly to a level-3 implementation and must then work backwards to detailed higher-levels of understanding.

This explanatory inversion, I want to suggest, actually constitutes one of the major *advantages* of the connectionist approach over traditional cognitive science. It is an advantage because it provides a means by which to avoid the *ad-hoc* generation of axioms and principles. Instead of having to decide on a rather arbitrary set of symbolic, language-based axioms to organize some cognitive task (recall naïve physics) the connectionist can let the task itself organize the network, and only *then* attempt to formulate various higher-level pictures of its activity. Such pictures, moreover, may depart (in ways we have yet to fully imagine) from the traditional picture of a theory as a set of propositions. Instead, they may be more geometric, or pictorial, or may use language in unexpected, apparently clumsy ways (see Churchland forthcoming: 1989).

There is, we may finally conjecture, a fairly deep reason why attitudes to competence polarize connectionists and classicists. It is that a competence model is a traditional *theory*, expressed in propositional or logical form. Classicists believe that thinking just *is* the manipulation of items having propositional or logical form; connectionists insist that this is just the icing on the cake and that *thinking* ('deep' thinking, rather than just sentence rehearsal) depends on the manipulation of quite different kinds of structure. As a result, the classicist attempts to give a level-2 processing model which is defined over the very same kinds of structure as figure in her level-1 theory. Whereas the connectionist insists on dissolving that structure and replacing it with something quite different.

A curious irony emerges. In the early days of Artifical Intelligence, the rallying cry was 'Computers *don't* crunch numbers, they *manipulate symbols!*' This was meant to inspire a doubting public by showing how much computation was like thinking. Now the wheel has come full circle. The virtue of connectionist systems, it seems, is that 'they don't manipulate symbols, they *crunch numbers*'. And nowadays we all know (don't we?) that thinking *isn't* mere symbol manipulation! So the wheel turns.[1]

[1] This paper was developed as part of the *Franco–British Programme of Philosophy and Cognitive Science*. I am especially indebted to Martin Davies for conversations and suggestions relating to the idea of a rogue model of competence developed in §4, and to Margaret Boden and Terence Sejnowski for discussion of the material contained in §5.

Chomsky, N. (1986). *Knowledge of Language: Its Nature, Origin and Use*. New York: Praeger Publishers.

Churchland, P. (forthcoming: 1989). 'On the Nature of Theories: A Neurocomputational Perspective.' In P. M. Churchland (ed.), *The Neurocomputational Perspective*. Cambridge, Mass.: MIT Press.

Clark, A. (1988). 'Thoughts, Sentences and Cognitive Science.' *Philosophical Psychology* 1: 263–78.

—— (1989). *Microcognition: Philosophy, Cognitive Science and Parallel Distributed Processing*. Cambridge, Mass.: MIT/Bradford Books.

Davies, M. (1987). 'Tacit Knowledge and Semantic Theory: Can a Five Per Cent Difference Matter?' *Mind* 96: 441–62.

—— (forthcoming). 'Connectionism, Modularity and Tacit Knowledge.' In *British Journal for the Philosophy of Science*.

Dennett, D. (1987). *The Intentional Stance*. Cambridge, Mass.: MIT/Bradford Books.

—— (1988a). 'The Evolution of Consciousness.' Jacobsen Lecture, University of London, May 1988. *Tufts University Current Circulating Manuscript* CCM-88-1.

—— (1988b). 'Review of Psychosemantics.' *Journal of Philosophy* 85: 384–9.

Fodor, J., and Pylyshyn, Z. (1988). 'Connectionism and Cognitive Architecture.' *Cognition* 28: 3–71.

Hayes, P. J. (1985a), 'The Second Naïve Physics Manifesto.' In J. R. Hobbs and R. C. Moore (ed.), *Formal Theories of the Commonsense World*, pp. 1–36. Norwood, NJ: Ablex.

—— (1985b). 'The Ontology of Liquids.' In J. R. Hobbs and R. C. Moore (eds.), *Formal Theories of the Commonsense World*, pp. 71–108. Norwood, NJ: Ablex.

Karmiloff-Smith, A. (1987). 'Beyond Modularity: A Developmental Perspective on Human Consciousness.' Transcript of talk given to the Annual Meeting of the British Psychological Society, Sussex, April 1987.

Marr, D. (1977). 'Artificial Intelligence—A Personal View.' Repr. in J. Haugeland (ed.), *Mind Design*, pp. 37–47. Cambridge, Mass.: MIT/Bradford Books.

Peacocke, C. (1986). 'Explanation in Computational Psychology: Language, Perception and Level 1.5.' *Mind and Language*, 1 (2): 101–23.

Pinker, A., and Prince, S. (1988). 'On Language and Connectionism.' *Cognition* 28: 73–193.

Ridley, M. (1985). *The Problems of Evolution*. Oxford: Oxford University Press.

Rosenberg, C., and Sejnowski, T. (1987). 'Parallel Networks That Learn to Pronounce English Text.' *Complex Systems* 1: 145–68.

Rumelhart, D., and McClelland, J. (1986a), 'PDP Models and General Issues in Cognitive Science.' In McClelland, Rumelhart, and the PDP Research Group (eds.), *Parallel Distributed Processing: Explorations in the Microstructure of Cognition*, Vol. 1, pp. 110–46. Cambridge, Mass.: MIT/Bradford Books.

—— (1986b), 'On Learning the Past Tenses of English Verbs.' In McClelland, Rumelhart, and the PDP Research Groups (eds.), *Parallel Distributed Processing: Explorations in the Microstructure of Cognition*, Vol. 2, pp. 216–27. Cambridge, Mass.: MIT/Bradford Books.

Rumelhart, D., Smolensky, P., McClelland, J., and Hinton, G. (1986). 'Schemata and Sequential Thought Processes in PDP Models'. In McClelland, Rumelhart, and the PDP Research Group (eds.), *Parallel Distributed Processing: Explorations in the Microstructure of Cognition*, Vol. 2, pp. 7–57. Cambridge, Mass.: MIT/Bradford Books.

Sejnowski, T., and Rosenberg, C. (1986). 'NETtalk: A Parallel Network That Learns to Read Aloud.' John Hopkins University Electrical Engineering and Computer Science Technical Report, JHU/EEC-86/01.

Smolensky, P. (1986). 'Information Processing in Dymanical Systems: Foundations of Harmony Theory.' In McClelland, Rumelhart, and the PDP Research Group (eds.), *Parallel Distributed Processing: Explorations in the Microstructure of Cognition*, Vol. 2, pp. 194–281. Cambridge, Mass.: MIT/Bradford Books.

—— (1988). 'On the Proper Treatment of Connectionism.' *Behavioral and Brain Sciences* 11: 1–74.

13

MAKING A MIND VERSUS MODELLING THE BRAIN: ARTIFICIAL INTELLIGENCE BACK AT A BRANCH-POINT

HUBERT L. DREYFUS and STUART E. DREYFUS

> [N]othing seems more possible to me than that people some day will come to the definite opinion that there is no copy in the . . . nervous system which corresponds to a *particular* thought, or a *particular* idea, or memory.
>
> Ludwig Wittgenstein (1948: i. 504(66e))

> [I]nformation is not stored anywhere in particular. Rather it is stored everywhere. Information is better thought of as 'evoked' than 'found'.
>
> David Rumelhart and Donald Norman (1981: 3)

In the early 1950s, as calculating machines were coming into their own, a few pioneer thinkers began to realize that digital computers could be more than number crunchers. At that point two opposed visions of what computers could be, each with its correlated research-programme, emerged and struggled for recognition. One faction saw computers as a system for manipulating mental symbols; the other, as a medium for modelling the brain. One sought to use computers to instantiate a formal representation of the world; the other, to simulate the interactions of neurones. One took problem-solving as its paradigm of intelligence; the other, learning. One utilized logic; the other, statistics. One school was the heir to the rationalist, reductionist tradition in philosophy; the other viewed itself as idealized, holistic neuroscience.

The rallying cry of the first group was that both minds and digital computers are physical-symbol systems. By 1955 Allen Newell and

Herbert L. Dreyfus and Stuart E. Dreyfus, 'Making a Mind Versus Modeling the Brain: Artificial Intelligence Back at a Branchpoint' from *Artificial Intelligence* 117, no. 1 (Winter 1988), Cambridge, Mass. Reprinted by permission of *Daedalus*, Journal of the American Academy of Arts and Sciences.

Herbert Simon, working at the Rand Corporation, had concluded that strings of bits manipulated by a digital computer could stand for anything—numbers, of course, but also features of the real world. Moreover, programs could be used as rules to represent relations between these symbols, so that the system could infer further facts about the represented objects and their relations. As Newell put it recently in his account of the history of issues in AI,

The digital-computer field defined computers as machines that manipulated numbers. The great thing was, adherents said, that everything could be encoded into numbers, even instructions. In contrast, the scientists in AI saw computers as machines that manipulated symbols. The great thing was, they said, that everything could be encoded into symbols, even numbers (Newell 1983: 196).

This way of looking at computers became the basis of a way of looking at minds. Newell and Simon hypothesized that the human brain and the digital computer, while totally different in structure and mechanism, had at a certain level of abstraction a common functional description. At this level both the human brain and the appropriately programmed digital computer could be seen as two different instantiations of a single species of device—a device that generated intelligent behaviour by manipulating symbols by means of formal rules. Newell and Simon stated their view as a hypothesis:

The Physical Symbol System Hypothesis. A physical symbol system has the necessary and sufficient means for general intelligent action.

By 'necessary' we mean that any system that exhibits general intelligence will prove upon analysis to be a physical symbol system. By 'sufficient' we mean that any physical symbol system of sufficient size can be organized further to exhibit general intelligence (Newell and Simon 1981: 41).

Newell and Simon trace the roots of their hypothesis back to Gottlob Frege, Bertrand Russell, and Alfred North Whitehead (1981: 42), but Frege and company were of course themselves heirs to a long, atomistic, rationalist tradition. Descartes had already assumed that all understanding consisted of forming and manipulating appropriate representations, that these representations could be analysed into primitive elements (*naturas simplices*), and that all phenomena could be understood as complex combinations of these simple elements. Moreover, at the same time, Hobbes had implicitly assumed that the elements were formal components related by purely syntactic operations, so that reasoning could be reduced to calculation. 'When a man *reasons*, he does nothing else but conceive a sum total from addition of parcels,' Hobbes wrote, 'for REASON . . . is nothing but reckoning . . . ' (1958: 45). Finally, Leibniz, working out the classical idea of mathesis—the formalization of every-thing—sought support to develop a universal symbol system so that 'we can assign to every object its determined characteristic number (1951: 18). According to Leibniz, in understanding we analyse concepts into more simple elements. In order to avoid a regress to simpler and simpler elements, there must be ultimate simples in terms of which all complex concepts can be understood. Moreover, if concepts are to apply to the world, there must be simple features that these elements represent. Leibniz envisaged 'a kind of alphabet of human thoughts' (1951: 20) whose 'characters must show, when they are used in demonstrations, some kind of connection, grouping and order which are also found in the objects' (1951: 10).

Ludwig Wittgenstein, drawing on Frege and Russell, stated in his *Tractatus Logico-Philosophicus* the pure form of this syntactic, representa-tional view of the relation of the mind to reality. He defined the world as the totality of logically independent atomic facts:

1.1 The world is the totality of facts, not of things.

Facts in turn, he held, could be exhaustively analysed into primitive objects.

2.01. An atomic fact is a combination of objects. . . .
2.0124. If all objects are given, then *thereby* all atomic facts are given.

These facts, their constituents, and their logical relations, Wittgenstein claimed, were represented in the mind.

2.1. We make to ourselves pictures of facts.
2.15. That the elements of the picture are combined with one another in a definite way, represents that the things are so combined with one another (1960).

AI can be thought of as the attempt to find the primitive elements and logical relations in the subject (man or computer) that mirror the primitive objects and their relations that make up the world. Newell and Simon's physical-symbol system hypothesis in effect turns the Wittgensteinian vision (which is itself the culmination of the classical rationalist philosophical tradition) into an empirical claim and bases a research-programme on it.

The opposed intuition, that we should set about creating artificial intelli-gence by modelling the brain rather than the mind's symbolic representa-tion of the world, drew its inspiration not from philosophy but from what was soon to be called neuroscience. It was directly inspired by the work of D. O. Hebb, who in 1949 suggested that a mass of neurones could learn if when neurone A and neurone B were simultaneously excited, that excitation increased the strength of the connection between them.

This lead was followed by Frank Rosenblatt, who reasoned that since intelligent behaviour based on our representation of the world was likely to be hard to formalize. AI should instead attempt to automate the procedures by which a network of neurones learns to discriminate patterns and respond appropriately. As Rosenblatt put it,

The implicit assumption [of the symbol manipulating research program] is that it is relatively easy to specify the behavior that we want the system to perform, and that the challenge is then to design a device or mechanism which will effectively carry out this behavior . . . [I]t is both easier and more profitable to axiomatize the *physical system* and then investigate this system analytically to determine its behavior, than to axiomatize the *behavior* and then design a physical system by techniques of logical synthesis (1962b: 386).

Another way to put the difference between the two research-programmes is that those seeking symbolic representations were looking for a formal structure that would give the computer the ability to solve a certain class of problems or discriminate certain types of patterns. Rosenblatt, on the other hand, wanted to build a physical device, or to simulate such a device on a digital computer, that could then generate its own abilities:

Many of the models which we have heard discussed are concerned with the question of what logical structure a system must have if it is to exhibit some property, X. This is essentially a question about a static system. . . .

An alternative way of looking at the question is: what kind of a system can *evolve* property X? I think we can show in a number of interesting cases that the second question can be solved without having an answer to the first (1962b: 387).

Both approaches met with immediate and startling success. By 1956 Newell and Simon had succeeded in programming a computer using symbolic representations to solve simple puzzles and prove theorems in the propositional calculus. On the basis of these early impressive results it looked as if the physical-symbol systems hypothesis was about to be confirmed, and Newell and Simon were understandably euphoric. Simon announced:

It is not my aim to surprise or shock you . . . But the simplest way I can summarize is to say that there are now in the world machines that think, that learn and that create. Moreover, their ability to do these things is going to increase rapidly until — in a visible future — the range of problems they can handle will be coextensive with the range to which the human mind has been applied (1958: 6).

He and Newell explained:

[W]e now have the elements of a theory of heuristic (as contrasted with algorithmic) problem solving; and we can use this theory both to understand human heuristic processes and to simulate such processes with digital computers. Intuition, insight, and learning are no longer exclusive possessions of humans: any large high-speed computer can be programmed to exhibit them also (1958: 6).[1]

Rosenblatt put his ideas to work in a type of device that he called a perceptron.[2] By 1956 Rosenblatt was able to train a perceptron to classify certain types of patterns as similar and to separate these from other patterns that were dissimilar. By 1959 he too was jubilant and felt his approach had been vindicated:

It seems clear that the . . . perceptron introduces a new kind of information processing automaton: For the first time, we have a machine which is capable of having original ideas. As an analogue of the biological brain, the perceptron, more precisely, the theory of statistical separability, seems to come closer to meeting the requirements of a functional explanation of the nervous system than any system previously proposed. . . . As concept, it would seem that the perceptron has established, beyond doubt, the feasibility and principle of non-human systems which may embody human cognitive functions . . . The future of information processing devices which operate on statistical, rather than logical, principles seems to be clearly indicated (1958: i. 449).

In the early sixties both approaches looked equally promising, and both made themselves equally vulnerable by making exaggerated claims. Yet

[1] Heuristic rules are rules that when used by human beings are said to be based on experience or judgement. Such rules frequently lead to plausible solutions to problems or increase the efficiency of a problem-solving procedure. Whereas algorithms guarantee a correct solution (if there is one) in a finite time, heuristics only increase the likelihood of finding a plausible solution.

[2] Rumelhart and McClelland (1986) describe the perceptron as follows: 'Such machines consist of what is generally called a *retina*, an array of binary inputs sometimes taken to be arranged in a two-dimensional spatial layout; a set of *predicates*, a set of binary threshold units with fixed connections to a subset of units in the retina such that each predicate computes some local function over the subset of units to which it is connected; and one or more decision units, with modifiable connections to the predicates' (i. 111). They contrast the way a parallel distributed processing (PDP) model like the perceptron stores information with the way information is stored by symbolic representation: 'In most models, knowledge is stored as a static copy of a pattern. Retrieval amounts to finding the pattern in long-term memory and copying it into a buffer or working memory. There is no real difference between the stored representation in long-term memory and the active representation in working memory. In PDP models, though, this is not the case. In these models, the patterns themselves are not stored. Rather, what is stored is the *connection strengths* between units that allow these patterns to be re-created (i. 31). . . . [K]nowledge about any individual pattern is not stored in the connections of a special unit reserved for that pattern, but is distributed over the connections among a large number of processing units' (i. 33). This new notion of representation led directly to Rosenblatt's idea that such machines should be able to acquire their ability through learning rather than by being programmed with features and rules: [I]f the knowledge is [in] the strengths of the connections, learning must be a matter of finding the right connection strengths so that the right patterns of activation will be produced under the right circumstances. This is an extremely important property of this class of models, for it opens up the possibility that an information processing mechanism could learn, as a result of tuning its connections, to capture the interdependencies between activations that it is exposed to in the course of processing' (i. 32).

the results of the internal war between the two research-programmes were surprisingly asymmetrical. By 1970 the brain simulation research, which had its paradigm in the perceptron, was reduced to a few lonely, under-funded efforts, while those who proposed using digital computers as symbol manipulators had undisputed control of the resources, graduate programmes, journals, and symposia that constitute a flourishing research-programme.

Reconstructing how this change came about is complicated by the myth of manifest destiny that any ongoing research-programme generates. Thus, it looks to the victors as if symbolic information processing won out because it was on the right track, while the neural network or connectionist approach lost because it simply didn't work. But this account of the history of the field is a retrospective illusion. Both research-programmes had ideas worth exploring, and both had deep, unrecognized problems.

Each position had its detractors, and what they said was essentially the same: each approach had shown that it could solve certain easy problems but that there was no reason to think either group could extrapolate its methods to real-world complexity. Indeed, there was evidence that as problems got more complex, the computation required by both approaches would grow exponentially and so would soon become intractable. In 1969 Marvin Minsky and Seymour Papert said of Rosenblatt's perceptron:

Rosenblatt's schemes quickly took root, and soon there were perhaps as many as a hundred groups, large and small, experimenting with the model. . . .
The results of these hundreds of projects and experiments were generally disappointing, and the explanations inconclusive. The machines usually work quite well on very simple problems but deteriorate very rapidly as the tasks assigned to them get harder (19).

Three years later, Sir James Lighthill, after reviewing work using heuristic programs such as Simon's and Minsky's, reached a strikingly similar negative conclusion:

Most workers in AI research and in related fields confess to a pronounced feeling of disappointment in what has been achieved in the past 25 years. Workers entered the field around 1950, and even around 1960, with high hopes that are very far from having been realized in 1972. In no part of the field have the discoveries made so far produced the major impact that was then promised. . . .
[O]ne rather general cause for the disappointments that have been experienced: failure to recognize the implications of the 'combinatorial explosion'. This is a general obstacle to the construction of a . . . system on a large knowledge base which results from the explosive growth of any combinatorial expression, representing numbers of possible ways of grouping elements of the knowledge base according to particular rules, as the base's size increases.

As David Rumelhart and David Zipser have succinctly summed it up, 'Combinatorial explosion catches you sooner or later, although sometimes in different ways in parallel than in serial (Rumelhart and McClelland 1986: i. 158). Both sides had, as Jerry Fodor once put it, walked into a game of 3-dimensional chess, thinking it was tick-tack-toe. Why then, so early in the game, with so little known and so much to learn, did one team of researchers triumph at the total expense of the other? Why, at this crucial branch-point, did the symbolic representation project become the only game in town?

Everyone who knows the history of the field will be able to point to the proximal cause. About 1965, Minsky and Papert, who were running a laboratory at MIT dedicated to the symbol-manipulation approach and therefore competing for support with the perceptron projects, began circulating drafts of a book attacking the idea of the perceptron. In the book they made clear their scientific position:

Perceptrons have been widely publicized as 'pattern recognition' or 'learning' machines and as such have been discussed in a large number of books, journal articles, and voluminous 'reports.' Most of this writing . . . is without scientific value (1969: 4).

But their attack was also a philosophical crusade. They rightly saw that traditional reliance on reduction to logical primitives was being challenged by a new holism:

Both of the present authors (first independently and later together) became involved with a somewhat therapeutic compulsion: to dispel what we feared to be the first shadows of a 'holistic' or 'Gestalt' misconception that would threaten to haunt the fields of engineering and artificial intelligence as it had earlier haunted biology and psychology (1969: 19).

They were quite right. Artificial neural nets may, but need not, allow an interpretation of their hidden nodes[3] in terms of features a human being could recognize and use to solve the problem. While neural network modelling itself is committed to neither view, it can be demonstrated that association does not *require* that the hidden nodes be interpretable. Holists like Rosenblatt happily assumed that individual nodes or patterns of nodes were not picking out fixed features of the domain.

Minsky and Papert were so intent on eliminating all competition, and so secure in the atomistic tradition that runs from Descartes to early Wittgenstein, that their book suggests much more than it actually

---

[3] Hidden nodes are nodes that neither directly detect the input to the net nor constitute its output. They are, however, either directly or indirectly linked by connections with adjustable strengths to the nodes detecting the input and those constituting the output.

demonstrates. They set out to analyse the capacity of a one-layer perceptron,[4] while completely ignoring in the mathematical portion of their book Rosenblatt's chapters on multilayer machines and his proof of the convergence of a probabilistic learning algorithm based on back-propagation[5] of errors (1962a: 292).[6] According to Rumelhart and McClelland,

Minsky and Papert set out to show which functions can and cannot be computed by [one-layer] machines. They demonstrated, in particular, that such perceptrons are unable to calculate such mathematical functions as parity (whether an odd or even number of points are on in the retina) or the topological function of connectedness (whether all points that are on are connected to all other points that are on either directly or via other points that are also on) without making use of absurdly large numbers of predicates. The analysis is extremely elegant and demonstrates the importance of a mathematical approach to analysing computational systems (1986: i. 111).

But the implications of the analysis are quite limited. Rumelhart and McClelland continue:

Essentially . . . although Minsky and Papert were exactly correct in their analysis of the *one-layer perceptron*, the theorems don't apply to systems which are even a little more complex. In particular, it doesn't apply to multilayer systems nor to systems that allow feedback loops (1986: i. 112).

Yet in the conclusion to *Perceptrons*, when Minsky and Papert ask themselves the question, 'Have you considered perceptrons with many layers?' they give the impression, while rhetorically leaving the question open, of having settled it:

Well, we have considered Gamba machines, which could be described as 'two layers of perceptron.' We have not found (by thinking or by studying the literature) any other really interesting class of multilayered machine, at least none whose principles seem to have a significant relation to those of the perceptron . . . [W]e consider it to be an important research problem to elucidate (or reject) our intuitive judgment that the extension is sterile (1969: 231–2).

Their attack on gestalt thinking in AI succeeded beyond their wildest dreams. Only an unappreciated few, among them Stephen Grossberg,

---

[4] A one-layer network has no hidden nodes, while multilayer networks do contain hidden nodes.

[5] Back-propagation of errors requires recursively computing, starting with the output nodes, the effects of changing the strengths of connections on the difference between the desired output and the output produced by an input. The weights are then adjusted during learning to reduce the difference.

[6] See also: 'The addition of a fourth layer of signal transmission units, or cross-coupling the A-units of a three-layer perceptron, permits the solution of generalization problems, over arbitrary transformation groups. . . . In back-coupled perceptrons, selective attention to familiar objects in a complex field can occur. It is also possible for such a perceptron to attend selectively to objects which move differentially relative to their background' (Rosenblatt 1962a: 576).

James A. Anderson, and Teuvo Kohonen, took up the 'important research problem'. Indeed, almost everyone in AI assumed that neural nets had been laid to rest forever. Rumelhart and McClelland note:

Minsky and Papert's analysis of the limitations of the one-layer perceptron, coupled with some of the early successes of the symbolic processing approach in artificial intelligence, was enough to suggest to a large number of workers in the field that there was no future in perceptron-like computational devices for artificial intelligence and cognitive psychology (1986: i. 112).

But why was it enough? Both approaches had produced some promising work and some unfounded promises.[7] It was too early to close accounts on either approach. Yet something in Minsky and Papert's book struck a responsive chord. It seemed AI workers shared the quasi-religious philosophical prejudice against holism that motivated the attack. One can see the power of the tradition, for example, in Newell and Simon's article on physical-symbol systems. The article begins with the scientific hypothesis that the mind and the computer are intelligent by virtue of manipulating discrete symbols, but it ends with a revelation: 'The study of logic and computers has revealed to us that intelligence resides in physical-symbol systems' (1981: 64).

Holism could not compete with such intense philosophical convictions. Rosenblatt was discredited along with the hundreds of less responsible network research groups that his work had encouraged. His research money dried up, and he had trouble getting his work published. By 1970, as far as AI was concerned, neural nets were dead. In his history of AI, Newell says the issue of symbols versus numbers 'is certainly not alive now and has not been for a long time' (1983: 10). Rosenblatt is not even mentioned in John Haugeland's (1985) or Margaret Boden's (1977) histories of the AI field.[8]

[7] For an evaluation of the symbolic representation approach's actual successes up to 1978, see Dreyfus (1979).

[8] Work on neural nets was continued in a marginal way in psychology and neuroscience. James A. Anderson at Brown University continued to defend a net model in psychology, although he had to live off other researchers' grants, and Stephen Grossberg worked out an elegant mathematical implementation of elementary cognitive capacities. For Anderson's position see Anderson (1978). For examples of Grossberg's work during the dark ages, see his (1982) book. Kohonen's early work is reported in *Associative Memory—A System-Theoretical Approach* (Berlin: Springer-Verlag, 1977). At MIT Minsky continued to lecture on neural nets and to assign theses investigating their logical properties. But according to Papert, Minsky did so only because nets had interesting mathematical properties, whereas nothing interesting could be proved concerning the properties of symbol systems. Moreover, many AI researchers assumed that since Turing machines were symbol manipulators and Turing had proved that Turing machines could compute anything, he had proved that all intelligibility could be captured by logic. On this view a holistic (and in those days statistical) approach needed justification, while the symbolic AI approach did not. This confidence, however, was based on confusing the uninterpreted symbols of a Turing machine (zeros and ones) with the semantically interpreted symbols of AI.

But blaming the rout of the connectionists on an anti-holistic prejudice is too simple. There was a deeper way philosophical assumptions influenced intuition and led to an overestimation of the importance of the early symbol-processing results. The way it looked at the time was that the perceptron people had to do an immense amount of mathematical analysis and calculating to solve even the most simple problems of pattern recognition, such as discriminating horizontal from vertical lines in various parts of the receptive field, while the symbol-manipulating approach had relatively effortlessly solved hard problems in cognition, such as proving theorems in logic and solving combinational puzzles. Even more important, it seemed that given the computing power available at the time, the neural-net researchers could do only speculative neuroscience and psychology, while the simple programs of symbolic representationists were on their way to being useful. Behind this way of sizing up the situation was the asumption that thinking and pattern recognition are two distinct domains and that thinking is the more important of the two. As we shall see later in our discussion of the common-sense knowledge problem, to look at things this way is to ignore both the pre-eminent role of pattern discrimination in human expertise and also the background of common-sense understanding that is presupposed in everyday real-world thinking. Taking account of this background may well require pattern recognition.

This thought brings us back to the philosophical tradition. It was not just Descartes and his descendants who stood behind symbolic information-processing, but all of Western philosophy. According to Heidegger, traditional philosophy is defined from the start by its focusing on facts in the world while 'passing over' the world as such (Heidegger 1962: §14–21; Dreyfus 1988). This means that philosophy has from the start systematically ignored or distorted the everyday context of human activity.[9] The branch of the philosophical tradition that descends from Socrates through Plato, Descartes, Leibniz, and Kant to conventional AI takes it for granted, in addition, that understanding a domain consists in having a *theory* of that domain. A theory formulates the relationships among objective, *context-free* elements (simples, primitives, features, attributes, factors, data points, cues, etc.) in terms of abstract principles (covering laws, rules, programs, etc.).

Plato held that in theoretical domains such as mathematics and perhaps ethics, thinkers apply explicit, context-free rules of theories they have learned in another life, outside the everyday world. Once learned, such theories function in this world by controlling the thinker's mind, whether

---

[9] According to Heidegger, Aristotle came closer than any other philosopher to understanding the importance of everyday activity, but even he succumbed to the distortion of the phenomenon of the everyday world implicit in common sense.

he or she is conscious of them or not. Plato's account did not apply to everyday skills but only to domains in which there is a priori knowledge. The success of theory in the natural sciences, however, reinforced the idea that in any orderly domain there must be some set of context-free elements and some abstract relations among those elements that account for the order of that domain and for man's ability to act intelligently in it. Thus, Leibniz boldly generalized the rationalist account to all forms of intelligent activity, even everyday practice:

[T]he most important observations and turns of skill in all sorts of trades and professions are as yet unwritten. This fact is proved by experience when passing from theory to practice we desire to accomplish something. *Of course, we can also write up this practice, since it is at bottom just another theory more complex and particular* . . . [italics added] (1951: 48).

The symbolic information-processing approach gains its assurance from this transfer to all domains of methods that have been developed by philosophers and that are successful in the natural sciences. Since, in this view, any domain must be formalizable, the way to do AI in any area is obviously to find the context-free elements and principles and to base a formal, symbolic representation on this theoretical analysis. In this vein Terry Winograd describes his AI work in terms borrowed from physical science:

We are concerned with developing a formalism, or 'representation,' with which to describe . . . knowledge. We seek the 'atoms' and 'particles' of which it is built and the 'forces' that act on it (1976: 9).

No doubt theories about the universe are often built up gradually by modelling relatively simple and isolated systems and then making the model gradually more complex and integrating it with models of other domains. This is possible because all the phenomena are presumably the result of the lawlike relations between what Papert and Minsky call 'structural primitives'. Since no one *argues* for atomistic reduction in AI, it seems that AI workers just implicitly *assume* that the abstraction of elements from their everyday context, which defines philosophy and works in natural science, must also work in AI. This assumption may well account for the way the physical-symbol system hypothesis so quickly turned into a revelation and for the ease with which Papert and Minsky's book triumphed over the holism of the perceptron.

Teaching philosophy at MIT in the mid-sixties, one of us—Hubert—was soon drawn into the debate over the possibility of AI. It was obvious that researchers such as Newell, Simon, and Minsky were the heirs to the philosophical tradition. But given the conclusions of the later Wittgenstein and the early Heidegger, that did not seem to be a good omen for the reductionist research-programme. Both these thinkers had called into question the very tradition on which symbolic information-processing was based. Both were holists, both were struck by the import-ance of everyday practices, and both held that one could not have a theory of the everyday world.

It is one of the ironies of intellectual history that Wittgenstein's devast-ating attack on his own *Tractatus*, his *Philosophical Investigations*, was published in 1953, just as AI took over the abstract, atomistic tradition he was attacking. After writing the *Tractatus*, Wittgenstein spent years doing what he called phenomenology (1975) — looking in vain for the atomic facts and basic objects his theory required. He ended by abandoning his *Tractatus* and all rationalistic philosophy. He argued that the analysis of everyday situations into facts and rules (which is where most traditional philosophers and AI researchers think theory must begin) is itself only meaningful in some context and for some purpose. Thus, the elements chosen already reflect the goals and purposes for which they are carved out. When we try to find the ultimate context-free, purpose-free elements, as we must if we are going to find the primitive symbols to feed a computer, we are in effect trying to free aspects of our experience of just that pragmatic organization which makes it possible to use them intelli-gently in coping with everyday problems.

In the *Philosophical Investigations* Wittgenstein directly criticized the logical atomism of the *Tractatus*:

'What lies behind the idea that names really signify simples'? — Socrates says in the *Theaetetus*: 'If I make no mistake, I have heard some people say this: there is no definition of the primary elements — so to speak — out of which we and everything else are composed . . . But just as what consists of these primary elements is itself complex, so the names of the elements become descriptive language by being compounded together.' Both Russell's 'individuals' and my 'objects' (*Tractatus Logico-Philosophicus*) were such primary elements. But what are the simple constituent parts of which reality is composed? . . . It makes no sense at all to speak absolutely of the 'simple parts of a chair' (1953: 21).

Already, in the 1920s, Martin Heidegger had reacted in a similar way against his mentor, Edmund Husserl, who regarded himself as the culmina-tion of the Cartesian tradition and was therefore the grandfather of AI (Dreyfus 1982). Husserl argued that an act of consciousness, or noesis, does not on its own grasp an object; rather, the act has intentionality (directedness) only by virtue of an 'abstract form', or meaning, in the noema correlated with the act.[10]

---

[10] 'Der Sinn . . . so wie wir ihn bestimmt haben, ist nicht ein konkretes Wesen im Gesamtbestande des Noema, sondern eine Art ihm einwohnender abstrackter Form.' See Husserl (1950). For evidence that Husserl held that the noema accounts for the intentionality of mental activity, see Hubert Dreyfus, 'Husserl's Perceptual Noema', in Dreyfus (1982).

This meaning, or symbolic representation, as conceived by Husserl, is a complex entity that has a difficult job to perform. In *Ideas Pertaining to a Pure Phenomenology* (1982), Husserl bravely tried to explain how the noema gets the job done. Reference is provided by 'predicate-senses', which, like Fregean *Sinne*, just have the remarkable property of picking out objects' atomic properties. These predicates are combined into complex 'descriptions' of complex objects, as in Russell's theory of descriptions. For Husserl, who was close to Kant on this point, the noema contains a hierarchy of strict rules. Since Husserl thought of intelligence as a context-determined, goal-directed activity, the mental representation of any type of object had to provide a context, or a 'horizon' of expectations or 'predelineations' for structuring the incoming data: 'a rule governing *possible* other consciousness of [the object] as identical—possible, as exemplifying essentially predelineated types' (1960: 45). The noema must contain a rule describing all the features that can be expected with certainty in exploring a certain *type* of object—features that remain 'inviolably the same: as long as the objectivity remains intended as *this* one and of this kind' (1960: 53). The rule must also prescribe predelineations of properties that are possible, but not necessary, features of this type of object: 'Instead of a completely determined sense, there is always, therefore, a *frame of empty sense. . . .*' (1960: 51).

In 1973 Marvin Minsky proposed a new data-structure, remarkably similar to Husserl's, for representing everyday knowledge:

A *frame* is a data-structure for representing a stereotyped situation, like being in a certain kind of living room, or going to a child's birthday party. . . .

We can think of a frame as a network of nodes and relations. The top levels of a frame are fixed, and represent things that are always true about the supposed situation. The lower levels have many *terminals*—slots that must be filled by specific instances or data. Each terminal can specify conditions its assignments must meet. . . .

Much of the phenomenological power of the theory hinges on the inclusion of expectations and other kinds of presumptions. A *frame's terminals are normally already filled with 'default' assignments* (1981: 96).

In Minsky's model of a frame, the 'top level' is a developed version of what, in Husserl's terminology, remains 'inviolably the same' in the representation, and Husserl's predelineations have become 'default assignments'—additional features that can normally be expected. The result is a step forward in AI techniques from a passive model of information-processing to one that tries to take account of the interactions between a knower and the world. The task of AI thus converges with the task of transcendental phenomenology. Both must try in everyday domains to find frames constructed from a set of primitive predicates and their formal relations.

Heidegger, before Wittgenstein, carried out, in response to Husserl, a phenomenological description of the everyday world and everyday objects like chairs and hammers. Like Wittgenstein, he found that the everyday world could not be represented by a set of context-free elements. It was Heidegger who forced Husserl to face precisely this problem by pointing out that there are other ways of 'encountering' things than relating to them as objects defined by a set of predicates. When we use a piece of equipment like a hammer, Heidegger said, we actualize a skill (which need not be represented in the mind) in the context of a socially organized nexus of equipment, purposes, and human roles (which need not be represented as a set of facts). This context, or world, and our everyday ways of skillful coping in it, which Heidegger called 'circumspection', are not something we *think* but part of our socialization, which forms the way we *are*. Heidegger concluded:

The context . . . can be taken formally in the sense of a system of relations. But . . . [t]he phenomenal content of these 'relations' and 'relata' . . . is such that they resist any sort of mathematical functionalization; nor are they merely something thought, first posited in an 'act of thinking'. They are rather relationships in which concernful circumspection as such already dwells (1962: 121–1).

This defines the splitting of the ways between Husserl and AI on the one hand and Heidegger and the later Wittgenstein on the other. The crucial question becomes, 'Can there be a theory of the everyday world as rationalist philosophers have always held?' Or is the common-sense background rather a combination of skills, practices, discrimination, and so on, which are not intentional states and so, *a fortiori*, do not have any representational content to be explicated in terms of elements and rules?

By making a move that was soon to become familiar in AI circles, Husserl tried to avoid the problem Heidegger posed. Husserl claimed that the world, the background of significance, the everyday context, was merely a very complex system of facts correlated with a complex system of beliefs, which, since they have truth conditions, he called validities. One could, in principle, he held, suspend one's dwelling in the world and achieve a detached description of the human belief system. One could thus complete the task that had been implicit in philosophy since Socrates: one could make explicit the beliefs and principles underlying all intelligent behaviour. As Husserl put it,

[E]ven the background . . . of which we are always concurrently conscious but which is momentarily irrelevant and remains completely unnoticed, still functions according to its implicit validities (1970: 149).

Since he firmly believed that the shared background could be made explicit as a belief system, Husserl was ahead of his time in raising the question of the possibility of AI. After discussing the possibility that a formal axiomatic system might describe experience and pointing out that such a system of axioms and primitives—at least as we know it in geometry—could not describe everyday shapes such as 'scalloped' and 'lens-shaped', Husserl left open the question whether these everyday concepts could none the less be formalized. (This was like raising and leaving open the AI question whether one can axiomatize common-sense physics.) Taking up Leibniz's dream of a mathesis of all experience, Husserl added:

The pressing question is . . . whether there could not be . . . an idealizing proced-ure that substitutes pure and strict ideals for intuited data and that would . . . serve . . . as the basic medium for a mathesis of experience (1952: v. 134).

But, as Heidegger predicted, the task of writing out a complete theoret-ical account of everyday life turned out to be much harder than initially expected. Husserl's project ran into serious trouble, and there are signs that Minsky's has too. During twenty-five years of trying to spell out the components of the subject's representation of everyday objects, Husserl found that he had to include more and more of the subject's common-sense understanding of the everyday world:

To be sure, even the tasks that present themselves when we take single types of objects as restricted clues prove to be extremely complicated and always lead to extensive disciplines when we penetrate more deeply. That is the case, for example, with . . . spatial objects (to say nothing of a Nature) as such, of psycho-physical being and humanity as such, culture as such (1960: 54–5).

He spoke of the noema's 'huge concreteness' (1969: 244) and of its 'tremendous complication' (1969: 246), and he sadly concluded at the age of seventy-five that he was a perpetual beginner and that phenomenology was an 'infinite task' (1970: 291).

There are hints in his paper 'A Framework for Representing Knowledge' that Minsky has embarked on the same 'infinite task' that eventually overwhelmed Husserl:

Just constructing a knowledge base is a major intellectual research problem . . . We still know far too little about the contents and structure of common-sense knowledge. A 'minimal' common-sense system must 'know' something about cause–effect, time, purpose, locality, process, and types of knowledge. . . . We need a serious epistemological research effort in this area (1981: 124).

To a student of contemporary philosophy, Minsky's naïveté and faith are astonishing. Husserl's phenomenology *was* just such a research effort. Indeed, philosophers from Socrates through Leibniz to early Wittgenstein carried on serious epistemological research in this area for two thousand years without notable success.

In the light of Wittgenstein's reversal and Heidegger's devastating critique of Husserl, one of us—Hubert—predicted trouble for symbolic information-processing. As Newell notes in his history of AI, this warning was ignored:

> Dreyfus's central intellectual objection . . . is that the analysis of the context of human action into discrete elements is doomed to failure. This objection is grounded in phenomenological philosophy. Unfortunately, this appears to be a nonissue as far as AI is concerned. The answers, refutations, and analyses that have been forthcoming to Dreyfus's writings have simply not engaged this issue—which indeed would be a novel issue if it were to come to the fore (1983: 222–3).

The trouble was, indeed, not long in coming to the fore, as the everyday world took its revenge on AI as it had on traditional philosophy. As we see it, the research-programme launched by Newell and Simon has gone through three ten-year stages. From 1955 to 1965 two research themes, representation and search, dominated the field then called 'cognitive simulation'. Newell and Simon showed, for example, how a computer could solve a class of problems with the general heuristic search principle known as means–end analysis—namely, to use any available operation that reduces the distance between the description of the current situation and the description of the goal. They then abstracted this heuristic technique and incorporated it into their General Problem Solver (GPS).

The second stage (1965–75), led by Marvin Minsky and Seymour Papert at MIT, was concerned with what facts and rules to represent. The idea was to develop methods for dealing systematically with knowledge in isolated domains called 'micro-worlds'. Famous programs written around 1970 at MIT include Terry Winograd's SHRDLU, which could obey commands given in a subset of natural language about a simplified 'blocks-world', Thomas Evan's analogy problem program, David Waltz's scene analysis program, and Patrick Winston's program, which could learn concepts from examples.

The hope was that the restricted and isolated micro-worlds could be gradually made more realistic and combined so as to approach real-world understanding. But researchers confused two domains, which, following Heidegger, we shall distinguish as 'universe' and 'world'. A set of interrelated facts may constitute a *universe*, like the physical universe, but it does not constitute a *world*. The latter, like the world of business, the world of theatre, or the world of the physicist, is an organized body of objects, purposes, skills, and practices on the basis of which human activities have meaning or make sense. To see the difference, one can contrast the *meaningless* physical universe with the *meaningful* world of the discipline of physics. The world of physics, the business world, and the theatre world make sense only against a background of common human concerns. They are local elaborations of the one common-sense world we all share. That is, subworlds are not related like isolable physical systems to the larger systems they *compose* but rather are local elaborations of a whole that they *presuppose*. Micro-worlds are not worlds but isolated meaningless domains, and it has gradually become clear that there is no way they could be combined and extended to arrive at the world of everyday life.

In its third stage, roughly from 1975 to the present, AI has been wrestling with what has come to be called the common-sense knowledge problem. The representation of knowledge was always a central problem for work in AI, but the two earlier periods—cognitive simulation and micro-worlds—were characterized by an attempt to avoid the problem of common-sense knowledge by seeing how much could be done with as little knowledge as possible. By the middle 1970s, however, the issue had to be faced. Various data-structures, such as Minsky's frames and Roger Schank's scripts, have been tried without success. The common-sense knowledge problem has kept AI from even beginning to fulfill Simon's prediction of twenty years ago that 'within twenty years machines will be capable of doing any work a man can do' (1965: 96).

Indeed, the common-sense knowledge problem has blocked all progress in theoretical AI for the past decade. Winograd was one of the first to see the limitations of SHRDLU and all script and frame attempts to extend the micro-worlds approach. Having 'lost faith' in AI, he now teaches Heidegger in his computer science course at Stanford and points out 'the difficulty of formalizing the common-sense background that determines which scripts, goals and strategies are relevant and how they interact' (1984: 142).

What sustains AI in this impasse is the conviction that the common-sense knowledge problem must be solvable, since human beings have obviously solved it. But human beings may not normally use common-sense *knowledge* at all. As Heidegger and Wittgenstein pointed out, what common-sense *understanding* amounts to might well be *everyday know-how*. By 'know-how' we do not mean procedural rules but knowing what to do in a vast number of special cases.[11] For example, common-sense physics has turned out to be extremely hard to spell out in a set of facts and rules. When one tries, one either requires more common sense to understand the facts and rules one finds or else one produces formulas of such complexity that it seems highly unlikely they are in a child's mind.

Doing theoretical physics also requires background skills that may not be formalizable, but the domain itself can be described by abstract laws that make no reference to these background skills. AI researchers mistakenly conclude that common-sense physics too must be expressible as a

[11] This account of skill is spelled out and defended in Dreyfus and Dreyfus (1986).

set of abstract principles. But it just may be that the problem of finding a *theory* of common-sense physics is insoluble because the domain has no theoretical structure. By playing with all sorts of liquids and solids every day for several years, a child may simply learn to discriminate prototypical cases of solids, liquids, and so on and learn typical skilled responses to their typical behaviour in typical circumstances. The same might well be the case for the social world. If background understanding is indeed a skill and if skills are based on whole patterns and not on rules, we would expect symbolic representations to fail to capture our common-sense understanding.

In the light of this impasse, classical, symbol-based AI appears more and more to be a perfect example of what Imre Lakatos (1978) has called a degenerating research-programme. As we have seen, AI began auspiciously with Newell and Simon's work at Rand and by the late 1960s turned into a flourishing research-programme. Minsky predicted that 'within a generation the problem of creating "artificial intelligence" will be substantially solved' (1977: 2). Then, rather suddenly, the field ran into unexpected difficulties. It turned out to be much harder than one expected to formulate a theory of common sense. It was not, as Minsky had hoped, just a question of cataloguing a few hundred thousand facts. The common-sense knowledge problem became the centre of concern. Minsky's mood changed completely in five years. He told a reporter that 'the AI problem is one of the hardest science has ever undertaken' (Kolata 1982: 1237).

The rationalist tradition had finally been put to an empirical test, and it had failed. The idea of producing a formal, atomistic theory of the everyday common-sense world and of representing that theory in a symbol manipulator had run into just the difficulties Heidegger and Wittgenstein had discovered. Frank Rosenblatt's intuition that it would be hopelessly difficult to formalize the world and thus to give a formal specification of intelligent behaviour had been vindicated. His repressed research-programme (using the computer to instantiate a holistic model of an idealized brain), which had never really been refuted, became again a live option.

In journalistic accounts of the history of AI, Rosenblatt is vilified by anonymous detractors as a snake-oil salesman:

Present-day researchers remember that Rosenblatt was given to steady and extravagant statements about the performance of his machine. 'He was a press agent's dream,' one scientist says, 'a real medicine man. To hear him tell it, the Perceptron was capable of fantastic things. And maybe it was. But you couldn't prove it by the work Frank did' (McCorduck 1979: 87).

In fact, he was much clearer about the capacities and limitations of the various types of perceptrons than Simon and Minsky were about their symbolic programs.[12] Now he is being rehabilitated. David Rumelhart, Geoffrey Hinton, and James McClelland reflect this new appreciation of his pioneering work:

Rosenblatt's work was very controversial at the time, and the specific models he proposed were not up to all the hopes he had for them. But his vision of the human information processing system as a dynamic, interactive, self-organizing system lies at the core of the PDP approach (1986: i. 45).

The studies of perceptrons . . . clearly anticipated many of the results in use today. The critique of perceptrons by Minsky and Papert was widely misinterpreted as destroying their credibility, whereas the work simply showed limitations on the power of the most limited class of perceptron-like mechanisms, and said nothing about more powerful, multiple layer models (1986: ii. 535).

Frustrated AI researchers, tired of clinging to a research-programme that Jerry Lettvin characterized in the early 1980s as 'the only straw afloat', flocked to the new paradigm. Rumelhart and McClelland's book

---

[12] Some typical quotations from Rosenblatt's *Principles of Neurodynamics*: 'In a learning experiment, a perceptron is typically exposed to a sequence of patterns containing representatives of each type or class which is to be distinguished, and the appropriate choice of a response is "reinforced" according to some rule for memory modification. The perceptron is then presented with a test stimulus, and the probability of giving the appropriate response for the class of the stimulus is ascertained. . . . If the test stimulus activates a set of sensory elements which are entirely distinct from those which were activated in previous exposures to stimuli of the same class, the experiment is a test of "pure generalization." The simplest of perceptrons . . . have no capability for pure generalization, but can be shown to perform quite respectably in discrimination experiments particularly if the test stimulus is nearly identical to one of the patterns previously experienced (p. 68). . . . Perceptrons considered to date show little resemblance to human subjects in their figure-detection capabilities, and gestalt-organizing tendencies (p. 71). . . . The recognition of sequences in rudimentary form is well within the capability of suitably organized perceptrons, but the problem of figural organization and segmentation presents problems which are just as serious here as in the case of static pattern perception (p. 72). . . . In a simple perception, patterns are recognized before "relations"; indeed, abstract relations, such as "A is above B" or "the triangle is inside the circle" are never abstracted as such, but can only be acquired by means of a sort of exhaustive rote-learning procedure, in which every case in which the relation holds is taught to the perceptron individually (p. 73). . . . A network consisting of less than three layers of signal transmission units, or a network consisting exclusively of linear connected in series, is incapable of learning to discriminate classes of patterns in an isotropic environment (where any pattern can occur in all possible retinal locations, without boundary effects) (p. 575). . . . A number of speculative models which are likely to be capable of learning sequential programs, analysis of speech into phonemes, and learning substantive "meanings" for nouns and verbs with simple sensory referents have been presented in the preceding chapters. Such systems represent the upper limits of abstract behaviour in perceptrons considered to date. They are handicapped by a lack of satisfactory 'temporary memory,' by an inability to perceive abstract topological relations in a simple fashion, and by an inability to isolate meaningful figural entities, or objects, except under special conditions (p. 577). . . . The applications most likely to be realizable with the kinds of perceptrons described in this volume include character recognition and "reading machines", speech recognition (for distinct, clearly separated words), and extremely limited capabilities for pictorial recognition, or the recognition of objects against simple backgrounds. "Perception" in a broader sense may be potentially within the grasp of the descendants of our present models, but a great deal of fundamental knowledge must be obtained before a sufficiently sophisticated design can be prescribed to permit a perceptron to compete with a man under normal environmental conditions' (p. 583).

Parallel Distributed Processing sold six thousand copies the day it went onto the market, and thirty thousand are now in print. As Paul Smolensky put it,

In the past half-decade the connectionist approach to cognitive modeling has grown from an obscure cult claiming a few true believers to a movement so vigorous that recent meetings of the Cognitive Science Society have begun to look like connectionist pep rallies (forthcoming).

If multilayered networks succeed in fulfilling their promise, researchers will have to give up the conviction of Descartes, Husserl, and early Wittgenstein that the only way to produce intelligent behaviour is to mirror the world with a formal theory in the mind. Worse, one may have to give up the more basic intuition at the source of philosophy that there must be a theory of every aspect of reality—that is, there must be elements and principles in terms of which one can account for the intelligibility of any domain. Neural networks may show that Heidegger, later Wittgenstein, and Rosenblatt were right in thinking that we behave intelligently in the world without having a theory of that world. If a theory is not *necessary* to explain intelligent behaviour, we have to be prepared to raise the question whether in everyday domains such a theoretical explanation is even *possible*.

Neural net modellers, influenced by symbol-manipulating AI, are expending considerable effort, once their nets have been trained to perform a task, in trying to find the features represented by individual nodes and sets of nodes. Results thus far are equivocal. Consider Hinton's (1986) network for learning concepts by means of distributed representations. The network can be trained to encode relationships in a domain that human beings conceptualize in terms of features, without the network being given the features that human beings use. Hinton produces examples of cases in which some nodes in the trained network can be interpreted as corresponding to the features that human beings pick out, although these nodes only roughly correspond to those features. Most nodes, however, cannot be interpreted semantically at all. A feature used in a symbolic representation is either present or not. In the net, however, although certain nodes are more active when a certain feature is present in the domain, the amount of activity not only varies with the presence or absence of this feature but is affected by the presence or absence of other features as well.

Hinton has picked a domain—family relationships—that is constructed by human beings precisely in terms of the features that human beings normally notice, such as generation and nationality. Hinton then analyses those cases in which, starting with certain random initial-connection strengths, some nodes can, after learning, be interpreted as representing those features. Calculations using Hinton's model show, however, that even his net seems to learn its associations for some random initial-connection strengths without any obvious use of these everyday features.

In one very limited sense, any successfully trained multilayer net can be interpreted in terms of features—not everyday features but what we shall call highly abstract features. Consider the simple case of layers of binary units activated by feed-forward, but not lateral or feedback, connections. To construct such an account from a network that has learned certain associations, each node one level above the input nodes could, on the basis of the connections to it, be interpreted as detecting when one of a certain set of input patterns is present. (Some of the patterns will be the ones used in training, and some will never have been used.) If the set of input patterns that a particular node detects is given an invented name (it almost certainly won't have a name in our vocabulary), the node could be interpreted as detecting the highly abstract feature so named. Hence, every node one level above the input level could be characterized as a feature detector. Similarly, every node a level above those nodes could be interpreted as detecting a higher-order feature, defined as the presence of one of a specified set of patterns among the first level of feature detectors. And so on up the hierarchy.

The fact that intelligence, defined as the knowledge of a certain set of associations appropriate to a domain, can always be accounted for in terms of relations among a number of highly abstract features of a skill domain does not, however, preserve the rationalist intuition that these explanatory features must capture the essential structure of the domain so that one could base a theory on them. If the net were taught one more association of an input–output pair (where the input prior to training produced an output different from the one to be learned), the interpretation of at least some of the nodes would have to be changed. So the features that some of the nodes picked out before the last instance of training would turn out not to have been invariant structural features of the domain.

Once one has abandoned the philosophical approach of classical AI and accepted the atheoretical claim of neural net modelling, one question remains: how much of everyday intelligence can such a network be expected to capture? Classical AI researchers are quick to point out—as Rosenblatt already noted—that neural net modellers have so far had difficulty dealing with stepwise problem-solving. Connectionists respond that they are confident that they will solve that problem in time. This response, however, reminds one too much of the way that the symbol manipulators in the sixties responded to the criticism that their programs were poor at the perception of patterns. The old struggle continues between intellectualists, who think that because they can do context-free logic they have a handle on everyday cognition but are poor at understanding perception, and gestaltists, who have the rudiments of an account of perception but no account of everyday cognition.[13] One might think, using the metaphor of the right and the left brain, that perhaps the brain or the mind uses each strategy when appropriate. The problem would then be how to combine the strategies. One cannot just switch back and forth, for as Heidegger and the gestaltists saw, the pragmatic background plays a crucial role in determining relevance, even in everyday logic and problem-solving, and experts in any field, even logic, grasp operations in terms of their functional similarities.

It is even premature to consider combining the two approaches, since so far neither has accomplished enough to be on solid ground. Neural network modelling may simply be getting a deserved chance to fail, as did the symbolic approach.

Still, there is an important difference to bear in mind as each research-programme struggles on. The physical-symbol system approach seems to be failing because it is simply false to assume that there must be a theory of every domain. Neural network modelling, however, is not committed to this or any other philosophical assumption. Nevertheless, building an interactive net sufficiently similar to the one our brain has evolved may be just too hard. Indeed, the common-sense knowledge problem, which has blocked the progress of symbolic representation techniques for fifteen years, may be looming on the neural net horizon, although researchers may not yet recognize it. All neural net modellers agree that for a net to be intelligent it must be able to generalize; that is, given sufficient examples of inputs associated with one particular output, it should associate further inputs of the same type with that same output. The question arises, however: what counts as the same type? The designer of the net has in mind a specific definition of the type required for a reasonable generalization and counts it a success if the net generalizes to other instances of this type. But when the net produces an unexpected association, can one say it has failed to generalize? One could equally well say that the net has all along been acting on a different definition of the type in question and that that difference has just been revealed. (All the

[13] For a recent influential account of perception that denies the need for mental representation, see Gibson (1979). Gibson and Rosenblatt collaborated on a research paper for the US Air Force in 1955; see Gibson, Olum, and Rosenblatt (1955).

'continue this sequence' questions found on intelligence tests really have more than one possible answer, but most human beings share a sense of what is simple and reasonable and therefore acceptable.)

Neural network modellers attempt to avoid this ambiguity and make the net produce 'reasonable' generalizations by considering only a prespecified allowable family of generalizations—that is, allowable transformations that will count as acceptable generalizations (the hypothesis space). These modellers then attempt to design the architecture of their nets so that they transform inputs into outputs only in ways that are in the hypothesis space. Generalization will then be possible only on the designer's terms. While a few examples will be insufficient to identify uniquely the appropriate member of the hypothesis space, after enough examples only one hypothesis will account for all the examples. The net will then have learned the appropriate generalization principle. That is, all further input will produce what, from the designer's point of view, is the appropriate output.

The problem here is that the designer has determined, by means of the architecture of the net, that certain possible generalizations will never be found. All this is well and good for toy problems in which there is no question of what constitutes a reasonable generalization, but in real-world situations a large part of human intelligence consists in generalizing in ways that are appropriate to a context. If the designer restricts the net to a predefined class of appropriate responses, the net will be exhibiting the intelligence built into it by the designer for that context but will not have the common sense that would enable it to adapt to other contexts, as a truly human intelligence would.

Perhaps a net must share size, architecture, and initial-connection configuration with the human brain if it is to share our sense of appropriate generalization. If it is to learn from its own 'experiences' to make associations that are humanlike rather than be taught to make associations that have been specified by its trainer, a net must also share our sense of appropriateness of output, and this means it must share our needs, desires, and emotions and have a humanlike body with appropriate physical movements, abilities, and vulnerability to injury.

If Heidegger and Wittgenstein are right, human beings are much more holistic than neural nets. Intelligence has to be motivated by purposes in the organism and goals picked up by the organism from an ongoing culture. If the minimum unit of analysis is that of a whole organism geared into a whole cultural world, neural nets as well as symbolically programmed computers still have a very long way to go.

REFERENCES

Anderson, J. A. (1978). 'Neural Models with Cognitive Implications.' In D. LaBerse and S. J. Samuels (eds.), *Basic Processing in Reading*, Hillsdale, NJ: Erlbaum.

Boden, M. (1977). *Artificial Intelligence and Natural Man*. New York: Basic Books.

Dreyfus, H. (1979). *What Computers Can't Do*, 2nd edn. New York: Harper & Row.

—— (1988). *Being-in-the-World: A Commentary on Division I of 'Being and Time'*. Cambridge, Mass.: MIT Press.

—— (ed.) (1982). *Husserl, Intentionality and Cognitive Science*. Cambridge, Mass.: MIT Press.

—— and Dreyfus, S. (1986). *Mind Over Machine*. New York: Macmillan.

Gibson, J. J. (1979). *The Ecological Approach to Visual Perception*. Boston: Houghton-Mifflin.

—— Olum, P., and Rosenblatt, F. (1955). 'Parallax and Perspective During Aircraft Landing.' *American Journal of Psychology* 68: 372–85.

Grossberg, S. (1982). *Studies of Mind and Brain: Neural Principles of Learning, Perception, Development, Cognition and Motor Control*. Boston: Reidel Press.

Haugeland, J. (1985). *Artificial Intelligence: The Very Idea*. Cambridge, Mass.: MIT Press.

Hebb, D. O. (1949). *The Organization of Behavior*. New York: Wiley.

Heidegger, M. (1962). *Being and Time*. New York: Harper & Row.

Hinton, G. (1986). 'Learning Distributed Representations of Concepts.' In *Proceedings of the Eighth Annual Conference of the Cognitive Science Society*. Amherst, Mass.: Cognitive Science Society.

Hobbes, T. (1958). *Leviathan*. New York: Library of Liberal Arts.

Husserl, E. (1950). *Ideen Zu Einer Reinen Phänomenologie und Phänomenologischen Philosophie*. The Hague: Nijhoff.

—— (1952). *Ideen Zu Einer Reinen Phänomenologie und Phenomenologischen Philosophie*, bk. 3 in vol. 5, *Husserliana*. The Hague: Nijhoff.

—— (1960). *Cartesian Meditations*, trans. D. Cairns. The Hague: Nijhoff.

—— (1969). *Formal and Transcendental Logic*, trans. D. Cairns. The Hague: Nijhoff.

—— (1970). *Crisis of European Sciences and Transcendental Phenomenology*, trans. D. Carr. Evanston: Northwestern University Press.

—— (1982). *Ideas Pertaining to a Pure Phenomenology and to a Phenomenological Philosophy*, trans. F. Kersten. The Hague: Nijhoff.

Kohonen, T. (1977). *Associative Memory: A System-Theoretical Approach*. Berlin: Springer-Verlag.

Kolata, G. (1982). 'How Can Computers Get Common Sense?' *Science* 217 (24 Sept.): 1237.

Lakatos, I. (1978). *Philosophical Papers*, ed. J. Worrall. Cambridge: Cambridge University Press.

Leibniz, G. (1951). *Selections*, ed. P. Wiener. New York: Scribner.

Lighthill, Sir James (1973). 'Artificial Intelligence: A General Survey.' In *Artificial Intelligence: A Paper Symposium*. London: Science Research Council.

McCorduck, P. (1979). *Machines Who Think*. San Francisco: W. H. Freeman.

Minsky, M. (1977). *Computation: Finite and Infinite Machines*. New York: Prentice-Hall.

—— (1981). 'A Framework for Representing Knowledge.' In J. Haugeland (ed.), *Mind Design*, pp. 95–128. Cambridge, Mass.: MIT Press.

—— and Papert, S. (1969). *Perceptrons: An Introduction to Computational Geometry*. Cambridge, Mass.: MIT Press.

Newell, A. (1983). 'Intellectual Issues in the History of Artificial Intelligence.' In F. Machlup and U. Mansfield (eds.), *The Study of Information: Interdisciplinary Messages*, pp. 196–227. New York: Wiley.

—— and Simon, H. (1958). 'Heuristic Problem Solving: The Next Advance in Operations Research.' *Operations Research* 6 (Jan.–Feb.): 6.

—— (1981). 'Computer Science as Empirical Inquiry: Symbols and Search.' In J. Haugeland (ed.), *Mind Design*, pp. 35–66. Cambridge, Mass.: MIT Press.

Rosenblatt, F. (1958). *Mechanisation of Thought Processes: Proceedings of a Symposium Held at the National Physical Laboratory*, Vol. 1. London: HMS Office.

—— (1962*a*). *Principles of Neurodynamics: Perceptrons and the Theory of Brain Mechanisms*. Washington, DC: Spartan Books.

—— (1962*b*). 'Strategic Approaches to the Study of Brain Models.' In H. von Foerster (ed.), *Principles of Self-Organization*, Elmsford, NY: Pergamon Press.

Rumelhart, D. E., and McClelland, J. L. (1986). *Parallel Distributed Processing: Explorations in the Microstructure of Cognition*, 2 vols. Cambridge, Mass.: MIT Press.

—— and Norman, D. A (1981). 'A Comparison of Models.' In G. Hinton and J. Anderson (eds.), *Parallel Models of Associative Memory*, pp. 3–6. Hillsdale, NJ: Erlbaum.

Simon, H. (1965). *The Shape of Automation for Men and Management*. New York: Harper & Row.

Smolensky, P. [1988]. 'On the Proper Treatment of Connectionism.' *Behavioral and Brain Sciences* [11: 1–74].

Winograd, T. (1976). 'Artificial Intelligence and Language Comprehension.' In *Artificial Intelligence and Language Comprehension*, Washington, DC: National Institute of Education.

—— (1984). 'Computer Software for Working with Language. *Scientific American* (Sept.): 142ff.

Wittgenstein, L. (1948). *Last Writings on the Philosophy of Psychology*, Vol. 1, trans. corrected. Chicago: University of Chicago Press, 1982.

—— (1953). *Philosophical Investigations*. Oxford: Basil Blackwell.

—— (1960). *Tractatus Logico-Philosophicus*. London: Routledge & Kegan Paul.

—— (1975). *Philosophical Remarks*. Chicago: University of Chicago.

14

SOME REDUCTIVE STRATEGIES IN COGNITIVE NEUROBIOLOGY

PAUL M. CHURCHLAND

1. INTRODUCTION

The aim of this paper is to make available to philosophers an intriguing theoretical approach to representation and computation currently under exploration in the neurosciences. The approach is intriguing for at least three reasons. Firstly, it provides a highly general answer to the question of how the brain might *represent* the many aspects of the world in which it lives. Later in the paper I shall explore that answer as it applies to a case familiar to philosophers: the case of the various subjective sensory qualia displayed in one's manifold of sensory intuition. There we will find the outlines of a genuine neurobiological reduction of the familiar sensory qualia. This application is but one among many, however, as I shall also try to show. One important result is that diverse cases of representation, cases which appear to common sense as being entirely distinct in character, emerge as being fundamentally the same. The approach thus finds unity in diversity.

The second intriguing aspect concerns the matter of *computation*. The style of representation to be outlined lends itself uniquely well to a powerful form of computation, a form well suited to the solution of a wide variety of problems. One of them is a problem relatively unfamiliar to philosophers: the problem of sensorimotor co-ordination. However unfamiliar it may be, this problem is of fundamental importance to cognitive theory, since the administration of appropriate behaviour in the light of current experience is where intelligence has its raw beginnings. Since sensorimotor co-ordination is the most fundamental problem that any animal must solve, a means of solution coupled to a general account of representation must surely arouse our curiosity.

Thirdly, the approach is intriguing in the way in which it embraces the mystery of the brain's *microphysical organization*, and the question of

Paul M. Churchland, 'Some Reductive Strategies in Cognitive Neurobiology' from *Mind* XCV, no. 379 (July 1986): 279–309. Reprinted with a few amendments by permission of Oxford University Press.

how its specific organization *implements* the representational and computational activities that the brain as a whole displays. Here, too, the approach finds empirical encouragement, for there are at least two major ways of physically implementing the abstract approach to representation and computation proposed, and each of them bears a suggestive resemblance to real neural structures displayed prominently throughout the empirical brain: the laminar organization of the cerebral cortex, and the dense orthogonal matrix of the cerebellar cortex.

Overall, the approach constitutes an unabashedly reductive strategy for the neuroscientific explanation of a variety of familiar cognitive phenomena. The propriety of such strategies is of course still a keenly debated issue within the philosophy of mind. Historically, this debate has been impoverished by the lack of any very impressive general *theories* from neurobiology, theories that promise actually to effect the neurobiological reduction of some familiar class of cognitive phenomena. At least, if such theories did exist in the outlying literature, they did not manage to make it into the philosophical debates. Given this absence of relevant theory, anti-reductionist arguments could and often did proceed simply by holding up some aspect of our cognitive life and asking the rhetorical question, 'How could *this* ever be accounted for, or even addressed, by any possible story about the nuts and bolts of neurones?'

Such rhetorical questions unfairly exploit the feebleness of our imaginations, since a reply even remotely adequate to the phenomena is not something we can reasonably be expected to think up on demand. As it happens, however, potentially adequate replies have indeed emerged from recent work in cognitive psychology and in cognitive neurobiology. Their existence, I believe, must soon shift our attention from the abstract issue of whether any such reduction is possible, to the concrete issue of which of various alternative neurobiological theories truly provides the right reduction, and to its long-term consequences for our overall self-conception.

The basic idea, to be explained as we proceed, is that the brain represents various aspects of reality by a *position* in a suitable *state space*; and the brain performs computations on such representations by means of general *co-ordinate transformations* from one state space to another. These notions may seem arcane and forbidding, but the graphics below will demystify them very swiftly. The theory is entirely accessible — indeed, in its simplest form it is *visually* accessible — even to the non-mathematical reader.

I was initially introduced to this theoretical approach by reading the provocative papers of the neuroscientists, Andras Pellionisz and Rodolfo Llinas (1979, 1982, 1984, 1985). Their presentations are much more general and more penetrating than the sketch to be provided here. But for didactic reasons, discussion of their ground-breaking work will be postponed until the later sections of the paper. At the outset, I wish to keep things as simple as possible.

We open our discussion, then, by confronting three apparently distinct puzzles:

1. the mystery of how the brain *represents* the world, and how it performs *computations* on those representations,
2. the mystery of *sensorimotor co-ordination,* and
3. the mystery of the brain's *microphysical organization.*

It is especially encouraging that these problems appear to admit of a simultaneous solution. Let us begin by addressing the third mystery, the microstructural mystery.

2. LAMINAR CORTEX, VERTICAL CONNECTIONS, AND TOPOGRAPHIC MAPS

The outer surface of the brain's great cerebral hemispheres consists of a thin layer, the classical 'grey matter', in which most of their neuronal cell bodies are located (see Fig. 1a). The remaining 'white matter' consists primarily of long axons projecting from the cells in this layer to other parts of the brain. If one examines the internal structure of this wrinkled layer, one finds that it subdivides into further layers (see Fig. 1b). Human cortex has six of these layers. Other creatures display a different number, but the laminar pattern is standard.

These further layers are distinguished by the type and concentration of cells within each sublayer, and by the massive intralayer or 'horizontal' projections within each sublayer. Moreover, these distinct layers are further distinguished by their proprietary inputs or outputs. The top several layers tend to have only inputs of one kind or another, from the sensory periphery, from other parts of the cortex, or from other parts of the brain. And the bottom layer seems invariably to be an output layer.

Finally, these distinct layers are systematically connected, in the fashion of nails struck through plywood, by large numbers of vertically oriented cells that permit communication between the several layers. These vertical cells conduct neuronal activity 'downwards', from the superficial input layers above to the output layer below.

If we now leave our microscopic edgewise perspective and look at the

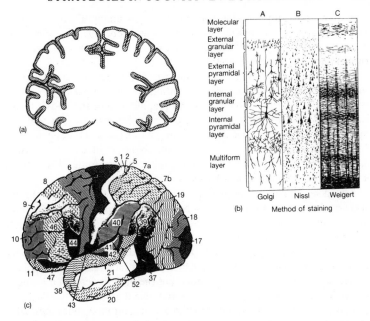

FIG. 1. (a) Cross-section of the cerebral hemispheres, showing the outer grey layer: the cerebral cortex. (b) The internal laminar structure of the cortical layer, as revealed by three different stains. (c) Brodmann's areas.

cortical sheet from the outside, we find that the cortical surface divides into a patchwork of smaller regions (see Fig. 1c). These areas are distinguished to some degree by differences in their laminar cytoarchitecture. An initial taxonomy, into what are called *Brodmann's areas* after their discoverer, is based on this criterion. These areas, or subareas of them, are of further interest because several of them plainly constitute *topographic maps* of some aspect of the sensory or motor periphery, or of some other area of the brain. For example, the neighbourhood relations holding between the cells in a given layer of the visual cortex at the rear of the brain correspond to the neighbourhood relations holding between the cells in the retina from which they receive inputs. The bundle of axonal projections from the retinal cells to the cortical cells preserves the topographic organization of the retinal cells. The surface of the primary visual cortex thus constitutes a topographic map of the retinal surface.

It is termed a 'topographic map', rather than simply a 'map', because the *distance* relations among retinal cells are generally *not* preserved. Typically, such maps are metrically deformed, as if they were made of rubber and then stretched in some fashion.

Many such maps have been identified. The so-called 'visual cortex' (areas 17, 18) has already been mentioned. The upper layer of the somatosensory cortex (area 3) is a topographic map of the body's tactile surface. The lower layer of the motor cortex (area 4) is a topographic map of the body's muscle system. The auditory cortex (areas 41, 42) contains a topographic map of frequency space. And there are many other cortical areas, less well understood as to exactly what they map, but whose topographical representation of distant structure is plain.

This general pattern of neural organization is not confined to the surface of the great cerebral hemispheres. As well, various nuclei of 'grey matter' in the more central regions of the brain—for example, the superior colliculus, the hippocampus, and the lateral geniculate nucleus—display this same multilayered, topographically organized, vertically connected structure. Not *everything* does (the cerebellum, for example, is rather different, of which more later), but the pattern described is one of the major organizational patterns to be found in the brain.

Why *this* pattern? What is its functional or cognitive significance? What do these structures do, and how do they do it? We can approach a possible answer to these questions by addressing the second mystery: the problem of sensorimotor co-ordination.

3. SENSORIMOTOR CO-ORDINATION

Let me begin by suggesting that vertically connected laminar structures are one of evolution's simplest solutions to a crucial type of problem, one that any sensorimotor system beyond the most rudimentary must somehow solve. In order to appreciate this type of problem, let us consider a schematic creature of a deliberately contrived simplicity.

Figure 2b is a plan view of a crablike schematic creature (2a) with two rotatable eyes and an extendable arm. If this equipment is to be useful to the crab, the crab must embody some functional relationship between its eye-angle pairs when an edible object is triangulated, and its subsequent shoulder and elbow angles, so that the arm can assume a position that makes contact with the edible target. Crudely, it must be able to grasp what it sees, wherever the seen object lies.

We can characterize the required arm/eye relationship as follows. First of all, let us represent the input (the pair of eye angles) by a point in a 2-dimensional sensory-system co-ordinate space or *state space* (Fig. 3a). The output (the pair of arm angles) can also be represented by an appropriate point in a separate 2-dimensional *motor* state space (Fig. 3b).

We now need a function to take us from any point in the sensory state

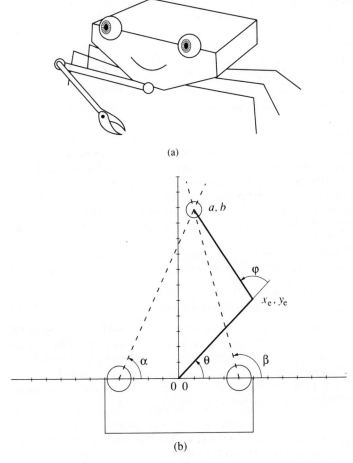

(a)

(b)

FIG. 2.

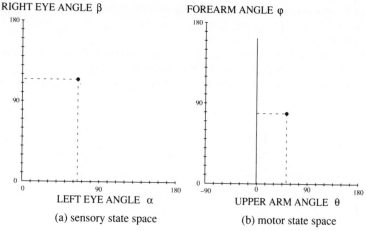

RIGHT EYE ANGLE β   FOREARM ANGLE φ

(a) sensory state space   (b) motor state space

LEFT EYE ANGLE α   UPPER ARM ANGLE θ

FIG. 3.

space to a suitable point in the motor state space, a function that will co-ordinate arm-position with eye-position in the manner described. (I here sketch the deduction of the relevant function so that its origin will not be a mystery, but the reader may leap past the algebra without any loss of comprehension. The only point to remember is that we are deducing a suitable function to take us from eye configurations to arm configurations.)

The two eye angles $\{\alpha, \beta\}$ determine two lines that intersect at the seen object. The co-ordinates (a, b) of that point (in *real* space) are given by

$$a = -4(\tan\alpha + \tan\beta)/(\tan\alpha - \tan\beta)$$
$$b = -8(\tan\alpha \cdot \tan\beta)/(\tan\alpha - \tan\beta)$$

The tip of the arm must make contact with this point. Assuming that both the forearm and the upper arm have a fixed length of 7 units, the elbow will therefore have to lie at the intersection of two circles of radius 7 units: one centred at (a, b), and the other centred at $(0, 0)$, where the upper arm projects from the crab's body. Solving for the relevant intersection, the real-space elbow co-ordinates (x_e, y_e) are given by

$$x_e = ((49 - ((a^2 + b^2)^2/4b^2) \cdot (1 - ((a^2/b^2)/((a^2/b^2) + 1))))^{1/2} + (((a/b) \cdot ((a^2 + b^2)/2b))/((a^2/b^2) + 1)^{1/2}))/((a^2/b^2) + 1)^{1/2}$$

$$y_e = (49 - x_e^2)^{1/2}$$

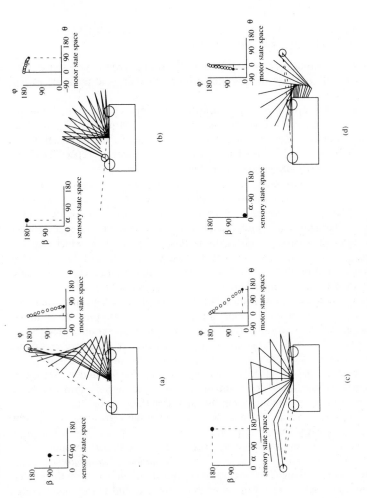

FIG. 4. The crab's arm in action

The three points in real space (a, b), (x_e, y_e), $(0, 0)$, determine the position of the arm, whose upper arm and forearm angles $\{\theta, \varphi\}$ are finally given by

$$\theta = \tan^{-1}(y_e/x_e)$$
$$\varphi = 180 - (\theta - \tan^{-1}((b - y_e)/(a - x_e)))$$

These are the desired co-ordinates of the arm in motor state space. The reader will note that the assembled functions that yield them are rather tangled ones.

Tangled or not, if the crab is drawn on a computer screen, such that its final arm position (drawn by the computer as output) is the specified function of its eye positions (entered by us as input), then it constitutes a very effective and well-behaved sensorimotor system, especially if we write the controlling program as follows.

Let the program hold the crab's arm folded against its chest (at $\theta = 0°$, $\varphi = 180°$), until some suitable stimulus registers on the fovea of both eyes. The arm is then moved from its initial state-space position (0°, 180°), along a straight line in motor state space, to its computed target position in motor state space. This is the state-space position at which, in *real* space, the tip of the arm contacts the triangulation point of the eyes. This arrangement produces a modestly realistic system that reaches unerringly for whatever it sees, anywhere within reach of its arm (Fig. 4a–c).

The algebraic representation of the crab's sensorimotor transformation, as represented in the six equations listed earlier, supplies no intuitive conception of its overall nature. A *geometrical* presentation is much more revealing. Let us therefore consider the projection of the active portion of the crab's sensory state space (Fig. 5a) into the orthogonal grid of its motor state space (Fig. 5b), as imposed by the function under discussion. That is to say, for *every* point in the displayed sensory grid, we have plotted the corresponding arm-position within the motor grid.

Here we can see at a glance the distortion of the vertical and horizontal lines of sensory space, as projected into motor space. The topological features of the sensory space are preserved, but its metrical properties are not. What we see is a systematic *transformation of co-ordinates*. (The heavy scored triangle and rectangle are drawn in solely to help the reader locate corresponding positions in the deformed and undeformed grids. Note also that the left border or β-axis of Fig. 5a shrinks to the left radial point in Fig. 5b, and that the top border of Fig. 5a shrinks to the right radial point in Fig. 5b.)

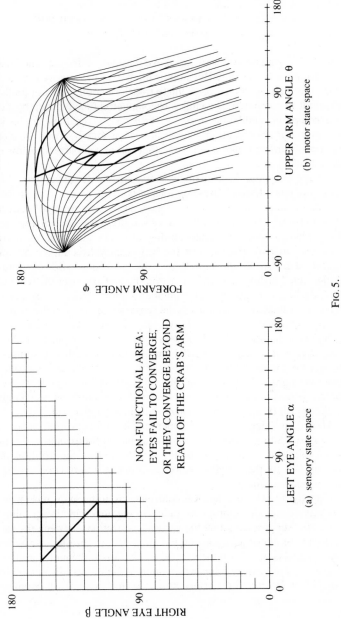

projection of sensory grid onto motor state space

UPPER ARM ANGLE θ

(b) motor state space

FOREARM ANGLE φ

NON-FUNCTIONAL AREA:
EYES FAIL TO CONVERGE,
OR THEY CONVERGE BEYOND
REACH OF THE CRAB'S ARM

LEFT EYE ANGLE α

(a) sensory state space

RIGHT EYE ANGLE β

Fig. 5.

4. CO-ORDINATE TRANSFORMATION: ITS PHYSICAL
IMPLEMENTATION

The transformation described above sustains effective and realistic sensorimotor behaviour. But how could a real nervous system possibly compute such a complex co-ordinate transformation? It is not realistic to expect it to compute this complex trigonometric function step by step, as our computer simulation does. Nevertheless, given their sophisticated sensorimotor co-ordination, biological systems somehow *must* be computing transformations like these, and others more complex still. How might they do it?

Figure 5 suggests a surprisingly simple means. If we suppose that the crab contains internal representations of both its sensory state space, and its motor state space, then the following arrangement will effect the desired transformation. Let the crab's sensory state space be represented by a physical grid of signal-carrying fibres, a grid that is metrically deformed in real space in just the way displayed in Figure 5b. Let its motor state space be represented by a second grid of fibres, in undeformed orthogonal array. Position the first grid over the second, and let them be connected by a large number of short vertical fibres, extending from co-ordinate intersections in the sensory grid down to the nearest co-ordinate intersection in the underlying motor grid, as in Figure 6.

Suppose that the fibres of the sensory grid receive input from the eyes' proprioceptive system, such that the position of each eye stimulates a unique fibre in the upper (deformed) grid. The left eye activates one fibre from the right radial point, and the right eye activates one from the left. Joint eye position will thus be represented by a simultaneous stimulation at the appropriate co-ordinate *intersection* in the upper grid.

Underneath that point in the upper map lies a unique intersection in the motor grid. Suppose that this intersecting pair of orthogonal motor fibres, when jointly activated, induces the arm to assume the position that is *appropriate* to the specific motor co-ordinate intersection where this motor signal originates.

Happily, the relative metrical deformations in the maps have placed in correspondence the appropriate points in the upper and lower maps. We need now suppose only that the vertical connections between the sensory grid and the motor grid function as 'and-gates' or 'threshold switches', so that a signal is sent down the vertical connection to the motor grid exactly if the relevant sensory intersection point is simultaneously stimulated by both of its intersecting sensory fibres. Such a system will compute the

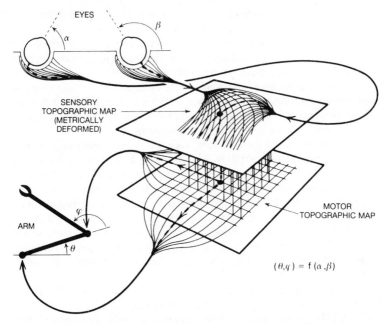

$$(\theta,\varphi) = f(\alpha,\beta)$$

FIG. 6. Co-ordinate transformation by contiguous topographic maps

desired co-ordinate transformations to a degree of accuracy limited only
by the grain of the two grids, and by the density of their vertical connections. I call such a system a *state-space sandwich*.

Three points are worth noting immediately about the functional properties of such an arrangement. Firstly, it will remain partially functional
despite localized damage. A small lesion in either grid will produce only
a partial dyskinesia (two permanent 'shadows' of fibre inactivity
downstream from the lesion), for which a shift of body position will
usually compensate (by bringing the target's state-space position out of
the shadow).

Indeed, we can do even better than this. Let the dimple on the back of
the crab's schematic eye (Fig. 6) be replaced by a bar, so as to stimulate
not one, but rather a set of adjacent proprioceptive cells. The position of
each eye will then be represented in the upper map by the activation of a
band of fibres centred on the 'correct' fibre. *Joint* eye position will then be
registered in the upper layer by a distributed *area* of stimulation, an area
centred on the 'correct' point. This will cause a corresponding area of

stimulation in the bottom grid, and thus a band of stimulation will be sent to each muscle. If the muscles are connected so as to assume a position appropriate to the 'mean' fibre within that distributed signal, even if that specific fibre happens to be inactive, then an appropriate motor response will be forthcoming even if the sandwich has suffered the scattered loss of a great many cells. Such a system will be functionally persistent despite widespread cell damage. The quality of the sensorimotor co-ordination would be somewhat degraded under cell damage, but a roughly appropriate motor response would still be forthcoming.

Seco; ᵈly, this system will be very, *very* fast, even with fibres of biological conduction velocities ($10 < v < 100$ m/s). In a creature the size of a crab, in which the total conduction path is less than 10 cm, this system will yield a motor response in well under 10 milliseconds. In the crab-simulation described earlier, my microcomputer (doing its trigonometry within the software) takes 20 times that interval to produce a motor response on-screen, and its conduction velocities are on the order of the speed of light. Evidently, the massively parallel architecture of the state-space sandwich buys it a large advantage in speed, even with vastly slower components.

Thirdly, the quality of the crab's co-ordination will not be uniform over its field of motor activity, since in the maximally deformed areas of the sensory grid, small errors in sensory registration produce large errors in the motor response (see again Fig. 5b). Accordingly, the crab is least well co-ordinated in the area close between its eyes, and to its extreme right and left.

All three of these functional properties are biologically realistic. And the sandwich appears biologically realistic in one further respect: it is relatively easy to imagine such a system being *grown*. Distinct layers can confine distinct chemical gradients, and can thus guide distinct morphogenetic processes. Accordingly, distinct topographical maps can appear in closely adjacent layers. But given that the maps are so closely contiguous, and assuming that they are appropriately deformed, the problem of connecting them up so as to produce a functional system becomes a trivial one: the solution is just to grow conductive elements that are roughly orthogonal to the layers.

Different creatures will have different means of locating objects, and different motor systems to effect contact with them, but all of them will face the same problem of co-ordinating positions in their sensory state space with positions in their motor state space, and the style of solution here outlined is evidently quite general in nature. In fact, the co-ordination of distinct biological subsystems by co-ordinate transformation is a matter that presumably extends far beyond the obvious case of basic sensorimotor co-ordination, and the same strategy may also be useful, or even essential, in the execution of higher cognitive activities, as we shall see later on. The point to be emphasized here is that a state-space sandwich constitutes a simple and biologically realistic means for effecting *any* 2D-to-2D co-ordination transformation, whatever its mathematical complexity, and whatever features—external or internal, abstract or concrete—that the co-ordinate axes may represent to the brain. If the transformation can be graphed at all, a sandwich can compute it. The sensorimotor problem solved above is merely a transparent example of the general technique at work.

Beyond its functional realism, the system of interconnected maps in Figure 6 is suggestively similar to the known *physical* structure of typical laminar cortex, including the many topographic maps distributed across the cerebral surface. In all of these areas, inputs address a given layer of cells, which frequently embodies a metrically deformed topographic map of something or other. And outputs leave the area from a different layer, with which the first layer enjoys massive vertical connections.

I therefore propose the hypothesis that the scattered maps within the cerebral cortex, and many sub-cerebral laminar structures as well, are all engaged in the co-ordinate transformation of points in one neural state space into points in another, by the direct interaction of metrically deformed, vertically connected topographic maps. Their mode of representation is state-space position; their mode of computation is co-ordinate transformation; and both functions are simultaneously implemented in a state-space sandwich.

[(Added in 1989.) The second and especially the third part of this hypothesis now seem almost certainly mistaken, at least as an account of the cerebral cortex. The cell population of a given layer in a given cortical area is indeed coding state-space positions, but by means of the overall *pattern* of activation-levels across the entire population of cells, rather than by the narrow spatial location of maximal cell activation. And the axonal projections from that layer into adjacent cell layers do indeed effect a transformation from one state-space to another, but by means of the 'matrix-multiplication' style of transformation outlined in §6 below, rather than by transferring an activational hot-spot between mutually deformed maps. The topographic mappings so characteristic of many cortical areas now appear as the occasional artefacts of a deeper coding strategy: the high-dimensional vector coding explained in §6. The discussion of this section remains instructive, however, as a possible account of the superior colliculus, and as an introduction to the relevant computational ideas.]

I can cite not a single cerebral area for which this functional hypothesis is known to be true. The decoding of cortical maps is a business that has only begun. Clear successes can be numbered on the fingers of at most three or four hands, and they are generally confined to the superficial cortical layers. There is a major subcortical area, however, whose upper-level and lower-level maps have been at least partially decoded, and which does display the general pattern depicted in Figure 6.

The *superior colliculus* is a phylogenetically very old laminar structure (Fig. 7a) located on the dorsal mid-brain. Among other things, it sustains the familiar reflex whereby the eye makes an involuntary saccade so as to foveate (= look directly at) any sudden change or movement that registers on the retina away from the foveal centre. We have all had the experience of being in a darkened movie theatre when someone down in the front row left suddenly ignites a match or lighter to light a cigarette. Every eye in the house makes a ballistic saccade to fixate this brief stimulus, before returning to the screen. This is the colliculus at work. Appropriately enough, this is sometimes called the 'visual grasp reflex'.

In humans and the higher mammals the superior colliculus is a visual centre secondary to the more important striate cortex (areas 17 and 18 on the Brodmann map) located at the rear of the cerebral hemispheres, but in lower animals such as the frog or snake, which lack any significant cortex, the superior colliculus (or *optic tectum*, as it is called in them) is their principal visual centre. It is an important centre even for mammals, however, and it works roughly as follows.

The topmost layer of the superior colliculus (hereafter, SC) receives projections directly from the retina, and it constitutes a metrically deformed topographic map of the retinal surface (Fig. 7b) (Schiller 1984; Goldberg and Robinson 1978; Cynader and Berman 1972; Gordon 1973). Vertical elements connect this layer to the deepest layer of the SC. These vertical connections appear to consist of a chain of two or three short interneurones descending stepwise through two intervening layers (Schiller 1984: 460, 466), of which more later. Also, the dendrites of some of the deep-layer neurones appear to ascend directly into the visual layer, to make synaptic connections with visual cells (Mooney *et al.* 1984: 185). The neurones of the deepest layer project their output axons via two distinct nervous pathways, one of which leads eventually to the pair of extra-ocular muscles responsible for vertical eye movements, and the

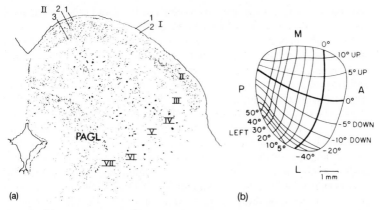

FIG. 7. (a) Projection drawing of Nissl-stained cross-section of cat superior colliculus illustrating laminar organization. Dots correspond to collicular neurones. (From Kanaseki and Sprague (1974); reprinted with permission) (b) Retinotopic map: a metrically deformed topographic map of the visual hemifield, in rectangular co-ordinates, on the superficial layer of the right colliculus of the cat. M = medial; L = lateral; A = anterior; P = posterior. (Adapted from Schiller 1984)

other to the pair responsible for horizontal eye movements (Huerta and Harting 1984: 287).

Intriguingly, this underlying motor layer also embodies a topographic map, a map of a state space that represents changes in the contractile position of the ocular muscles (Robinson 1972: 1800). Microstimulation by an electrode at a given point in this deepest layer causes the eye to execute a saccade of a characteristic size and direction, a saccade which moves the fovea into the position formerly occupied by that retinal cell which projects to *the immediately overlying visual cell in the topmost layer of the colliculus* (Robinson 1972; Schiller and Stryker 1972). In other words, the relative metrical deformations in the two maps have placed into correspondence the appropriate points in the upper and lower maps. (This means that the 'deformation' seen in Fig. 7b should not of itself be taken as evidence for the state-space sandwich hypothesis. What is crucial is the deformation of the maps relative to each other.)

Finally, any sufficiently strong retinally produced stimulation in that topmost visual map is conveyed downwards to the motor map by the appropriate vertical elements, where it produces a saccade of just the size and direction appropriate for the foveation of the external stimulus that provoked it. The SC thus appears to be an instance of both the structural and the functional pattern displayed in Figure 6.

One might expect a biological sandwich to code the position of retinal stimulations with an *area* of stimulation in the upper map, rather than a single point, so as to be functionally persistent in the face of small lesions and scattered cell death, as explained earlier. Activity in the SC does display this pattern (McIlwain 1975; 1984; 268). The schematic model of Figure 6 also predicts that the size and direction of the motor response induced by microstimulation at various points within the collicular sandwich will be a function solely of where in either map the stimulation occurs, and not of the magnitude of the stimulation, nor of its vertical position between the two maps. Experimentation has already yielded this result (Robinson 1972; Schiller and Stryker 1972). The SC, it appears, is a real sensorimotor co-ordinate transformer of roughly the kind at issue. It foveates on changing or moving visual targets by essentially the same means whereby the schematic cortex of the crab reaches out for triangulated objects.

A word of caution is in order here, since the account just offered does not do justice for the full complexity of the SC. In mammals, especially the higher mammals, the SC is a tightly integrated part of a larger modulating system that includes inputs from the visual cortex and the frontal eye fields, and outputs to the neck muscles. The functional properties of the entire system are more varied and more subtle than the preceding outline suggests, and the job of sorting them out is still underway (Mays and Sparks 1980; Schiller and Sandell 1983). The above discussion is submitted as an account of the central or more primitive functions of the SC, at best.

With these examples in mind—the crab's 'cortex', and the superior colliculus—it is appropriate to focus on the many other topographically organized multilayered cortical areas scattered throughout the brain, and ask what co-ordinate transformations *they* might be effecting. Here it is very important to appreciate that the topographic maps we seek to decode need not be, and generally will not be, maps of something anatomically obvious, such as the surface of the retina, or the surface of the skin. More often they will be maps of some *abstract state space*, whose dimensional significance is likely to be opaque to the casual observer, though of great functional importance to the brain. Two nice examples of such abstract maps are the map of *echo delays* in the bat's auditory cortex, and the map of *binaural disparities* in the owl's inferior colliculus (Konishi 1986).

In the case of the crab's schematic cortex, the angular state of an external system (the eyes) is directly mapped onto the angular state of another external system (the jointed arm). But in a creature of any complexity, we can expect a long chain or hierarchy of internal systems interacting with one another, systems that are maps of the output of other internal systems, and whose output drives the activities of further internal systems. To understand such maps will require that we understand the function of the other maps in the overall system.

All of this suggests that the brain may boast many more topographic maps than have so far been identified, or even suspected. Certainly the brain has a teeming abundance of topographically organized areas, and recent work has expanded the number of known sensory-related maps considerably (Merzenich and Kaas 1980; Allman *et al.* 1982). All of this further suggests that we will make better progress in trying to understand the significance of the many topographically organized cortical areas when we approach them as maps of abstract but functionally relevant state spaces.

There has been a tendency among neuroscientists to restrict the term 'topographic map' to neural areas that mirror some straightforward aspect of the physical world or sensory system, such as the retina, or the surface of the skin. This is unfortunate, since there is no reason for the brain to show any such preference in what it constructs maps of. Abstract state spaces are just as mappable as concrete physical ones, and the brain surely has no advance knowledge of which is which. We should expect it, rather, to evolve maps of what is functionally significant, and that will frequently be an abstract state space.

5. CORTEX WITH MORE THAN TWO LAYERS

While we are discussing the biological reality of the laminar mechanism proposed, consider the objection that our model cortex has only two layers, whereas typical human cortex has six layers, and, counting fine subdivisions, perhaps eight or nine in some areas. What are they for?

There is no difficulty in perceiving a function for such additional layers. Let us return again to the superior colliculus, which illustrates one of many possibilities here. Between the visual and motor maps of the SC there are, in some creatures, one or two intermediate layers (see again Fig. 7). These appear to constitute an auditory map and/or a somatosensory map (a facial or whisker map), whose function is again to orient the eye's fovea, this time towards the source of sudden *auditory* and/or *somatosensory* stimulation (Goldberg and Robinson 1978). Not surprisingly, these intervening maps are each metrically deformed in such a fashion as to be in rough co-ordinate 'register' with the motor map, and hence with each other. Altogether, this elegant three- or four-layer topographic sandwich constitutes a *multimodal* sensorimotor co-ordinate transformer.

Multilayered structures have further virtues. It is plain that maps of several distinct modalities, suitably deformed and placed in collective register within a 'club sandwich', provide a most effective means of cross-modal integration and comparison. In the SC, for example, this multimodal arrangement is appropriate to the production of a motor response to the *joint* receipt of faint but spatiotemporally coincident auditory and visual stimuli, stimuli which, in isolation, would have been *sub*threshold for a motor response. For example, a faint sound from a certain compass point may be too faint to prompt the eyes into the foveating saccade, and a tiny movement from a certain compass point may be similarly impotent; but if both the sound *and* the movement come from the same compass point (and are thus coded in the SC along the same vertical axis), then their simultaneous conjunction will indeed be sufficient to make the motor layer direct the eyes appropriately. This prediction is strongly corroborated by the recent results of Meredith and Stein (1985).

Further exploration reveals that multilayered sandwiches can subserve decidedly sophisticated cognitive functions. In an earlier publication on these matters (Churchland 1986), I have shown how a *three*-layer state-space sandwich can code, and project, the path of a moving object in such a fashion as to position the crab's arm to catch the moving target on the fly. Evidently, a multilayered cortex can offer considerable advantages.

6. BEYOND STATE-SPACE SANDWICHES

The examples studied above are uniform in having an input state space of only two dimensions, and an ouput state space of only two dimensions. It is because of this fact that the required co-ordinate transformation can be achieved by a contiguous pair of sheetlike maps. But what of cases where the subsystems involved each have more than two parameters? What of cases where the co-ordinate transformations are from an input space of n to an output space of m dimensions, where n and m are different, and both greater than 2? Consider, for example, the problem of co-ordinating the joint angles of a limb with three or more joints, and the problem of co-ordinating several such limbs with each other. Or consider the problem of co-ordinating the even larger number of muscles that collectively control such limbs. As soon as one examines the problems routinely faced, and solved, by real creatures, one appreciates that many of them are far more complex than can be represented by a simple 2D-to-2D transformation.

Perhaps some of these more complex problems might be solved by dividing them into a set of smaller ones, problems that can be managed after all by a set of distinct 2-D state-space sandwiches, each addressing some slice or aspect of the larger problem (for some specific suggestions in this vein, see Ballard 1986). The predominance of laminar cortex in the brain certainly encourages speculation along these lines. But such solutions, even approximate ones, cannot in general be guaranteed. The brain badly needs some mechanism beyond the state-space sandwich if it is routinely to handle these higher dimensional problems.

Pellionisz and Llinas have already outlined a mechanism adequate to the task, and have found impressive evidence of its implementation within the cerebellum. The cerebellum is the large structure at the rear of the brain, just underneath the cerebral hemispheres. Its principal function, divined initially from lesion studies, is the co-ordination of complex body movements, such as would be displayed in preparing a dinner or in playing basketball. It displays a neural organization quite different from that of the cerebral hemispheres, an organization whose significance may be rendered transparent by the Pellionisz–Llinas account.

To illustrate this more general mechanism for co-ordinate transformation, let us consider an input system of four dimensions whose inputs a, b, c, d, are transformed into the values x, y, z, of a 3-dimensional output system. As before, the inputs and outputs can each be regarded as points in a suitable state space. Since they are n-tuples, each can also be regarded as a vector (whose base lies at the origin of the relevant state space, and whose arrowhead lies at the point specified by the n-tuple).

A standard mathematical operation for the systematic transformation of vectors into vectors is matrix multiplication. Here it is the *matrix* that embodies or effects the desired co-ordinate transformation. To see how this works, consider the matrix of Figure 8, which has four rows and three columns. To multiply the input vector $\{\ a,\ b,\ c,\ d\ \}$ by this matrix we multiply a times p_1, b times p_2, c times p_3, d times p_4, and then sum the four results to yield x. We then repeat the process with the second column to yield y, and again with the third column to yield z. Thus results the output vector $\{\ x, y, z\ \}$.

This algebraic operation can be physically realized quite simply by the neural array of Figure 9. The parallel input fibres at the right each send a train of electrochemical 'spikes' towards the waiting dendritic trees. The numbers a, b, c, d represent the amount by which the momentary spiking frequency of each of the four fibres is above (positive number) or below

$$\{a, b, c, d\} \begin{bmatrix} p_1 & q_1 & r_1 \\ p_2 & q_2 & r_2 \\ p_3 & q_3 & r_3 \\ p_4 & q_4 & r_4 \end{bmatrix}$$

$$= \quad \{x, \quad y, \quad z\}$$

FIG. 8.

(negative number) a certain baseline spiking frequency. The topmost input fibre, for example, synapses onto each of the three output cells, making a stimulatory connection in each case, one that tends to depolarize the cell body and make it send a spike down its vertical output axon. The output frequency of spike emissions for each cell is determined by (1) the simple *frequency* of input stimulations it receives from all incoming synaptic connections, and (2) the *weight* or *strength* of each synaptic connection,

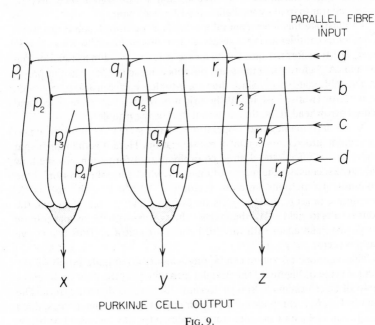

PURKINJE CELL OUTPUT

FIG. 9.

which is determined by the placement of the synapses and by their cross-sectional areas. These strength values are individually represented by the coefficients of the matrix of Figure 8. *The neural interconnectivity thus implements the matrix.* Each of the three cells of Figure 9 'sums' the stimulation it receives, and emits an appropriate train of spikes down its output axon. Those three output frequencies differ from the background or baseline frequencies of the three output cells by positive or negative amounts, and these amounts correspond to the output vector $\{ x, y, z \}$.

Note that with state-space sandwiches, the coding of information is a matter of the spatial location of neural events. By contrast, with the matrix-multiplication style of computation under discussion, input and output variables are coded by sets of spiking frequencies in the relevant pathways. The former system uses 'spatial coding'; the latter system uses 'frequency coding'. But both systems are engaged in the co-ordinate transformation of state-space positions.

The example of Figure 9 concerns a three-by-four matrix. But it is evident that neither the mathematical operation nor its physical realization suffers any dimensional limitations. In principle, a Pellionisz–Llinas connectivity matrix can effect transformations on state spaces of a dimensionality into the thousands and beyond.

The schematic architecture of Figure 9 corresponds very closely to the style of micro-organization found in the cerebellum (Fig. 10). (For an accessible summary of cerebellar architecture, see Llinas 1975). The horizontal fibres are there called *parallel fibres*, and they input from the higher motor centres. The bushy vertical cells are there called *Purkinje cells*, and they output through the cerebellar nucleus to the motor periphery. In fact, it was from the observation of the cerebellum's beautifully regular architecture, and from the attempt to re-create its functional properties by modelling its large-scale *physical* connectivity within a computer, that Pellionisz and Llinas were originally led to the view that the cerebellum's job is the systematic transformation of vectors in one neural hyperspace into vectors in another neural hyperspace (Pellionisz and Llinas 1979).

Given that view of the problem, the tensor calculus emerges as the natural framework with which to address such matters, especially since we cannot expect the brain to limit itself to Cartesian co-ordinates. In the examples discussed so far, variation in position along any axis of the relevant state space is independent of variation along any of the other axes, but this independence will not characterize state spaces with non-orthogonal axes. Indeed, this generalization of the approach, to include non-Cartesian hyperspaces, is regarded by Pellionisz and Llinas as one of the

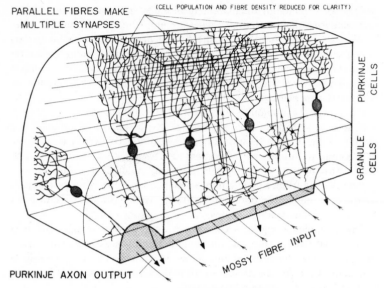

PARALLEL FIBRES MAKE
MULTIPLE SYNAPSES
(CELL POPULATION AND FIBRE DENSITY REDUCED FOR CLARITY)

PURKINJE CELLS

GRANULE CELLS

MOSSY FIBRE INPUT

PURKINJE AXON OUTPUT

FIG. 10. Schematic section: cerebellum

most important features of their account, a feature that is essential to understanding all but the simplest co-ordination problems. I cannot pursue this feature here.

Three final points about the neural matrix of Figure 9. Firstly, it need not be limited to computing linear transformations. The individual synaptic connections might represent any of a broad range of functional properties. They need not be simple multipliers. In concert then, they are capable of computing a large variety of non-linear transformations. Secondly, a neural matrix will have the same extraordinary speed displayed by a state-space sandwich. And thirdly, given large matrices and/or cell redundancy, such structures will also display a functional persistence despite the scattered loss of their cellular components.

These brief remarks do not do justice to the very extensive work of Pellionisz and Llinas, nor have I explored any criticisms. (For the latter, see Arbib and Amari 1985. For a reply, see Pellionisz and Llinas 1985.) The reader must turn to the literature for deeper instruction. The principal lesson of this section is that the general functional schema being advanced in this paper—the schema of representation by state-space position, and computation by co-ordinate transformation—does not encounter implementational difficulties when the representational and computational task exceeds the case of two dimensions. On the contrary, the brain boasts neural machinery that is ideally suited to cases of very high dimensionality. We have then at least two known brain mechanisms for performing co-ordinate transformations: the state-space sandwich specifically for 2-dimensional cases, and the neural matrix for cases of any dimensionality whatever.

7. THE REPRESENTATIONAL POWER OF STATE SPACES

Discussion so far has been concentrated on the impressive *computational* power of co-ordinate transformations of state spaces, and on the possible neural implementation of such activity. But it is important to appreciate fully the equally powerful *representational* capacity of neural state spaces. The global state of a complex system of n distinct variables can be economically represented by a single point in an abstract n-dimensional state space. And such a state-space point can be neurally implemented, in the simplest case, by a specific distribution of n spiking frequencies in a system of only n distinct fibres. Moreover, a state-space representation embodies the *metrical* relations between distinct possible positions within it, and thus embodies the representation of *similarity* relations between distinct items thus represented. Five examples will illustrate these claims, all of which may be real, and three of which pose problems familiar to philosophers.

The qualitative character of our sensations is commonly held to pose an especially intractable problem for any neurobiological reduction of mental states (see Nagel 1974; Jackson 1982; Robinson 1982). And it is indeed hard to see much room for reductive purchase in the subjectively discriminable but 'objectively uncharacterizable' qualia present to consciousness.

Even so, a determined attempt to find order rather than mystery in this area uncovers a significant amount of expressible information. For example, we will all agree that the 'colour' qualia of our visual sensations arrange themselves on a continuum. Within this continuum of properties there are similarity relations (orange is similar to red), relative similarity relations (orange is more similar to red than to purple), and betweenness relations (orange is between red and yellow). There are also an indefinite number of distinct 'paths' through continuously similar colours that will take us from any given colour to a different colour.

To this we can add that people who suffer one or another of the various types of colour blindness appear to embody a significantly *reduced* continuum of colour qualia, one reduced in at least partially specifiable ways (it fails to display a contrast between red and green, or between blue and yellow, etc.). This question of the relative variety of qualia displayed within a given modality raises the point that, across the familiar five modalities, there is noteworthy variation. For example, though the variety of discriminable colour sensations is large, the variety of discriminable taste sensations is even larger, and the variety of discriminable smell sensations is larger still. Such cross-modal variation reminds us further of the presumed variation across species, as instanced in the canine's extraordinary ability to discriminate, by smell alone, any one of the 3.5 billion people on the planet. One presumes that the canine's continuum of olfactory sensations is somehow much 'larger' than a human's, in the sense of containing a greater variety of discriminable types of sensation.

Here, then, are some humdrum facts about the manifold(s) of subjective sensory qualia, facts which a reductive account of mind might attempt to explain. It must do this by reconstructing these facts, in some revealingly systematic way, in neurobiological terms. (For a general account of the nature of cross-theoretic identities and intertheoretic reductions, see Churchland 1985, 1979.) This possibility will now be explored. For several of the relevant modalities, physiological and cognitive psychologists have already sketched the outlines of such an account, and state-space representations play a prominent role in all of them.

Consider first the abstract 3-dimensional 'colour cube' proposed by Edwin Land (Fig. 11), within which every one of the many hundreds of humanly discriminable colours occupies a unique position or small volume (Land 1977). Each axis represents the eye/brain's construction of the *objective* reflectance of the seen object at one of the three wavelengths to which our cones are selectively responsive. Two colours are closely similar just in case their state-space positions within this cube are close to one another. And two colours are dissimilar just in case their state-space positions are distant. We can even speak of the degree of the similarity, and of the dimensions along which it is reckoned (see also Zeki 1983).

If the human brain does possess an internal implementation of such a state space, it has purchased a great deal of representational power at a very low price. For example, if our native discrimination along each axis of Land's colour state space is only 10 distinct positions, then a ternary system should be able to represent fully 10^3 distinct colours. If anything, this underestimates our capacities, so the assumption of 10-unit axial discrimination is likely too low. In any case, there is no trouble accounting for our broad discriminatory powers: one's discrimination within Land's

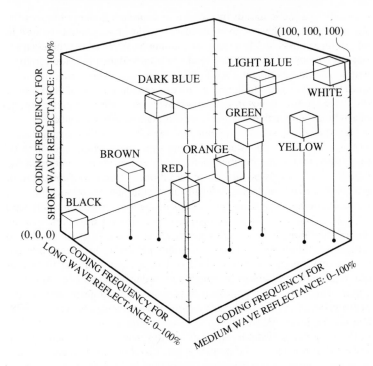

FIG. 11. Colour state space

state space explodes as the third power of one's discrimination along each axis. And certainly our peripheral machinery tends to bear out the general hypothesis. All colour perception arises from the inputs of exactly three kinds of retinal cones.

All of which suggests the hypothesis that a visual *sensation* of any specific colour is literally identical with a specific triplet of spiking frequencies in some triune brain system. If this is true, then the similarity of two colour sensations emerges as just the proximity of their respective state-space positions. Qualitative 'betweenness' falls out as state-space betweenness. And of course there are an indefinite number of continuous state-space paths connecting any two state-space points. Evidently we can reconceive the cube of Figure 11 as an internal 'qualia cube'. Just think of each axis as representing the instantaneous activity level or spiking frequency of one of the three internal pathways for reflectance information.

Finally, if genetic misadventure should deprive a human of one of the standard three pathways, then his or her qualia space should collapse to one of three possible 2-D spaces, depending on which of the three axes is rendered inoperative. A specific and predictable deficit in colour discrimination should accompany each loss. And thus it is. There are three principal types of colour blindness, each corresponding to the loss of one of the three types of cones in the retina. What we have here is the outline of a genuinely reductive account of one domain of sensory qualia.

Our gustatory system appears to exploit a similar arrangement, although here the dimensionality of the state space would appear to be four, since that is the number of distinct kinds of taste receptors in the mouth. Any humanly possible taste sensation, it is therefore conjectured, is a point somewhere within a 4-D gustatory state space. Or more literally, it is a quadruple of spiking frequencies in the four proprietary pathways carrying information from the gustatory receptors for distribution to the rest of the brain. If our discrimination along each axis is comparable to that within colour space (10 or more units per axis), this means that the variety of different taste sensations will be greater than the variety of different colour sensations by roughly an order of magnitude. And so it seems. This state-space approach to gustatory sensations appears in the neuroscience literature as the *across-fibre pattern theory* (Bartoshuk 1978; Smith 1983; Pfaff 1985).

This insight into gustatory space allows us to say something determinate about 'what it is like' to be a rat or a cat. Like humans, rats and cats are mammals, and they also possess a four-channelled gustatory system. One difference merits mention, however. One of the four pathways— sometimes labelled the 'bitter' pathway, since the 4-value code for bitterness requires a high level of activity in that pathway—shows a different sensitivity across the three species. In rats, this pathway shows a narrower range of evocable activity (is less discriminating) than it does in humans. In cats, it shows a wider range of activity (is more discriminating).

Consider now the slight contrast, in humans, between the taste of sugar (sucrose) and the taste of saccharin. Sugar is generally preferred because saccharin has a faintly bitter aspect to it. The preceding information about rats and cats suggests that in rats, this difference in respect of bitterness will be smaller than it is in humans, and that in cats the difference will be larger. Saccharin, that is, should taste rather more bitter to cats than it does to us. Or so the preceding would suggest. As it happens, the choice behaviour of rats does not discriminate between sugar and saccharin: they will eat either indiscriminantly. Cats, by contrast, will eat sugar, but reject saccharin (Bartoshuk 1978).

An account of this same general kind may hold for our olfactory system, which has six or more distinct types of receptor. A 6-D space has greater volume still, and will permit even greater feats of discrimination. A 6-D space, at 10-unit axial discrimination, will permit the discrimination of 10^6 odours. And if we imagine only a 7-D olfactory space, with only three times the human axial discrimination, which a dog almost certainly possesses, then we are contemplating a state space with 30^7, or 22 billion, discriminable positions! Given this, the canine's ability to distinguish, by smell, any one of the 3.5 billion people on the planet no longer presents itself as a mystery.

I have neither the space nor the understanding to discuss the complex case of auditory qualia, but here, too, a state-space approach is claimed to be illuminating (see Risset and Wessel 1982). Depending on the researchers and the modality involved, the state-space approach is variously called 'multivariate analysis', or 'multidimensional scaling', or 'across-fibre pattern coding', or 'vector coding', and so forth. But these are all alternative incarnations of the same thing: state-space representations.

Evidently this approach to understanding sensory qualia is both theoretically and empirically motivated, and it lends support to the reductive position advanced in an earlier paper (Churchland 1985) on the ontological status of sensory qualia. In particular, it suggests an effective means of expressing the allegedly inexpressible. The 'ineffable' pink of one's current visual sensation may be richly and precisely expressible as a '95 Hz/80 Hz/80 Hz chord' in the relevant triune cortical system. The 'unconveyable' taste sensation produced by the fabled Australian health tonic, Vegamite, might be quite poignantly conveyed as a '85/80/90/15 chord' in one's four-channelled gustatory system (a dark corner of taste-space that is best avoided). And the 'indescribable' olfactory sensation produced by a newly opened rose might be quite accurately described as a '95/35/10/80/60/55 chord' in some 6-D system within one's olfactory bulb.

This more penetrating conceptual framework might even displace the common-sense framework as the vehicle of intersubjective description and spontaneous introspection. Just as a musician can learn to recognize the constitution of heard musical chords, after internalizing the general theory of their internal structure, so may we learn to recognize, introspectively, the n-dimensional constitution of our subjective sensory qualia, after having internalized the general theory of *their* internal structure. This analogy has the further advantage of pre-empting a predictable response: that such a reconception of the 'internal world' would rob it of its beauty and peculiar identity. It would do so no more than reconceiving musical phenomena in terms of harmonic theory robs music of its beauty and peculiar identity. On the contrary, such reconception opens many aesthetic doors that would otherwise have remained closed.

The familiar 'ineffable qualia', I believe, are continuous with features that clearly do divide into components. Consider the human 'module' for facial recognition. We apparently have one, since the specific ability to recognize faces can be destroyed by specific right parietal lesions. Here it is plausible to suggest an internal state-space representation of perhaps twenty dimensions, each coding some salient facial feature such as nose length, facial width, etc. (Police 'Identi-kits' attempt to exploit such a system, with some success.) Even if discrimination along each axis were limited to only 5 distinct positions, such a high-dimensional space would still have an enormous volume (5^{20} positions), and it would permit the discrimination and recognition of billions of distinct faces. It would also embody similarity relations, so that close relatives could be successfully grouped, and so that the same person could be reidentified in photos taken at different ages. Consider two photos of the young and the old Einstein. What makes them similar? They occupy proximate positions in one's facial state space.

Finally, let us turn to a motor example, and let us consider one's 'body image': one's continuously updated sense of one's overall bodily configuration in space. That configuration is constituted by the simultaneous position and tension of several hundreds of muscles, and one monitors it all quite successfully, to judge from the smooth co-ordination of most of one's movements. How does one do it? With a high-dimensional state space, according to the theories of Pellionisz and Llinas, who ascribe to the cerebellum the job of computing appropriate transformations among high-dimensional codings of actual and intended motor circumstances, codings lodged in the input parallel fibres and the output Purkinje axons.

Some of the possibilities here can be evoked by a very simple example. Consider a highly complex and critically orchestrated periodic motion, such as occurs in feline locomotor activity (Fig. 12a). Consider now a three-dimensional joint-angle motor state space for the cat's hind limb, a space in which every possible configuration of that limb is represented by a point, and every possible movement is represented by a continuous path. The graceful step cycle of the galloping cat will be very economically represented by a closed loop in that joint-angle state space (Fig. 12b). If the relevant loop is specified or 'marked' in some way, then the awesome task of co-ordinated locomotion reduces to a simple tracking problem: make your motor state-space position follow the path of that loop.

Whether anything in the brain answers to this suggestion is moot. But exploration of this technique, with an ultimate aim of using portable

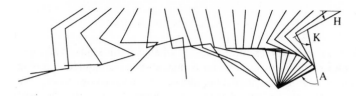

(a) STEP CYCLE: FELINE HIND LEG

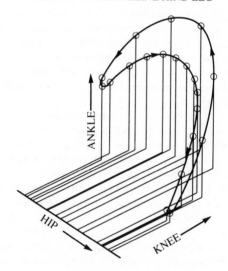

(b) SKELETAL STATE SPACE

FIG. 12.

microcomputers as an artificial means of generating effective locomotor activity in paraplegics, is under way in the CNS Lab of Larry Jordan at the University of Manitoba.

Once we have taken the step beyond the cognitive significance of points in 2-D state space, to the cognitive significance of *lines* and *closed loops* in *n*-dimensional state spaces, it seems possible that we will also find cognitive significance in *surfaces*, and *hypersurfaces*, and *intersections* of hypersurfaces, and so forth. What we have opening before us is a 'geometrical', as opposed to a narrowly syntactic, conception of cognitive activity.

8. CONCLUDING REMARKS

We have seen how a representational scheme of this kind can account, in a biologically realistic fashion, for a number of important features of motor control, sensory discrimination, and sensorimotor co-ordination. But has it the resources to account for the so-called higher cognitive activities, as represented by *language use*, for example, and by our propositional knowledge of the world in general?

Conceivably, yes. One might try to find, for example, a way of representing 'anglophone linguistic hyperspace' so that all grammatical sentences turn out to reside on a proprietary hypersurface within that hyperspace, with the logical relations between them reflected as spatial relations of some kind. I do not know how to do this, of course, but it holds out the possibility of an alternative to, or potential reduction of, the familiar Chomskyan picture.

As for the 'set of beliefs' that is commonly supposed to constitute a person's knowledge, it may be that a geometrical representation of sentences will allow us to solve the severe problem of 'tacit belief' (Dennett 1975; Lycan 1985). Just as a hologram does not 'contain' a large number of distinct 3-D images, curiously arranged so as to present a smoothly changing picture of a real object as the hologram is viewed from different positions; so may humans not 'contain' a large number of distinct beliefs, curiously arranged so as collectively to present a coherent account of the world.

Perhaps the truth is rather that, in both cases, a specific image or belief is just an arbitrary projection or 'slice' of a deeper set of data-structures, and the collective coherence of such sample slices is a simple consequence of the manner in which the global information is stored at the deeper level. It is not a consequence of, for example, the busywork of some fussy inductive machine applying inductive rules for the acceptance or rejection of discrete slices taken singly. Which means that, to understand learning, we may have to understand the forces that dictate directly the evolution of the global data-structures at the deeper level.

These highly speculative remarks illustrate one direction of research suggested by the theory outlined in this paper: just what are the abstract representational and computational capacities of a system of state spaces interacting by co-ordinate transformations? Can we use it to articulate models for the 'higher' forms of cognitive activity? The theory also begs research in the opposite direction, towards the neurophysiology of the brain. Since the brain is definitely not a 'general-purpose' machine in the way that a digital computer is, it may often turn out that, once we are primed to see them, the brain's localized computational tactics can simply be read off its microstructure. There is point, therefore, to studying that microstructure. (For an accessible review of cognitive neurobiology, see Patricia Churchland 1986.)

Taken jointly, the prodigious representational and computational capacities of a system of state spaces interacting by co-ordinate transformations suggest a powerful and highly general means of understanding the cognitive activities of the nervous system, especially since the physical mechanisms appropriate to implement such a system are widespread throughout the brain.

REFERENCES

Allman, J. M., *et al.* (1982). 'Visual Topography and Function.' In C. N. Woolsey (ed.), *Cortical Sensory Organization*, Vol. 2, pp. 171–86. Clifton, NJ: Humana Press.

Arbib, M., and Amari, S. (1985). 'Sensorimotor Transformations in the Brain.' *Journal of Theoretical Biology* 112: 123–55.

Ballard, D. H. (1986). 'Cortical Connections and Parallel Processing: Structure and Function.' *Behavioral and Brain Sciences* 9 (1): 67–90.

Bartoshuk, L. M. (1978). 'Gustatory System.' In R. B. Masterton (ed.), *Handbook of Behavioral Neurobiology*, Vol. 1: *Sensory Integration*, pp. 503–67. New York: Plenum Press.

Churchland, P. M. (1979). *Scientific Realism and the Plasticity of Mind.* Cambridge: Cambridge University Press.

—— (1985). 'Reduction, Qualia, and the Direct Introspection of Brain States.' *J. Philosophy* 82 (1): 8–28.

—— (1986). 'Cognitive Neurobiology: A Computational Hypothesis for Laminar Cortex.' *Biology and Philosophy* I (1): 25–51.

Churchland, P. S. (1986). *Neurophilosophy: Toward a Unified Understanding of the Mind–Brain.* Cambridge, Mass.: MIT Press.

Cynader, M., and Berman, N. (1972). 'Receptive Field Organization of Monkey Superior Colliculus.' *Journal of Neurophysiology* 35: 187–201.

Dennett, D. C. (1975). 'Brain Writing and Mind Reading.' In K. Gunderson (ed.), *Minnesota Studies in the Philosophy of Science*, Vol. VII, pp. 403–15. Minneapolis: University of Minnesota Press.

Goldberg, M., and Robinson, D. L. (1978). 'Visual System: Superior Colliculus.' In R. Masterson (ed.), *Handbook of Behavioral Neurobiology*, Vol. 1, pp. 119–64. New York: Plenum Press.

Gordon, B. (1973). 'Receptive Fields in Deep Layers of Cat Superior Colliculus.' *Journal of Neurophysiology* 36: 157–78.

Huerta, M. F., and Harting, J. K. (1984). 'Connectional Organization of the Superior Colliculus.' *Trends in Neuroscience* 7 (8): 286–9.

Jackson, F. (1982). 'Epiphenomenal Qualia.' *Philosophical Quarterly* 32 (127): 127–36.

Kanaseki, T., and Sprague, J. M. (1974). 'Anatomical Organization of Pretectal Nuclei and Tectal Laminae in the Cat.' *Journal of Comparative Neurology* 158: 319–37.

Konishi, M. (1986). 'Centrally Synthesized Maps of Sensory Space.' *Trends in Neuroscience* 9 (4): 163–8.

Land, E. (1977). 'The Retinex Theory of Color Vision.' *Scientific American* (Dec.): 108–28.

Llinas, R. (1975). 'The Cortex of the Cerebellum.' *Scientific American* 232 (1): 56–71.

—— (1986). ' "Mindness" as a Functional State of the Brain.' In C. Blakemore and S. Greenfield (eds.), *Mind and Matter*, pp. 339–60. Oxford: Blackwell.

Lycan, W. G. (1986). 'Tacit Belief.' In R. J. Bogdan (ed.), *Belief*, pp. 61–82. Oxford: Oxford University Press.

McIlwain, J. T. (1975). 'Visual Receptive Fields and their Images in the Superior Colliculus of the Cat.' *Journal of Neurophysiology*, 38: 219–30.

—— (1984). *Abstracts: Society for Neuroscience* 10 (Part I): 268.

Mays, L. E., and Sparks, D. L. (1980). 'Saccades are Spatially, Not Retinocentrically, Coded.' *Science* 208: 1163–5.

Meredith, M. A., and Stein, B. E. (1985). 'Descending Efferents from the Superior Colliculus Relay Integrated Multisensory Information.' *Science* 227 (4687): 657–9.

Merzenich, M., and Kaas, J. (1980). 'Principles of Organization of Sensory-Perceptual Systems in Mammals.' *Progress in Psychobiology and Physiological Psychology* 9: 1–42.

Mooney, R. D., *et al.* (1984). 'Dendrites of Deep Layer, Somatosensory Superior Collicular Neurons Extend into the Superficial Layer.' *Abstracts: Society for Neuroscience* 10 (Part I): 158.

Nagel, T. (1974). 'What Is It Like to Be a Bat?' *Philosophical Review* 83, (4): 435–50.

Pellionisz, A. (1984). 'Tensorial Aspects of the Multi-Dimensional Approach to the Vestibulo-Oculomotor Reflex.' In A. Berthoz and E. Melvill-Jones (eds.), *Reviews in Oculomotor Research*. New York: Elsevier.

—— and Llinas, R. (1979). 'Brain Modelling by Tensor Network Theory and Computer Simulation. The Cerebellum: Distributed Processor for Predictive Coordination.' *Neuroscience* 4: 323–48.

—— (1982). 'Space-Time Representation in the Brain: The Cerebellum as a Predictive Space-Time Metric Tensor.' *Neuroscience* 7 (12): 2949–70.

—— (1985). 'Tensor Network Theory of the Metaorganization of Functional Geometries in the Central Nervous System.' *Neuroscience* [16 (2): 245–74].

Pfaff, D. W. (ed.) (1985). *Taste, Olfaction, and the Central Nervous System*. New York: Rockefeller University Press.

Risset, J. C., and Wessel, D. L. (1982). 'Exporation of Timbre by Analysis and Synthesis.' In D. Deutsch (ed.), *The Psychology of Music*, pp. 26–58. New York: Academic Press.

Robinson, D. A. (1972). 'Eye Movement Evoked by Collicular Stimulation in the Alert Monkey.' *Vision Research* 12: 1795–1808.

Robinson, H. (1982). *Matter and Sense*. New York: Cambridge University Press.

Schiller, P., and Sandell, J. H. (1983). 'Interactions between Visually and Electrically Elicited Saccades before and after Superior Colliculus and Frontal Eye Field Ablations in the Rhesus Monkey.' *Experimental Brain Research* 49: 381–92.

—— (1984). 'The Superior Colliculus and Visual Function.' In I. Darian-Smith (ed.), *Handbook of Physiology*, Vol. III, pp. 457–504.

Schiller, P., and Stryker, M. (1972). 'Single-unit Recording and Stimulation in Superior Colliculus of the Alert Rhesus Monkey.' *Journal of Neurophysiology* 35: 915–24.

Smith, D. V., et al. (1983). 'Coding of Taste Stimuli by Hamster Brain Stem Neurons.' *Journal of Neurophysiology* 50 (2): 541–58.

Stein, B. E. (1984). 'Development of the Superior Colliculus.' In W. M. Cowan (ed.), *Annual Review of Neuroscience* 7: 95–126.

Zeki. S. (1983). 'Colour Coding in the Cerebral Cortex: The Reaction of Cells in Monkey Visual Cortex to Wavelengths and Colours.' *Neuroscience* 9 (4): 741–65.

15

THE CONNECTIONIST CONSTRUCTION OF CONCEPTS

ADRIAN CUSSINS

1. PREFACE[1]

Two Cognitive-Science Frameworks for a Solution to the Problem of Embodied Cognition

Cognitive-science theories are theories of how physical systems think. But a framework for cognitive-science theorizing must explain how it is *possible* for physical systems to think. How can intentional phenomena be part of the same world which is described by the natural sciences? How can there be organisms *in* the world which are capable of thinking *about* the world? How can the world include, as a part of itself, perspectives on the world? I shall call the problem of possibility introduced by these questions, 'the problem of embodied cognition'.

This article is about solutions to the problem of embodied cognition, which are both psychological and computational in character. The Language of Thought (LOT) framework (Fodor 1976, 1987, and see §3) is exhibited as a candidate solution, and a rival cognitive-science framework ('C3' for *Connectionist Construction of Concepts*) developed. Both LOT and C3 serve also as methodologies for work in cognitive science, helping to direct research and to understand its significance. Hence two contrasts emerge from the paper: a contrast between two general conceptions of the enterprise of cognitive science and a contrast between two ways of understanding how embodied cognition is possible.

This paper is published for the first time in this collection, by permission of the author. Copyright © 1990 Adrian Cussins.

[1] I use footnotes to give the article, to some degree, different levels at which it may be read depending on the background of the reader. Some footnotes are for non-philosophers only where I give a 'pocket definition' of a technical term used in the text. Other footnotes are for philosophers only where I use the footnote to indicate a connection to some established philosophical topic. And there are also ordinary footnotes.

A Theory of Representation as a Means for Deriving Psychological Explanations from Computational Models

A computational artefact which is held to have significance for psychological explanation is a 'model'. A model is just a physical object. How are psychological explanations to be extracted from it?

A cognitive-science theory ('a theory') is a structured articulation of psychological explanations based on the functioning of the model. Cognitive-science theorizing thus rests on a conception of the relation between computational artefacts and psychological explanations. This relation is mediated by a theory of representation.

A representation is itself a physical object which has two kinds of properties; properties of the representational 'vehicle' and properties of the representational 'content'. For example, a sequence of marks on a marking surface may be a representation. The alphanumeric letter sequencing that these marks instantiate is a property of the representational vehicle. And if the sequence happens to be the following, 'Stanford is warmer than Oxford', then the content of the representation is that Stanford is warmer than Oxford. The representational vehicle is the medium that carries the representational content as its message.

In a model, the properties of a representational *vehicle* are all properties which have computational impact (e.g. syntactic properties of LISP code). They are properties which affect the computational functioning of the model. And the properties which form the representational *content* are all properties which have psychological impact (e.g. the task-domain semantic properties of the LISP code). They are properties which affect the psychological explanations which can be derived from the model. So, on one side, the properties of a representation have a role in psychological explanation, and on the other side, they have a role in the computational functioning of the model. It is the theory of representation which must tie together these two sets of properties, and hence establish the connection between computational functioning and psychological explanation. It is a theory of representation which allows us to extract psychological import from computational physical objects; that gets a theory out of a model.

If cognitive science involves getting a psychological theory out of a computational model, and if a theory of representation is the way to do this, then in order to understand the nature of cognitive-science theorizing we need to understand the relation between computation, representational vehicles, representational content, and psychological explanation. The task is inherently multidisciplinary (see Fig. 1).

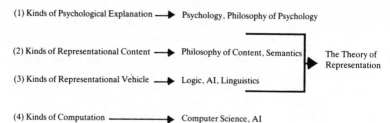

(1) Kinds of Psychological Explanation ——▶ Psychology, Philosophy of Psychology

(2) Kinds of Representational Content ——▶ Philosophy of Content, Semantics   The Theory of Representation

(3) Kinds of Representational Vehicle ——▶ Logic, AI, Linguistics

(4) Kinds of Computation ——————▶ Computer Science, AI

FIG. 1. Four Levels of Analysis in Cognitive Science

Cognitive-Science Frameworks

A cognitive-science framework consists of an analysis at each of the four levels in Figure 1. A central theme of this paper is that the analyses are not independent of each other. For example, given a von Neumann analysis of the computational architecture of a model, and a syntactic analysis of the model's representational vehicles, the theory of content for the model would have to be a semantic theory. And the choice of semantic contents entails that a particular kind of psychological explanation (conceptualist explanation) is derived from the model (as will be explained later). Or, if one chooses a connectionist computational architecture, one may be led as, for example, Smolenksy (1988a) has been, to reject syntactic representational vehicles. As is shown in this article, this consequence should itself have implications for the kinds of contents which connectionist representational vehicles can carry, and thus implications for the kinds of psychological explanation which can be extracted from connectionist models. The analysis at each level constrains the analysis at the adjacent levels, so consequences can also be traced in a top-down direction.

A cognitive-science framework, then, involves a decision at each of these levels, so that the decision at each level is compatible with the decisions at all of the other levels. Diagramming the possible choices at each level provides a representation of competing cognitive-science frameworks. The terms which denote the choices in Figure 2 are explained in the body of the article. But it may help to begin with the diagram.

The Strategy of the Paper

An alternative kind of content from that presupposed by LOT is suggested, and the consequences of its use are considered for psychological explanation, for theories of representation, and for computational implementation.

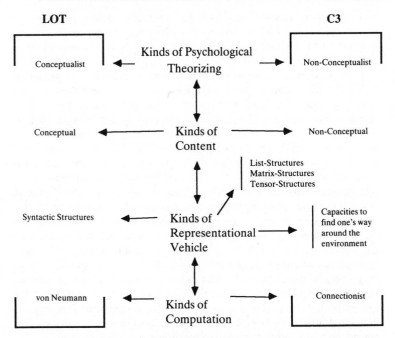

Fig. 2. LOT and C3 as Cognitive-Science Frameworks. Vertical arrows indicate constraints. Horizontal arrows indicate some possible kinds or choices at each level. The left-hand side of the diagram is labelled as the set of choices which constitute the Language of Thought framework for cognitive science, and the right-hand side is labelled as the set of choices which constitute the C3 framework for cognitive science. The unfamiliar terms in the diagram are explained in the body of the text.

If the alternative framework is to be a genuine alternative to LOT, then it must provide a solution to the problem of embodied cognition; it must indicate how the physical embodiment of cognition is possible. Section 2 explains why this problem is a problem and provides a necessary and sufficient condition for its solution.

In §3, I explain why the LOT interpretation of cognitive science offers a candidate solution to the problem of embodied cognition. I point out that this status depends on the computational use of the classical *syntax/semantics theory of representation (S/S theory)*.

The dependence of LOT theorizing on the S/S theory of representation entails that psychological modelling based on LOT employs *conceptual content*. So in §4, I explain the distinction between *conceptual content* and *non-conceptual content,* give several examples which indicate the psychological need for a notion of non-conceptual content, and introduce a particular kind of non-conceptual content: *construction-theoretic content (CTC).*

In §5, I explore the cognitive psychological consequences of modelling with conceptual content: psychological phenomena are modelled in terms of the relations that hold between concepts, and between the sensors/ effectors and concepts. This establishes the contrast for §6 to develop the idea that the *non-conceptualist* psychological task is to explain the *cognitive emergence of objectivity.* The psychologically fundamental structure of cognition is not the structure that holds *between* concepts, but, rather, the structure *within* concepts. Section 7 makes the notion of *objectivity* more precise, and provides a way of assessing any system for the degree to which it is a concept-exercising system. Section 8 develops the connection between objectivity and *perspective-independence.* It explains why certain forms of non-conceptual content do not present the world to a subject as the objective world, and it explains the conditions under which non-conceptual content would present the world objectively. Section 9 shows how a psycho-computational theory of transformations of non-conceptual content can decrease the perspective-dependence of the abilities of the system by the formation of a cognitive maplike structure. This explains how a cognitive science which models in terms of non-conceptual content can nevertheless satisfy conceptual constraints on cognition.

In conclusion, I suggest that the interesting cognitive employment of connectionism should not rest on the S/S theory, because the S/S theory entails conceptualist theorizing, and connectionist cognitive modelling is suited to non-conceptualist psychological modelling. I give some reason to think that C3 is as suited to (a way of cognitively using) connectionism as LOT is to classical AI. The potential for connectionism to use non-conceptual content shows why Fodor and Pylyshyn's (1988) criticism is misplaced. Connectionism can use the apparatus I have introduced to show how connectionist cognitive modelling can, in principle, respond to the problem of embodied cognition.

2. THE PROBLEM OF EMBODIED COGNITION AND THE CONSTRUCTION CONSTRAINT ON ITS SOLUTION

The Problem of Embodied Cognition

Consider the following way to bring out the problem of embodied cognition.

Suppose, for the purposes of this article, that there is an irreducible and indispensable scientific level of cognitive explanation of human behaviour and that, even by the end of the next millennium, cognitive science will not have been made redundant by neurophysiology, quantum mechanics, or some other non-cognitive level of explanation.[2]

Let us also accept naturalism: that all non-physical properties are either reducible[3] to or must be *realized*[4] or *implemented*[5] in physical properties. In other words, anything that has a causal power either has only physical causal powers or must be built out of physical components, so that it is possible, in principle, to understand why it is that something which is built physically like *that* (pointing to the physical science description), has *those* causal powers (pointing to the non-physical description). Naturalism does not require that non-physical properties be—despite appearances—really physical properties (naturalism does not require reduction), but it does require that if we knew all the science there could be, we should not find it coincidental that certain physical objects have the irreducible non-physical properties that they have.

Whether or not human behaviour can be explained physiologically, humans behave as they do because of neurophysiological properties of humans. But—given our first supposition—humans behave as they do because of certain irreducible cognitive properties. Neurophysiological explanation and cognitive explanation are independent of one another, and—apart from cognitive or physiological breakdowns—are each complete,[6] in their own terms. How then can it be that both of the following are true: (1) cognitive explanations of behaviour are not causally redundant, and (2) the physical causation of the behaviour of a person *marches in step* with the cognitive causation of the behaviour of a person, so that a person is not torn apart in a tug of war between the physical and

---

[2] For the purposes of this paper, I am simply assuming this as a premiss. There are many good reasons for taking cognitive explanation to be irreducible and indispensable. See e.g. Fodor in Block (1980: vol. i.), Fodor, 'Computation and Reduction' in Fodor (1981b), Fodor (1987: ch. 1), Pylyshyn (1984: chs. 1–2), Putnam (1973). I argue that there must be a *scientific* level of cognitive explanation in Cussins (Jan. 1987).

[3] One level of description and explanation reduces to another level only if all the explanation at the reduced level can be derived from explanation at the reducing level (thanks to David Charles). If all the properties at one level were identical to properties at another, then it would follow that one of the levels would reduce to the other; but reduction does not require property identity. It is important to see that the *construction* of cognition, which I favour (see below), does not entail the *reduction* of cognition, which I do not favour.

[4] In the sense in which digestion is *realized* in the stomach.   [5] See n. 71 below.

[6] That is, barring a physiological breakdown, a particular token effect characterized in physiological terms, can, in principle, be derived by wholly physiological means, without having to use, for example, psychological laws or quantum mechanical laws. There may be exceptions to this, but they are not the general case. This is not to deny, of course, that the particular effect, under other descriptions of it, may be more satisfactorily explained non-physiologically. And similarly for effects characterized in psychological terms.

cognitive causal powers in the person. I write as I do because of my beliefs about how best to communicate a philosophical problem to a readership that is partly non-philosophical, but it is also true that I behave as I do because of certain neurophysiological causes within me. How do we avoid the conclusion that there is a battle for the control of my hand?

This problem is not resolved by supposing that cognitive explanation is non-causal, for the problem will then re-emerge as: how is it possible that the behaviour of a person, which is physically caused, is *coherent* from a cognitive perspective? My writing as I am is cognitively predictable (whether or not it is cognitively caused), but had my neurophysiology been different in any of a very large number of imaginable ways, I would not be writing this at all; because, for example, my hand would be motionless, or stuck behind my back. How can my physiology keep on making my body do one of the limited range of things that it must do if it is to make cognitive sense?[7]

In short, how can we understand cognition naturalistically without either the reduction of cognitive properties to non-cognitive properties,[8] or the elimination of cognitive properties,[9] or the rejection of the scientific indispensability of cognitive properties?[10] To understand this is to understand how cognition can be physically embodied, and thus to understand how to solve the problem of embodied cognition.

The Construction Constraint

There is a naturalistic alternative to reduction, elimination, and explanatory dispensability: the *construction* of cognitive properties out of non-cognitive properties. This idea may be introduced by an example.

The notion of architectural functionality may be essential to the work of an architect, even though the notion cannot be reduced to any builder's notion of the arrangement of materials. For example, an architect may need to work with the notion of an efficient corporate headquarters. But this notion cannot be defined in terms of the spatial arrangement of

---

[7] I develop a thought-experiment to help make vivid this problem in Cussins (Jan. 1987).

[8] Cognitive reductionists include Smart (1970), Armstrong (1968), Place (1970), Lewis (1966).

[9] Cognitive eliminativists include Quine (1960). Churchland, P. M. (1979), Churchland, P. S. (1986), and Stich (1983) are often cited as eliminativists, but may be better thought of as recommending the elimination of conceptual content only, rather than every notion of content.

[10] Dennett (1987) occupies this dispensability position: there are irreducible cognitive properties, but they are not an essential part of the complete scientific explanation of human behaviour. Strictly speaking there is no scientific psychology for Dennett because there are no psychological natural kinds (see Cussins 1988). Stephen Schiffer (1987) is also a dispensability theorist. For Schiffer there are irreducible cognitive properties, but only in a pleonastic sense.

commercial sizes of bricks, stone, metal, glass, plastic, wood, and concrete. An unspecifiable infinite set of different arrangements of builders' materials will be sufficient for an efficient corporate headquarters, given a particular company at a particular stage of development and a particular technological and ethnographic context. Not only does an unspecifiable infinite set make reduction impossible, but which unspecifiable set this is will vary with the contextual parameters. What is efficient for a small company may not be efficient for IBM in the eighties. And what is efficient given telephones, electronic mail, and fax machines would not be efficient in a technological context which pre-dated these communication media.

So, the notions in terms of which a client would specify a building to an architect cannot be reduced to notions that a builder must work with. There are, thus, two distinct levels of notions (levels of description) which an architect must somehow bridge if he is to do his job. For a layman, the architect's ability to produce a builder's specification from an architectural specification, or to know which architectural properties would be instantiated by a building built to a builder's specification, may appear unintelligible. There is no further level of description that the architect employs. Rather, in learning his job, an architect has gained an understanding of the architectural notions and the builders' notions that allows him to move back and forth between descriptions at each of these levels. For the architect the relation between the two levels of description is Intelligible,[11] not coincidental, whereas for the layman the relation is not Intelligible and so may appear coincidental.

The architect's understanding may be more practical than theoretical. Finding Intelligible the gap between the architectural level of description, and the materials level of description may consist simply in the following skill: given any building specification, an architect should be able to tell[12] (and to know that he can tell) what architectural properties a building constructed like that would have, and given any architectural specification (e.g. to provide functional office accommodation which is appropriate to the context of St Paul's Cathedral), an architect should know how to put together building materials so as to satisfy the specification. For the skilled architect, the gap between the two levels of description is Intelligible not coincidental: we may say that the architect—but not the layman— has the ability to *construct* architectural notions from building notions.

---

[11] I capitalize the first letter of 'intelligible' to indicate that it expresses a semi-technical notion. There is more discussion of this notion in Cussins (1992).

[12] Perceive, not infer. Inferential connections are only ever within a level. Perception (whether sensory or not) is the cognitive means to cross between levels.

Thus the relation of *construction* is an explanatory relation between levels which differs from the relation of reduction and from the relations of elimination and dispensability. The *construction constraint*, applied generally, claims that any non-physical level of description and explanation should be constructable out of, ultimately, some physical level, in an analogous sense to the architect's construction of architectural notions out of building and materials notions. If we have to construct some notion φ out of physical notions then we need to be able to understand the nature of an object's being φ in terms of a sequence of levels of description which are such that the top level makes manifest the φ-ness of the object (as, for example, the architectural level makes manifest the corporate efficiency of a building), the bottom level is a physical level of description, and every two adjacent levels are such that the gap between them is Intelligible (not coincidental), as is the gap—for the architect—between the architectural and builders' levels.[13]

Relying on this intuitive idea of the distinction between *Intelligible* gaps between levels and gaps which are coincidental or miraculous,[14] the construction constraint can be stated more formally as follows:

A theory (or, rather, a framework of theories) of an ability φ meets the construction constraint with respect to φ if, and only if, it explains what it is for an organism to possess the ability in terms of the possession of a sequence of more than two levels of abilities, $L_1 - L_n$, such that:

1. the base level, L_1, abilities are such that we do not understand why it is that an organism which has these abilities is thereby an organism which has the ability φ, (that is, L_1 and L_n are so related as to generate a

[13] An Intelligible connection between two levels does not involve laws between the levels, nor does it involve a third level of description in terms of which the connection between the two principal levels is understood. This is why I emphasized the practical character of the architect's ability. The connection between two levels is Intelligible if the marching in step of the two levels does not appear as a miraculous coincidence (Cussins Jan. 1987). One aspect of this is that somebody who grasps the Intelligible connection should have a fallible, practical capacity to arrange a system at the lower level so as to satisfy upper-level constraints, and should have some idea about the circumstances under which upper-level performance will degrade. I can see no good reason why the distinction between Intelligible gaps and miraculous gaps being merely intuitive (depending on, for example, what we *recognize* as Intelligible) should be a problem. The construction constraint is a constraint on explanation. It may be the case that what it is to be explanatory can itself only be explained in terms of what the creatures, for whom it is an explanation, *find* explanatory. It would be very nice to be able to say what an Intelligible gap amounts to, in a way which goes beyond this, but for my purposes it is sufficient that we can tell of any gap between levels, whether it is Intelligible or miraculous. The construction constraint, like Tarski's Convention T, is a criterion for success, the satisfaction of which we are able to recognize. And, like Tarski's convention, it rests on an undefined notion. (This parallel was brought to my attention by David Charles.)

[14] I give a miraculous coincidence thought-experiment in Cussins (Jan. 1987).

miraculous coincidence problem about how they march in step), and,

2. the theory shows why it is that possession of the top level, L_n, abilities constitutes or manifests possession of φ, and,

3. for each pair of levels, L_i and L_{i+1} ($1 \leq i < n$), L_i and L_{i+1} are not related in such a way that they generate a miraculous coincidence problem about how they march in step (that is, the gap between L_i *and* L_{i+1} is Intelligible).[15]

The gap between the folk-psychological level and the neurophysiological level is *not* Intelligible in this way: I exploited this gap to make vivid the problem of embodied cognition as a tug of war for the control of my hand. Given only our folk-psychological and neurophysiological knowledge, the marching in step of these two levels appears to be a miraculous coincidence.[16] In contrast, because we know how to build machine languages out of electronic components and a high-level language like LISP out of a machine language, the fact that the behaviour of the computer is determined by two levels of description (LISP and electronics) each complete in its own terms, does not appear as a coincidence. The construction relation between LISP and electronics does not require that there be lawlike relations between any two adjacent levels: there are no known laws connecting electronic and computational descriptions. It requires just that somebody with knowledge of, for example, machine languages, LISP, and electronics, should know roughly how to go about putting together electronics so as to build a LISP machine. Not that one has to succeed every time; just that one shouldn't find it coincidental that what has been built functions as a computer.

[15] Philosophers of mind will be interested to note that the construction constraint is, in one way, weaker than supervenience (even supervenience on the entire physical world, rather than just the cranium), and, in another way, stronger. It is weaker in that it does not require that a lower level in a construction is sufficient for an upper level, as the subvenient level must be sufficient for the supervenient level. So, in connection with §2, we might note against LOT that syntax does not determine semantics, because there will always be distinct semantic interpretations of, for example, the connectives given the same syntax and proof theory (Williams unpublished). Nevertheless, the fact that syntax preserves semantic constraints, under any of the possible semantic interpretations, may be sufficient to account for the *marching in step* of explanations of behaviour which employ semantic notions (psychological explanation), with explanations of behaviour which are due to the computational implementation of syntax. Hence, although syntax does not determine semantics, this fact cannot rule out the use of the syntax/semantics theory of representation in a construction of cognition.

The construction constraint is stronger than supervenience in that it imposes demands on explanation that supervenience does not (supervenience does not require that there be an Intelligible connection between levels).

One advantage, to me, of the construction constraint over supervenience is that I know how to argue for the construction constraint, but I do not know how to argue for supervenience.

[16] For some writers, such as Swinburne (1986), the marching in step doesn't just appear to be, but *is* miraculous: God must be invoked in order to account for the behavioural coherence of a person.

There is an interpretation of cognitive science according to which it has been exploring a construction[17] of conceptual abilities in terms of interposing representational and computational levels between the folk-psychological and neurophysiological levels. In the next section I show how the LOT theory can be seen as offering an account along these lines. The rest of the paper develops an alternative account which is more appropriate to the use of connectionist computational architectures and is not subject to the same kind of difficulties as LOT.

3. THE LANGUAGE OF THOUGHT AS CONCEPT CONSTRUCTION

Remarkably, there has only been one serious attempt to solve the problem of embodied cognition by satisfying the construction constraint on the possession of concepts.[18] This is the model which Fodor has called the Language of Thought (LOT), and which he and others have defended at a philosophical level for over a decade. The possibility of LOT turns crucially on the possibility of both the computational and the psychological application of a theory of representation which has been developed in logic since Frege. This is the theory which characterizes a representational system in terms of its combinatorial syntax and combinatorial semantics.

The syntactic theory of a representational system provides a recursive specification of all and only the legal concatenations of the atomic representations of the system. The semantic theory of a representational system provides an axiomatizable, recursive specification of the interpretation of all the legal representations. If the representational system is to do any work it must be possible to define over the syntax either a proof theory for logical work, or a theory of procedural consequence for computational work, which specifies all of the legal transformations from each

---

[17] Throughout, I use the term 'construction' in the technical sense employed by the construction constraint. A relation between two adjacent levels may be a construction relation, even though the construction *constraint* required more than two levels. It will be a construction relation if it is Intelligible. The naturalistic constraint on a framework of theories is not to make the relation between every two levels a construction relation, but rather to satisfy the construction constraint. The relation between non-adjacent levels in a framework which satisfies the construction constraint may not be an Intelligible relation, for a grasp of it may depend on grasp of the theory of the intermediate levels.

[18] There are the beginnings of construction theories based on the communication theoretic notion of information (Dretske 1981). The trouble with these theories, as with behaviourism, is that although the notion of information is suited to a low level in a construction, nobody has yet shown how it is possible, even in principle, to construct a level of concept possession on top of the informational level. Dretske has a go in ch. 7 of his book, but unfortunately the attempt fails (Cussins forthcoming). Other non-reductionistic, non-eliminativist theoretical frameworks include Millikan (1984), but it is not yet clear how this framework could yield a construction.

legal representation. Much of the value of the logical tradition has rested on our being able to define purely syntactically a theory of legal transformation which nevertheless respects semantic constraints.

The syntactic and semantic theories must be explanatorily independent, yet linked. They must be independent in that it must be possible to understand how to apply the theory of legal transformation without understanding anything of the semantic theory; but they must be linked in that the syntactic application of the theory of legal transformation must not violate semantic constraints: traditionally, it must not transform a set of true premises into a false conclusion. And, the syntactic and semantic theories must be so related that even if the representational system is not complete, a useful proportion of all of the semantically coherent transformations must be capturable by the syntactic application of the theory of legal transformation.

What is so remarkable about LOT is its insight that the way to achieve the required Intelligible connection between the computational component and the psychological component of modelling in cognitive science is by developing a syntactic and semantic representational system for which the syntax is implemented computationally and the semantics[19] is appropriate for psychological explanation. If this can be done, then the way to achieve the required relation between the computation and the psychology will simply follow from our understanding how to establish representational systems for which it is possible to specify a syntactic theory of legal transformation which respects semantic constraints. What the S/S theory delivers is a semantic-independent level (syntax) that marches in step with the semantic level. So if the syntax can be implemented computationally and the semantics can provide the basis for psychological explanation, then LOT will have shown how computational transitions can march in step with psychological transitions.

The syntactic characterization of classical computational architectures is natural in part because these architectures themselves grew out of the logical tradition. And the psychological employment of semantics is natural because of a tradition that also goes back to Frege; the tradition of taking the meaning of a sentence to be the object of the propositional

---

[19] For LOT *aficionados*: a number of issues within LOT have revolved around whether a special notion of semantics is required for psychological interpretation. The special notion is generally called 'narrow semantics', which, under psychological interpretation, becomes 'narrow content' (Fodor 1987). Narrow content does not fully determine reference. Narrow contents may have to be general, while all the singular aspects of content are taken to be broad phenomena, and narrow content may be restricted to the presentation of observational properties to the exclusion of natural kind properties. But the point is that these innovations are heavily constrained: narrow content and narrow semantics must be *narrow*—they must form a subset of classical content and classical semantics.

attitude which is expressed by the concatenation of a psychological verb ('believes', 'desires', . . .), 'that' and the sentence itself.

Cognitive science, as interpreted by LOT, becomes distinct from both the theory of computation and the theory of psychology, because it is the attempt to establish empirically an intermediate level of explanation built on a system of cognitive representation which is such that (1) the syntax is computationally implementable, (2) the semantics captures the important psychological generalizations, and (3) the theory of procedural consequence is consistent and usefully complete. It is of course true that work in cognitive science will proceed without LOT, that many cognitive scientists disagree with LOT, and that numerous objections have been raised to LOT. But LOT remains the only theory which gives clear criteria for the evaluation of the success of cognitive science, which is plausibly workable, and which shows how cognitive science may achieve the significance it hopes for: to explain how cognition can be physically embodied by constructing the psychological component of a cognitive-science model out of the computational component.

My purpose here is not to criticize LOT, but to understand how to develop an alternative theory which has a comparable explanatory scope. LOT is so impressive because it rests on the remarkable tradition of S/S representational theory. To develop an alternative solution to the problem of embodied cognition, we need to develop an alternative theory of representation. An alternative theory of *computation* is not sufficient.

4. CONTENT, CONCEPTUAL CONTENT, AND NON-CONCEPTUAL CONTENT

I have suggested how it is, in the case of LOT, that providing the kind of explanation appropriate to the problem of embodied cognition depends on the theory of representation which LOT employs. But it is also true that the choice of a representational theory determines the kind of psychological explanation which a model can offer. This is so because the representational theory determines the kind of *content* which can be assigned to states of the model, and this, in turn, determines the kind of psychological explanation that the model can make available. The link between representation and psychological explanation is *content*.

Introducing Content

I will begin with a pocket account of the notions of content, conceptual content, and non-conceptual content, before presenting a more careful analysis of them.

Human persons act as they do, and thus often behave as they do, because some aspect of the world is presented to them in some manner. The term 'content', as I shall use it, refers in the first instance, to the way in which some aspect of the world is presented to a subject; the way in which an object or property or state of affairs is given in, or presented to, experience or thought. For example, I see the grey, plastic rectangular object in front of me as being a typing board, having the familiar Qwerty structure. I also see it *as being* in front of me, and these facts are responsible for my hands moving in a certain way. Representational states of mine have content in virtue of which they make the world accessible to me, guide my action, and (usually) are presented to me as something which is either correct or incorrect. I shall speak of *a representational state (or vehicle) having content*. It may be that a single representational vehicle carries more than one content, even more than one kind of content.

The theory of content—in terms of which we explain what content is—locates the notion with respect to our notions of experience, thought, and the world. But it is important to see that this is consistent with the notion of content being applied to (though not explained in terms of) states which are not states of an experiencing subject.[20] There are derivative uses of the notion in application to the communicative products of cognition, such as speech, writing, and other sign-systems, or to non-conscious states of persons such as subpersonal information-processing states, but these uses must ultimately be explained in terms of a theory of the primary application of content in cognitive experience.[21]

Conceptual content is content which presents the world to a subject *as* the objective, human world about which one can form true or false judgements. If there are other kinds of content, kinds of *non-conceptual* content, then that will be because there are ways in which the world can be presented to a subject of experience which do not make the objective, human world accessible to the subject. It is not unnatural to suppose that there must be non-conceptual forms of content, because this is the kind of thing that we want to say about very young human infants (before the acquisition of the object concept, say), or very senile people, or certain other animals. It is compelling to think of these beings as having experience, yet they are unable to communicate thoughts to us; we are unable to

---

[20] It is even consistent with this to suppose that there were simple kinds of content around in the world, before there were any experiencing subjects—but more of this later.

[21] In this context, this is just stipulation. Some people working in non-Fregean semantical traditions, rather than a Dummettian/Strawsonian tradition, will find my use odd, which is why I have begun with this stipulation. I need to have a notion like my notion of content—whatever it is called—because part of the problem of embodied cognition is to explain how there can be certain physical creatures, like us but unlike paramoecia, whose response to the world does not consist wholly in their response to physical stimulations of their sensory surfaces, but which rests, in part, on a conception of how the world is.

understand—from the inside—how they are responding to the world; we are unable to impose our world on them.

Conceptual content presents the world to a subject as divided up into objects, properties, and situations: the components of truth conditions. For example, my complex conceptual content (thought) that the old city wall is shrouded in mist today presents the world to me as being such that the state of affairs of the old city wall being shrouded in mist obtains today. To understand this content I have to think of the world as consisting of the object, the old city wall, the property of being shrouded in mist, and the former satisfying the latter. The possession of *any* content will involve carving up the world in one way or another. There will be a notion of *non-conceptual* content if experience provides a way of carving up the world which is not a way of carving it up into objects, properties, or situations (i.e. the components of truth conditions).[22]

It is natural to say that the possession of content consists in having a *conception* of the world as being such and such. But the word 'conception' is too closely related to 'concept' for it to function neutrally as between conceptual and non-conceptual presentations of the world. I shall say[23] that a content *registers* the world as being some way, and so ask, is there a way of registering the world which does not register it into objects, properties, or situations?

Definitions of Conceptual and Non-Conceptual Properties

I will begin a more careful analysis of these notions by introducing definitions of conceptual and non-conceptual *properties*, and then show how these definitions can be applied within the theory of content.[24]

A property is a **conceptual property** if, and only if, it is canonically characterized,[25] relative to a theory, only by means of concepts which

---

[22] How could there be such a way? Well, this is what I am devoting much of the paper to trying to explain. So far in this section I take myself to have given only a pocket account of the notions of content, conceptual, and non-conceptual. The rest of this section begins on an analysis, and gives an argument for the existence on non-conceptual content, while in §7 I make more precise the notion of a content's presenting the world objectively as consisting of objects, properties, and situations. The claims in this section don't tell us what content is; they are intended just to give an intuitive feel for the notions. Later, we will see that conceptual content is the availability in experience of a task-domain, and non-conceptual content is the availability in experience of substrate-domain abilities.

[23] Following, with some differences, the usage in Bennett (1976) and by Brian Smith (1987).

[24] NB: these are not yet definitions of two kinds of **content**.

[25] Something is canonically characterized (within a theory) if, and only if, it is characterized in terms of the properties which the theory takes to be essential to it. A game of football, for example, is canonically characterized, in the Football Association, in terms of the notions employed in the rules of the game, not in terms of temporal patterns of disruption to the playing

are such that an organism *must have* those concepts in order to satisfy the property.

A property is a **non-conceptual property** if, and only if, it is canonically characterized, relative to a theory, by means of concepts which are such that an organism *need not have* those concepts in order to satisfy the property.

Notice that the difference between these two definitions lies principally in the difference between the italicized 'must have' in the first definition, and 'need not have' in the second definition.

Consider the property of thinking of someone as a bachelor. A specification of what this property is will use the concepts *male*,[26] *adult*, and *unmarried*. But nothing could satisfy the property unless it possessed[27] these concepts, since nothing would count as thinking of someone as a bachelor, unless he or she was able to think of the person as being male, adult, and unmarried. So the property of thinking of someone as a bachelor (unlike the property of being a bachelor) is a conceptual property.

Or consider the belief property of believing that the Stanford Campus is near here (where I think of the Stanford Campus *as* the Stanford Campus, rather than as the campus of the richest university in the West, and I think of here as here, rather than as 3333 Coyote Hill Road). Given this, nothing could satisfy the property unless it possessed the concept of the Stanford Campus *qua* Stanford Campus. Thus the property is canonically characterized only by means of concepts which an organism must have in order to satisfy the property, and is therefore a conceptual property. Contrast the property of having an active hypothalamus. Such a property is characterized by means of the concept *hypothalamus*, but an organism may satisfy the property without possessing this concept. Therefore the property of having an active hypothalamus is a non-conceptual property.[28]

Formally, the idea is that conceptual content is content which consists of conceptual properties, while non-conceptual content is content which consists of non-conceptual properties. Can we give any substance to this formal idea?

---

field. A *content* is canonically characterized by a specification which reveals the way in which it presents the world. See below.

[26] I use asterisks, like quotation marks, to indicate that the enclosed words do not have their normal reference. But asterisks indicate that the words refer to the concept or concepts, or other kind of content, that the words express, rather than to the linguistic items themselves.

[27] Notice the difference between instantiating or satisfying or falling under a concept, on the one hand, and possessing a concept on the other. I possess the concept *bachelor*, but I don't fall under that concept.

[28] Not a *content* property, obviously.

*The Application of the Definitions of Conceptual and Non-Conceptual
Properties within the Theory of Content*

In order to show that there is a notion of non-conceptual content we need
to show that the definition of non-conceptual properties can be applied
within the theory of content. What does this mean?

The definitions of conceptual and non-conceptual properties use the
notion of canonical specification, for otherwise every property would be a
non-conceptual property, since, trivially, every property—including
conceptual properties—can be specified by means of concepts that the
subject need not possess. So we need to employ the notion of *canonical*
specification. If we are to apply these definitions within the theory of
content then the notion of canonicality that we are interested in is the
notion of being a canonical specification within the theory of content.
Certain specifications of a state or an activity are identified within a theory
of content as being canonical when they are specifications generated by
the theory in order to capture the distinctive way in which some aspect of
the world is given to the subject of the state or activity. So, as brought out
by McDowell (1977), '"aphla" refers to aphla' would be canonical, but
'"aphla" refers to ateb' would not be, even though both would be true,
because aphla is ateb. The notion of being canonical within the theory of
content is parallel to the notion of being canonical in the theory of
number, where the canonical specification of the number nine is not 'the
number of planets', but 'nine'.

The Case of Conceptual Content

The Notion of a Task-Domain. In order to understand how conceptual
content works we need the notion of a task-domain for a behaviour. *A*
task-domain *is a bounded domain of the world which is taken as already
registered into a given organization of a set of objects, properties, or
situations,*[29] *which contains no privileged point or points of view, and with
respect to which the behaviour is to be evaluated.*[30]

[29] The task-domain objects, properties, and situations are presumed to be fully objective in
the sense that it is presumed that it is, in principle, possible to explain what it is for them to exist
in a way which is independent of any explanation of what it is for organisms to recognize or
perceive or act on them. (It will then turn out that the notion of a task-domain is an idealization.) It is important to see that a task-domain is entirely abstracted from any perceiver or
subject. There is no point of view in a task-domain, no essentially indexical elements.

[30] One might suppose that a task-domain is simply a part of the world. But this is not so,
because a task-domain is a part of the world *under a given conceptualization*. Not only does the
world permit of many different true conceptualizations, it also permits of registrations which
are NOT conceptualizations (I shall argue).

SHRDLU's[31] blocks micro-world was SHRDLU's task-domain. The notion of a Model in formal semantics, and (often) the notion of a possible world in logic are notions of task-domains. Likewise, the performance of a chess computer is evaluated with respect to a chess task-domain which consists of 64 squares categorized into two types, 32 pieces—each with an ownership property—, a legal starting position, three types of legal ending position, and a set of transformations from each legal position to all of the legal continuations from that position. The computer's task-domain excludes, for example, human emotions and plans, lighting conditions, reasons for, and the point of, winning . . . What this means is that the performance of a chess-playing computer is evaluated with respect to transformations of chess tokens on a 64-square board, but not with respect to its response to human emotions, the lighting conditions, the historical pattern of the game, or 'its reasons for winning'. Moreover, because the domain is fixed so that certain situations are registered as wins for White, and certain others as wins for Black, the performance of the computer is not assessed with respect to its ability to transfer its knowledge to a different game, chess*, which is identical to chess except that those situations which are wins for White in chess, are wins for Black in chess*, and those situations which are wins for Black in chess are wins for White in chess*.[32]

A task-domain, then, is a conceptualized region of the world which provides the context of evaluation (true/false, win/lose, true-in-a-model/ false-in-a-model, adaptive/non-adaptive, successful/unsuccessful . . .) for the performance of some system. How is the notion of a task-domain connected to the notion of conceptual content?

The Specification of α Content by Concepts of the Task-Domain. Consider again the cognitive occurrence in me that we express in words as, 'I am thinking that the Stanford Campus is near here.' This is a representational state of mine, and may possess more than one kind of content.[33] What kind of content does the state carry? There is a type of content (let us call it 'α content') which is stipulated within the theory of content[34] to

---

[31] See Winograd (1973).

[32] For such a game to be playable, it would have to be supplemented with new rules, such as the rule of obligatory capture: if, on a turn, a player can capture an opponent's piece, then he must do so. But this does not alter the point that an intelligent capacity to play a game (unlike a conventional computer's capacity) entails the capacity to adapt to be able to play related games, whose task-domains may differ from each other and from that of the original game.

[33] We shouldn't assume that because the state has a linguistic expression, that therefore it has only one kind of content: linguistic content isn't a kind of content, only a kind of *expression* of, or vehicle for, content. It turns out that we need more than one kind of content to do justice to our language use. At this stage in the paper, I am trying to be neutral on this point.

[34] See e.g. Dummett (1975) and (1976), Davidson (1967), Evans (1982: chs. 1–4).

be a kind of content that has determinate[35] truth conditions;[36] that is, whose evaluation as correct imposes a determinate condition on the world. It follows that the linguistic expression, 'that the Stanford Campus is near here' cannot fully capture the α content of the representational state, since this requires a fixed interpretation for 'near' and 'here'. (In order for the state to be a state with α content, we need to know what truth condition it imposes on the world. But the words 'here' and 'near' do not tell us.)

Now suppose that this state occurs as part of a project of mine in which I am planning how best to eat lunch given various parameters and constraints on me: time, money, hunger, distance to eating locations, speed of transport available to me, cost of food at various locations. These parameters and constraints establish a task-domain which fixes an interpretation for the terms 'near' and 'here': suppose that it follows from the time constraints on me, and my hunger, that I need to be eating within fifteen minutes. Then 'near' means: can be reached by a mode of transport available to me within fifteen minutes. Likewise 'here' will mean something like: the region between the spot on which I am standing and a line joining the embarkation points for all the modes of transport which are part of my planning domain.

The interpretation of my cognitive occurrence as having α content depends on specifying the content by means of concepts of a task-domain; in this case, the domain of my planning to eat lunch under various constraints and given various parameters. In other words, the provision of determinate truth conditions for my cognitive state, required by the interpretation of it as having α content, entails that the content is canonically specified by means of concepts which reflect the objective structure of the task-domain: its organization into objects, properties, and situations. Since an organism can only grasp an α content if it grasps its truth conditions (or its contribution to the truth conditions of contents containing it), it follows that an organism which grasps such a content

[35] Having probabilistic truth conditions is one way to have determinate truth conditions. When Quine argued that the reference of 'gavagai' was indeterminate (Quine 1960), he did not mean that it referred with a certain probability to rabbit, and with a certain probability to rabbit-stage, and with a certain probability to connected-rabbit-parts. From my perspective, fuzzy set theory and probabilistic emendations of semantic theories do not offer us a notion of content different from conceptual content. Rather, they provide a way in which a state or item, etc., may have its conceptual content probabilistically. A coin tossed in a task-domain may come up heads with probability 1/2. Task-domains are fully determinate, not deterministic.

[36] I am not prejudicing the issue of whether there is more than one kind of content. I am noticing a certain constraint within the theory of content and calling 'α content' the content which satisfies this constraint. Later, I introduce a different constraint within the theory of content, and call 'β content' the content which satisfies this new constraint. This leaves it open that β content may be identical to α content.

must know what the (relevant part of the) t-domain of the content is. But a t-domain (unlike the world) is essentially conceptually structured, so there is no way of knowing what the t-domain of a content is without possessing the concepts in terms of which the t-domain is structured. Hence possession of an α content requires possession of the concepts in terms of which it is canonically specified. It follows that α content is a kind of content which consists of conceptual properties, as defined above. That is, α content is conceptual content.

The process of identification of α content as conceptual content may be mimicked in order to demonstrate a notion of non-conceptual content. We must ask, is there a way to motivate in a similar fashion the application of the definition of non-conceptual properties within the theory of content? In asking this, I am asking whether non-conceptual specifications of states or activities can ever be canonical within the theory of content. Thus I am asking whether non-conceptual specifications of an activity can ever be *required* by a correct theory of content in order to capture the distinctive way in which some aspect of the world is given to the subject of the activity.

We can clarify what is involved in doing this by setting out, as a summary of the above discussion, the different elements that I have used in motivating the definition of conceptual properties within the theory of content:

1. The definition of conceptual *properties* (by stipulation);
2. The claims that there is a constraint within the theory of *content* which requires determinate truth conditions,[37] and that possession of content which satisfies this constraint requires knowledge (grasp) of its truth conditions (these claims are given by the theory of content, and are constitutive of this notion of content);
3. A psychological state expressed linguistically as 'thinking[38] that the Stanford Campus is near here', not yet analysed with respect to the kind of content that it has;

[37] Or, a determinate contribution to determinate truth conditions. I shan't continue to make this qualification.

[38] It has been a philosophical convention since Frege (cf. Frege 1977) that *thinking* is a psychological concept, whereas *thought* is a logical or philosophical notion. *Concept*, like *thought*, is, in the first instance, a logical notion: concepts are thought constituents. So saying 'my *thought* that p' within the convention, entails that the content of the state is conceptual. Saying, merely, 'my *thinking* that p' does not entail any consequence about the kind of content that the state has.

Part of what I am addressing in the paper is the question whether *concept* should, as well as being a logical notion, also be a psychological notion. Psychology, I am assuming, must employ some notion of content, but I will suggest that the kind of content which is conceptual content has only a logico-philosophical role; psychology requires a different kind of content—non-conceptual content.

4. The claim, argued in the text, that the interpretation of (3) under (2) requires the notion of a task-domain and the specification of the content of (3) by means of concepts of the task-domain.
5. (4) results in the satisfaction of (1), hence the identification of content which satisfies the constraint in (2) as *conceptual* content.

The notion of a task-domain provides the link between the philosophical notion of α content, and my stipulative definition of conceptual properties; a link which is needed to show that the analysis of a psychological state in terms of α content entails satisfaction of the definition of conceptual properties.

The case of Non-Conceptual Content

I can show the need for non-conceptual content by showing that there are psychological states the full understanding of which requires a notion of content which cannot be analysed in this way; that is, which must be canonically specified by means of concepts that the subject need not have. The discussion will have to parallel the discussion for the case of conceptual content, so we need a parallel for (1) – (5):

1′. The definition of non-conceptual properties (by stipulation);
2′. Some constitutive conditions on a kind of content, β content, which are provided by the theory of content, but which are different from the conditions in (2).
3′. Some psychological or representational state as yet unanalysed with respect to the kind of content it has.
4′. An argument for the claim that the interpretation of (3′) under (2′) requires the notion of some domain other than the task-domain and the specification of the content of (3′) by means of concepts of this domain.
5′. A demonstration that (4′) results in satisfaction of (1′), hence the identification of β content as non-conceptual content.

We already have (1′). What about (2′)?

Cognitive Significance. A good theory of content is answerable to various constraints. For example, a good theory of content should be appropriate for use within a content-based scientific psychology, it should have resources to explain how certain contents have determinate truth conditions and a good theory of content should also capture *cognitive significance*, that is, the role that content plays with respect to perception, judgement, and action.

How can the theory of content accommodate cognitive significance? Frege's notion of sense[39] was introduced, in the first instance, to explain how certain identity statements could be informative. For example, to learn that Hesperus = Hesperus is not to learn anything new, but to learn that Hesperus = Phosphorus may be to learn something of considerable significance, yet Hesperus *is* Phosphorus. It follows that possession of the content expressed here by the word 'Phosphorus' cannot consist just in the ability to think of the planet Venus (specified no further than this), because just the same ability is associated with 'Hesperus'. There is here a motivation for introducing a notion of content (sense) which differs from a purely referential notion of content (reference). There is a content expressed by 'Hesperus' which is different from the content expressed by 'Phosphorus' because the former content plays a different role from the latter content in a person's judgements of the truth value of contents of the form '. . . = Hesperus'. Frege generalized this motivation into a criterion of identity for such contents (senses).[40] We may generalize it still further to yield a generalized notion of sense which I call 'β content', whose identity conditions are fixed, not just by its constitutive connections to judgement, but by its constitutive connections to perception, action, and judgement.[41] Possession of a particular β content requires possession of a contentful state which plays that role in the psychological economy of the subject which is constitutive of the β content.

A major success within recent work in the theory of content has been to show that there are indexical and demonstrative β contents that cannot be canonically specified, in the way appropriate to conceptual content, by means of any description.[42] This has been achieved by showing that were a description—*per impossibile*—to provide canonical specification of the content, in the way appropriate to conceptual content, it would alter the cognitive significance of the content, that is, the character of its constitutive connections to action and judgement. Since cognitive significance is constitutive of β content, it follows that this form of specification cannot canonically capture β contents.

For example, Perry (1979) shows this for the indexical 'I' and connections to action, and Peacocke (1986) shows it for demonstrative perceptual contents and connections to perception and judgement. Perry's point is that the conceptual use of any descriptive canonical specification—*the

[39] Frege (1891).
[40] See n. 43 below.
[41] See e.g. Peacocke (1986). Frege added the further condition on sense, that it *determine* reference. This is not, however, a condition on ß content. Only certain ß contents (those that are senses) determine reference.
[42] Of course, the people I cite don't put their conclusions this way!

x such that φx^*—for the indexical content $*I^*$, will alter the cognitive significance of the thought $*I$ am ψ^* by altering its constitutive connections to action. The reason for this is that it is always possible that one may not realize that I am the x such that φx, so that even if one would act immediately on the basis of judging $*I$ am ψ^* (e.g. $*I$ am spilling sugar all over the supermarket floor*), one might not act on the basis of judging $*$the x such that φx is ψ^*.

Peacocke contrasts what a person knows when he or she knows the length of a wall in virtue of just having read an estate agent's handout, and what a person knows when he or she knows the length of a wall just in virtue of looking at it. Frege's intuitive criterion of difference[43] for contents can be used to show that although both people know the length of the wall, neither knows what the other knows. Thus suppose that my wife's and my cognitive states were identical except for the fact that I know what the length of the wall is just in virtue of having read the handout, and she knows what the length of the wall is just in virtue of having seen it. But then, thinking of the length of the wall in only that way which is available to each of us, I may be agnostic about the thought $*$that length is greater than the length of our piano* (because, for example, we don't know how long in feet our piano is), whereas my wife will judge this thought to be true because, simply by looking, she can see that our piano will fit against the wall. Therefore, the perceptual demonstrative β content differs from any descriptively specified conceptual content, and so cannot be canonically specified, in the way appropriate to conceptual contents, by means of any specification such as 'the person sees that distance-in-feet $(a,b) = n$' where a and b are the end-points of the wall.

We could treat examples such as Perry's and Peacocke's in a way which was similar to my treatment of thinking that the Stanford Campus is near here—that is conceptually—by means of concepts of the respective task-domains. That would be, in effect, to characterize these indexical contents in a descriptive, conceptual fashion.[44] But Perry's and Peacocke's arguments show that justice cannot be done, in such a way, to the *cognitive significance* of these contents. So we have only to recognize a notion of content for which cognitive significance is essential, to see that there is

---

[43] Frege's intuitive criterion of difference: the thought grasped in one cognitive act, x, is different from the thought grasped in another cognitive act, y, if, and only if, it is possible for some rational person at a time to take incompatible attitudes to them; i.e. accepting (rejecting) one while rejecting (accepting) or being agnostic about the other.

[44] Cf. n. 29 above where I say that the notion of a task-domain prescinds from any notion of indexicality. There is no point of view in a task-domain, so if point of view is essential to indexicality, a notion of content for which indexicality is essential cannot be captured by means of concepts of the task-domain.

a kind of content which cannot be canonically specified by means of concepts of the task-domain.

The argument so far shows that there is a very large class of cognitive states (all states which contain indexical or demonstrative elements[45]) which have a kind of content (β content) for which the only canonical conceptual specification is the use of a simple demonstrative or indexical under the conditions of a shared perceptual environment or shared memory experience. Such a specification is evidently useless for the construction-theoretic purposes of a scientific psychology since the only way the theorist can have to understand the nature of the content is either to share the experiential environment of the content, or draw on similar experiential environments available to the theorist in memory experience.[46] (Scientific psychology, here, is psychology which is aiming to solve the problem of embodied cognition, and which therefore is aiming to construct any explanatorily indispensable notion of content out of non-content involving levels of description.[47]) Yet this class of contents is particularly important for psychology, at least because of its direct connections to action and its crucial role in learning. Is the theoretical psychologist therefore incapable of capturing those contents which are basic to our ability to act in the world and to learn from it?

Only if the psychologist assumes that he or she must work with conceptual content. The problem arises because there is no conceptual structure within the demonstrative or the indexical or the observational content which can be exploited to yield a canonical conceptual specification of the

---

[45] Are there any representational states which don't contain, either explicitly or implicitly, indexical or demonstrative elements? Perhaps *God is good*, because it is part of the essence of God that He is unique. (Just about every definite description contains an implicit indexical reference to, for example, *our* earth). But does 'good' mean 'good to us', 'good from our point of view, rather than say, the Devil's'?

[46] The theorist can *refer* to the mode of presentation in question without *employing* it, but this doesn't help. What is in question is the kind of explanation of the nature of these contents that a scientific psychologist can give or use, if the psychologist is restricted to conceptual kinds of specification, and accepts as the explanatory task the need to construct any psychologically indispensable notion of content. The trouble is that if the specification is canonical, the theorist's capacity to understand the nature of the content in question depends, ineliminably, on his or her having had similar experiences. Thus, conceptual specification of these contents which is both canonical and *theoretically* adequate fails because there are only two ways to conceptually specify such contents: by means of concepts of the task-domain, or by use of the indexical or demonstrative term where the understanding of the use of the term depends on either sharing the experiential environment, or having had similar experiences. Perry's and Peacocke's arguments show that the first method of specification cannot be canonical for β contents, while the ineffable dependence on having had certain sorts of experiences shows that the second method of specification cannot be theoretically adequate. (Thanks to Christopher Peacocke for pointing out this worry to me.)

[47] It may be objected that I am imposing overly strict explanatory demands on a theory of content. I consider this objection in Cussins (1992).

content which would be appropriate for the purposes of a scientific psychology. But this doesn't exclude there being any *non-conceptual* structure within the content. If we can make sense of this notion, then there is here an argument to show that much of the psychological life can only be captured by means of, and should, therefore, only be modelled in terms of, non-conceptual content.

The Notion of a Substrate-Domain.   Abandon, then, the demand that every content must have its theoretical specification given in the way which is constitutive of conceptual content; that is, by means of concepts of the task-domain. What other theoretically adequate method of speci-fication could there be? I introduce below one kind of canonical non-conceptual specification. It is not necessarily the only kind,[48] although I believe that it is the only kind in terms of which we can solve the problem of embodied cognition.

It will help to consider the operation of an autonomous, mobile robot known as 'Flakey' which lives at a research institute, SRI, in California.[49] Flakey navigates the corridors of SRI. His task is to move up and down the corridors, avoiding hitting the walls, and to turn into particular doorways.

In order to be able to behave flexibly in a range of task-domains a system must be able to employ representations[50] of features which are special to the domain in which it happens to find itself. For example, if the width of corridors varies in Flakey's environment, then Flakey will need to respond differentially to corridor width. Given the kind of system that Flakey is, this will mean that Flakey will have to represent this variable. The system need only not represent that which does not change through-out the career of the system. So the greater the system's representational capacity, the greater its potential flexibility. Should we suppose, there-fore, that the cognitively ideal system would computationally represent— in the traditional AI style—all the facts there are? That although nothing

[48] E.g. Dretske's notion of information (Dretske 1981) would be a notion of non-conceptual content, were it to be a notion of content, because one does not need to possess the concept of information in order to be in information carrying states. (Evidently so, since even trees — for that matter, anything at all — carry information.) The trouble comes when Dretske tries to justify the notion of information as a notion of content. Peacocke (1989) develops a different notion of non-conceptual content.

[49] See Reifel (1987).

[50] I don't want to beg the question as to what kind of representational system is sufficient for the possession of concepts, so, in the discussion of systems such as Flakey, I use a general notion of representation, which is neutral with respect to whether its significance (e.g. its semantics) is only extrinsically attributed, or whether its significance (like that of content) is intrinsically available. I consider the conditions for a physical system to have states whose significance is intrinsically available in §7.

achieves this ideal, the closer one comes to it, the better one's cognitive capacities will be?

To suppose this[51] is to miss an important distinction between two kinds of fact. What I want to show is that computational representation of only one of these kinds of fact is required for the ideal Artificial Intelligence system. Flakey is sometimes imagined to deliver pizza throughout SRI. It might be that only one weight of pizza is allowed through the extensive security system, and that Flakey would therefore be built on the assumption that if something is recognized as a pizza, then the mobile arm needs to exert a certain force to lift it. This would have the effect of 'unburdening'[52] the representational capacities of Flakey, with respect to having to work out each time it was about to lift a pizza, how much force was required to lift it. This connection could simply be built into the hardware. However, the folks down at Hewlett Packard, intrigued by Flakey's growing reputation, might want to try him out on delivering pizza for them. They would be sorely disappointed because, unfortunately for Flakey, the security system at HP labs lets all weights of pizza through. Flakey was discovered to be throwing pizza around in a way not likely to impress DARPA.[53]

Indeed, DARPA could reasonably argue that this was a *cognitive* defect of Flakey's. We treat intelligence in an open-ended way: so-and-so may be great at chess, but if he can't learn to play Go, then we think him the less intelligent for it. For Flakey, representation of pizza weight is required for acceptable, let alone ideal, cognition.

But we shouldn't conclude therefore that to be truly intelligent Flakey must represent all the facts there are. For example, it would be surprising if Flakey were to represent the distance between the sonar sensors at its base. This is not only for the reason that this distance is a constant throughout Flakey's career, but, more importantly, because Flakey's own structure is not part of Flakey's task-domain. Flakey never has to manipulate the distance between his sonar sensors; this distance is not something with respect to which Flakey's performance will be evaluated. Rather, it is part of Flakey's substrate of abilities in virtue of which Flakey has those corridor-movement behavioural capacities which he in fact has. This distinction between task-domain ('t-domain') and the domain of the system's substrate of abilities ('s-domain') is essential to understanding what a flexible system is required to represent. To be able to operate flexibly in a range of t-domains a system must be able to represent those

---

[51] As certain theorists in AI do; see e.g. Lenat and Feigenbaum (1987).

[52] See Barwise (1987).

[53] The American defence department funding agency for 'advanced research projects'.

features of a t-domain which vary, or may vary, within the range of t-domains. But so long as the s-domain is outside this range, as it usually will be, a flexible system has no need to represent aspects of its s-domain.[54]

My visual capacity may be quite superb and open-ended: I can visually discriminate any kind of object, in an extensive range of conditions of illumination, and distances from me, and so forth. But nobody would suggest that it is a defect of my visual capacity, that I am ignorant of the algorithms employed by my visual information-processing system. With respect to my personal level visual capacity, my subpersonal information-processing capacities are part of the s-domain.[55] Given a division between t-domain and s-domain in a particular case, performance in the task-domain—even fully conceptual performance—does not require the possession of any concepts of the s-domain.

Specifying β Contents by Concepts of the Substrate-Domain.   As we saw, the notion of a t-domain provided the link between α content and the definition of conceptual properties. Can the notion of the s-domain provide a parallel link between β content and the definition of non-conceptual properties? An intelligent agent does not need to have concepts of its s-domain, so if β content can be canonically specified by reference to the objects and properties of the s-domain, we will have motivated a kind of content which is specified by means of concepts that the system or organism need not have.

Consider the following quotation from Evans (1982: ch. 6):

What is involved in a subject's hearing a sound as coming from such and such a position in space? . . . When we hear a sound as coming from a certain direction, we do not have to think or calculate which way to turn our heads (say) in order to look for the source of the sound. If we did have to do so, then it ought to be possible for two people to hear a sound as coming from the same direction and yet to be disposed to do quite different things in reacting to the sound, because of differences in their calculations. Since this does not appear to make sense, we must say that having spatially significant perceptual information consists at least partly in being disposed to do various things.

When Evans asks, 'what is involved in a subject's hearing a sound as coming from such and such a position in space?', he is asking about the

[54] See Cussins (May 1987).

[55] Evidently, what abilities are part of the s-domain will be relative to a particular task-domain. Knowledge of visual information-processing algorithms is not part of many people's task-domain, but it was part of David Marr's. Hence what kind of content some state possesses, will be relative to the kind of evaluation which is appropriate to it: the particular way of dividing up t- and s-domains for the particular case. There may be more than one task-domain for a single state at a time.

nature of the content by which the subject is presented in experience with this aspect of the world. Evidently the content is indexical or demonstrative since, were we to express the content in words, we would say that perception presents the sound as coming from 'that location', or 'from over there'. The conclusion drawn on the basis of Perry's and Peacocke's examples applies: there is no way to canonically specify this content as a conceptual content, if we wish to do theoretical justice to the cognitive significance of the content; in particular its direct connection to action. What Evans adds, is, first, a further reason why this kind of content cannot be captured conceptually (no conceptual content can be necessarily linked to action as directly as certain β contents require), and, secondly, the suggestion that the way to capture the cognitive significance of the content is by reference to a way of moving in the world; the subject's ability to reach out and locate the object, or walk to the source of the sound, which the perceptual experience makes available. At the place in the argument which we have now reached, it is this second idea which is important, because, for Evans's content, a way of moving in the world is part of the s-domain.

Given our usual views about consciousness, the idea here can seem quite strange: it is the idea that certain contents consist in a means of finding one's way in the world (tracking the object, say) being available to the subject in his or her experience, even though it may not be available to the subject conceptually, and, indeed, the subject may be incapable of expressing in words what this way of moving is.[56] My knowledge of where the sound is coming from consists in, say, knowledge of how I would locate the place: knowledge which is exhausted by what is available to me directly—without depending on any concepts—in experience. I may have that knowledge even though I am unable to entertain any thoughts about the way of moving in question; I require no concepts of my ability to find my way in the environment, in order to have an experience whose content consists in presenting to me a way of moving.

It may help to consider one of the most extreme cases of non-conceptual content (to which I will return in §8): the case of pain-experience. We have been taught in the philosophical tradition not to view pain-experience as experience with any *content* at all; its function isn't to represent the world, we are told. But the reason for this is not because pain-experience isn't phenomenologically very similar to the

---

[56] For that matter, the subject may also be incapable of moving. A way of moving may be available in my experience, even though I am incapable of acting on the basis of it. (The content would still be canonically characterized in terms of its constitutive connections to perception and action.)

experience of colour or shape of objects, but, rather, because we do not view the world as possessing various paining properties. We say that the edge of my desk is coloured brown on the basis of a visual experience as of brownness, but we do not say that the edge of my desk has a sharp paining on the basis of a tactile experience of a sharp pain. I will give a reason in §8 as to why this is the case, but for now the point is to think of experience as a spectrum of kinds of experience ranging from pain-experience where we are not remotely inclined to attribute the experienced property to the world, through colour-experience where we do attribute the experienced property to the world (but we get into some trouble for doing so—§6), to shape- or motion-experience. Pain-experience is just much less objective[57] than shape-experience. This will show up in the kinds of content that pain-experience can have, as against the kinds of content that shape-experience can have. Pain-experience never has conceptual content, but it doesn't follow that it has no content at all. Pain-experience presents the world as being painful; paining is made available to one in pain-experience. But we don't suppose that we need concepts of pain for this to be the case; we just have to *be* in pain, or to remember being in pain. In a similar way, experience can present a way of moving in the world, even though the subject of the experience has no concepts of ways of moving.

Our kinaesthetic sense provides another example. On the basis of kinaesthetic experience the subject knows how his body is arranged; how his hands are in relation to each other and to his head, for example. But the person need have no concepts of this spatial arrangement in order to have this knowledge. Rather, the knowledge consists in an experiential sensitivity to, for example, moving one's hands closer together, or bringing one's hands next to one's torso. The capacities one has to rearrange one's body are directly present in kinaesthetic experience without having to possess any concepts of the arrangement of a body.

Returning to the example from auditory experience, Evans's idea is that the spatial content of the auditory perception has to be specified in terms of a set of conceptually unmediated abilities to form judgements and to move in the egocentric space[58] around the organism. This is because the content consists in the experiential availability to the subject of a dispositional ability to move. The experiential content of perception is specified in terms of certain fundamental skills which the organism possesses, 'the ability to keep track of an object in a visual array, or to follow an instrument in a complex and evolving pattern of sound'. These are skills which

---

[57] In the sense of 'objective' which I make clear in §§ 7 and 8. Bascially, pain-experience is less objective because it is less perspective-independent.

[58] For some exposition of this term, see the discussion of the map-maker in §9.

belong to the subject's s-domain. So, if Evans is right, this class of contents is canonically specified by reference to abilities which are part of the s-domain, and therefore by means of concepts which a subject need not have in order to grasp any member of this class of contents. So the structure of the (conceptually atomic) indexical, demonstrative, and observational contents of experience is the structure of their non-conceptual content. ß content is non-conceptual content.

People often misunderstand this as a behaviouristic theory, so let me emphasize again that the claim is not, in the first instance, about the characterization of a *subpersonal*[59] perceptual state of the organism. The aim is to capture how the person's perceptual experience presents the world as being (i.e. a genuine notion of personal level content). The notion of non-conceptual content is a notion which must ultimately be explained in terms of what is available in *experience*. If the content is canonically characterized as a complex disposition of some specified sort, then the claim is that this disposition is directly available to the person in his or her experience, and that the content of the experience consists in this availability. But for a behaviourist, the notion of experience can have no explanatory role.[60]

In summary, then, I have discerned a constraint on content in terms of cognitive significance, rather than in terms of truth conditions; I have suggested in a Fregean spirit that we need to introduce a kind of content which is answerable to this constraint; I have shown that this kind of content cannot be canonically specified in any way which is appropriate to conceptual content, and that it is therefore not a type of conceptual content; we have seen that we need to employ this kind of content to do full justice to any cognitive psychological state with indexical or demonstrative elements (most of our cognitive life); that a plausible suggestion for how to canonically capture the content is by means of concepts of the s-domain; and that since a cognitive creature does not need to have concepts of its s-domain, I have shown that this kind of content satisfies the definition of non-conceptual properties, and is therefore a kind of non-conceptual content.[61] I will call the kind of non-conceptual content which I have introduced, 'construction-theoretic content (CTC)',

---

[59] See Dennett (1969: 93–4) and Dennett (1978: 101–2, 153–4, 219).

[60] Not just for a behaviourist, actually. The cognitive revolution may have reinstated the notion of representation, but it hasn't yet reinstated experience (my notion of content). I hope by this article to push us a little way towards doing that.

[61] It should be noted that there may be kinds of ß content which are not canonically specified by means of concepts of the s-domain, or, more narrowly, by means of concepts of ways of finding one's way in the environment; if so, these will be kinds of content which are not kinds of conceptual content. Their status will depend on how it is proposed to canonically capture them.

because I shall go on to show how this kind of content can form the basis for a construction of conceptual capacities.

How Widespread is the Phenomenon of Non-Conceptual Content?

The content of certain conceptual states has only the structure of their non-conceptual content, and so can only be psychologically analysed in terms of their non-conceptual structure. There are two levels of analysis of content, conceptual and non-conceptual, and it has been demonstrated that the psychological explanation of a certain portion of our cognitive life can only be given in terms of its non-conceptual structure. It is irresistible to wonder, how widespread is this phenomenon? Could it be that even for those areas of cognition where there is conceptual structure, the correct level of scientific psychological analysis is still in terms of its non-conceptual structure? *Is the psychological structure of cognition its non-conceptual structure?* I believe that the hypothesis that it is is the basis for a connectionist alternative to LOT. But this is to run ahead of ourselves.[62]

It will help to consider some other examples. Evans quotes Charles Taylor as follows:

Our perceptual field has an orientational structure, a foreground and a background, an up and down . . . This orientational structure marks our field as essentially that of an embodied agent. It's not just that the field's perspective centres on where I am bodily—this by itself doesn't show that I am essentially agent. But take the up-down directionality of the field. What is it based on? Up and down are not simply related to my body—up is not just where my head is and down where my feet are. For I can be lying down, or bending over, or upside down; and in all these cases 'up' in my field is not the direction of my head. Nor are up and down defined by certain paradigm objects in the field, such as the earth or sky: the earth can slope for instance . . . Rather, up and down are related to how one would move and act in the field.

Taylor is here asking what the significance of our concept *up* consists in. He considers three answers, two of which are: up is where my head is, and, up is where the sky is. But the significance of our notion of up cannot consist in our grasp of the direction of our head or the direction of the sky, because, for example, we can perfectly correctly employ the concept *up* when we are lying down. And so on. Then Taylor offers a third answer, 'up and down are related to how one would move and act in the field'. This immediately strikes one as a very different sort of answer from the first two answers that Taylor considers. In the first two cases what is being offered is a traditional conceptual analysis; a definition, as we might

---

[62] The argument given here can be extended from the case of indexical and demonstrative senses to all senses.

define 'bachelor' to refer to an unmarried adult male. Where it is proper to give a traditional conceptual analysis, a person's understanding of the left-hand side of the definition must consist in the cognitive availability of the conceptual structure which is displayed on the right-hand side. But Taylor's third answer is not a definition; it simply states that our possession of the concept *up* must be analysed in terms of certain basic, non-conceptual abilities that we possess, such as our ability to move and act in a co-ordinated way. These basic abilities may be characterized by means of technical concepts (such as concepts of the way in which the gravitational force structures our field) which an organism need not possess in order to possess these basic abilities. Taylor has hit upon the analysis of a concept in terms of its non-conceptual content.

Or consider recognitional abilities in those cases (the majority) where recognition does not depend on the recognition of the object as *the x such that φx and ψx* (for any concepts of properties *φ* and *ψ*). For example, my ability to recognize my wife's face as Charis's face is not an ability (even a subpersonal ability) to recognize the unique face with certain conceptual features (e.g. Roman nose, distance between eyes being n inches . . .). When I think a perceptual-demonstrative thought of the form *That is Charis*, my cognitive state is not correctly reconstructable as involving the inference, that is the φ person, the φ person is Charis, so that is Charis. In fact, my ability to recall a person's features (even a person whom I know very well) when not in their presence is extremely limited, but this in no way diminishes my ability to hold a particular person in memory. (In an extreme case, I might not be able to recall a single perceptual feature of my wife, and yet be unrivalled in my ability to think singular thoughts about her.) So it cannot be that the capacity for me to hold someone in memory in the way required for me to have a singular thought about the person consists in my storing some set of conceptual features which, as it so happens, are uniquely satisfied in the world.[63]

What this suggests is that although there will be mental features in our theory of recognition, they won't be features whose analysis depends on a *semantic* account, i.e. a semantic relation between the feature and some objective element of the appropriate ready registered task-domain. Our

---

[63] Remember that *conceptual* does not equal *conscious*. Of course recognition of Charis does not depend on matching to consciously stored features; this is not the point I am making. In claiming that the psychological structure of recognition of individuals is its non-conceptual structure, rather than its conceptual structure, I am claiming that a computational model of individual recognition must be suited to transforming representations which have non-conceptual, rather than conceptual, content. Much of this computational transformation of representations will normally, of course, be quite unconscious.

ability to recognize massively outstrips our ability to recall, and cannot be analysed in terms of it.[64] The suggested alternative is that our ability to recall objective features of the world is dependent on the structure of the non-conceptual content of our recognitional capacities; content specified in terms of basic spatial and temporal tracking and discriminatory skills which are required to find our way around the environment.

Non-Conceptual Content and Representational Vehicles

Equipped with the distinction between conceptual and non-conceptual content, we can return to the general argument that the kind of representational theory that a computational model employs determines the kind of content appropriate for that model, which in turn determines the kind of psychological explanation that the model can provide. In the next section, I consider the kind of psychological explanation that requires conceptual content, and in §6 I consider the kind of psychological explanation that non-conceptual content can make available. But it will help first to move down the psycho-computational hierarchy one step to see the connection between the two kinds of content and two kinds of representational theory.

I explained in §3 that LOT's capacity to respond to the problem of embodied cognition depended on its use of the S/S theory of representation. We are now in a position to add a feature to those features discerned in §3 which are constitutive of S/S theory. I have already pointed out that the level of semantics and the level of syntax are explanatorily independent of each other, in the sense that one does not have to know the semantic theory in order to understand what the theory of the syntax is saying, and vice versa. Syntax must respect semantic constraints, but the operations of proof theory, or procedural consequence, defined over syntax, are formal in that they are independent of semantic features. All that I have said so far about semantics is that we often want a semantic theory to be a finitely axiomatizable recursive theory of semantic properties. We can now see that a semantic theory is a theory of the relation between syntactic items and *conceptual* contents.[65] A semantic theory has the form that it has because the base axioms specify relations of reference or denotation to objects, properties, or situations of the task-domain. S/S theory is committed to conceptual content.

[64] See Evans (1982: ch. 8).

[65] A typical S/S system will not, strictly speaking, have conceptual contents, since it won't, strictly speaking, be a concept-exercising system. Such a use of the notion of conceptual content is a derivative use, which must ultimately be explained in terms of the paradigmatic use of the notion with respect to a subject of experience and thought (see n. 20). Nevertheless, derivative uses of the notions of content may have considerable utility.

What form would a representational theory for non-conceptual content take? Notice, first, that the relation between vehicle and content will not be given by a semantic theory, since the contents are non-conceptual. The contents carried by the vehicles cannot be given by referential or satisfaction (i.e. semantic) relations between the vehicles and the elements of the task-domain. Notice, secondly, that the relation between the representational vehicles of CTC, and CTC itself cannot be a semantic relation, because the level of CTC is *not* explanatorily independent of the level of the s-domain abilities of an organism in virtue of which the organism is able to find its way in its environment. These s-domain abilities are the vehicles which carry CTC,[66] but we saw that we could not understand how experience non-conceptually presents the world, without specifying such contents in terms of these very abilities. The 'syntax' and the 'semantics' of non-conceptual content are not explanatorily independent, so they are not, strictly speaking, syntax and semantics. The notion of non-conceptual content shows how there can be a radical alternative to S/S representational theory. The value of this point is that when we come to consider Fodor and Pylyshyn's (1988) criticism of connectionism to the effect that connectionist causation is not syntactically systematic, we can agree with Fodor and Pylyshyn, but show how this fact redounds to the advantage of connectionism.

5. CONCEPTUALIST THEORIZING AND PSYCHOLOGICAL EXPLANATION

The computational use of S/S theory entails the use of conceptual content, which in turn entails conceptualist theorizing. Conceptualist theorizing is theorizing about cognition, at the level of *psychological* explanation (see Fig. 2), in terms of conceptual properties. The application of conceptual properties, as defined in §4, to theorizing about an organism's cognitive functioning requires that the organism be assumed to possess at least that set of basic concepts on which the analytic hierarchy of concepts[67] is to be

---

[66] The theory of the vehicles of CTC might take many different forms: a cybernetic account, an information-processing account, an ecological account (Gibson 1986), or a tensor theoretic account (Pellionisz and Llinas 1979, 1980, 1982, 1985), and Smolensky (1988b). But the theory of the representational vehicles of content should not be confused with a theory of content. We should not speak of cybernetic, or information-theoretic contents, only of cybernetic or information-theoretic vehicles of content. In §9 I draw a distinction between representational vehicles of non-conceptual contents which are abilities, and connectionist computational vehicles of non-conceptual contents which are patterns of activation distributed over processing units.

[67] The 'analytic hierarchy of concepts' just means the ordering of concepts due to the relations of one-way logical dependency between concepts. Thus the concept *spectacles*

founded. This set of presupposed basic concepts will be taken to be either innate or acquired by non-psychological (e.g. neurophysiological) processes, or both.

Logic has given us a way of understanding the atemporal logical relations which obtain between constituent concepts and between complex thoughts. Work on knowledge representation in cognitive science[68] has thrown up numerous formalisms for extending this to capture the dynamic temporal relations that constitute actual theoretical and practical reasoning. All of psychological activity will then be taken to consist in various manipulations of the relations between the basic concepts (central processes of reasoning and learning), or in establishing the connection between the central concept manipulating system and the sensory and effector systems of the organism (a connection achieved by the peripheral, perception, and action modules; cf. Fodor 1983).

By presupposing the possession by the organism of a set of basic conceptual contents, the theorist is presupposing the availability to the organism of a ready registered world (a task-domain), consisting of that set of objects, properties, and situations which are taken in the semantics to be the referents of the basic concepts and their complexes. The connection between this system of basic concepts and the world is held to be achieved by peripheral modules which, apart from some parameter setting, are essentially innate. Thus the *cognitive* starting-point, according to conceptualist theorizing, is already a point at which an objective (if, yet, simple) world is available to the subject of cognition: objectivity is presupposed in order that cognitive theorizing can begin. If there is a problem about how physical organisms can acquire capacities of registration which register the world objectively (unlike capacities based on pain-experience,[69] say), it is taken to be not a psychological problem, but rather a problem for neurophysiology or the theory of natural selection.

By contrast, because the possession of CTC does not depend on the possession of the concepts by means of which the CTC is canonically specified, non-conceptualist theorizing need not presuppose the organism's possession of basic concepts, and so need not presuppose the availability to the organism, of a ready registered world. Rather, non-conceptualist theorizing (as we shall see) is theorizing about those processes that *give rise* to the availability to the organism of an objective,[70]

logically or analytically depends on the concept *eye*, but not vice versa. The concept *bachelor* depends on the concept *adult*, but not vice versa.

[68] See e.g. Brachman and Levesque (1985).

[69] See §4 and the next footnote.

[70] The reader will have noticed that I am using the word 'objective' more and more. A very rough idea is that something is objective if it is independent of a subject's grasp of it. Paining

registered world. The problem of how basic concepts are acquired is not therefore cast aside for other disciplines to deal with, but is treated directly as the central problem for psychology. As I explain in §6, non-conceptualist theorizing takes cognition to be the *emergence of objectivity*, not the inferential manipulation of a ready-made registration of objectivity. One result of this is that the perception and action systems are not peripheral modules whose function is to establish a connection between surface impingements and a central reasoning system, but are the core of cognition. High-level reasoning forms merely a peripheral structure built around the core.

Another consequence of conceptualist theorizing is that beneath the explanatory level of the basic concepts is nothing psychological; just an implementation[71] theory. This is a theory of how the conceptual activities are instantiated within non-conceptual processes, e.g. neurophysiological processes. The conceptual theory within cognitive psychology may be realized within a theory of computational procedural consequence as, within logic, the semantics (entailment relations) is realized within syntax (proof-theory). The point is that, for conceptualist theorizing, what lies beneath the basic concepts is *explanatorily independent* of the conceptual level, in the sense that one can understand fully what is happening at the conceptual level without, necessarily, understanding anything about what is happening beneath this level.[72] The slogan of conceptualist theorizing is that what is beneath the concepts is beneath explanatory bedrock.

By restricting attention to the conceptual structure of cognition, the conceptualist is forced to model all cognitive processes as processes of inference, either demonstrative or non-demonstrative (even learning is modelled as hypothesis formation, and hypothesis verification or

properties are not objective in this sense, neither are university degrees (?), but being triangular is. I shall be less rough about the notion in §7.

[71] x implements y if x is a substrate for y, but the explanation of what it is to be y is independent of (does not in any way rely on) the explanation of what it is to be x. A basic dogma of cognitive science (rejected by e.g. Churchland 1986) has been that neurophysiology implements cognition: cognitive facts may be true in virtue of neurophysiological facts, but the explanation of the nature of cognitive facts is quite independent of the explanation of neurophysiological facts. People speak of a 'mere' implementation theory, not because its provision would not be both major and desirable, but because it would not shed any light on the psychological nature of cognition.

[72] For philosophers: notice that the conceptualist's explanatory independence between levels is compatible with supervenience. A much more difficult question is whether it is compatible with the construction constraint. Ultimately I think not, but showing this depends on showing that all conceptualist attempts to construct concepts, such as LOT and functionalism, fail. Fodor rejects total explanatory independence of the conceptual and non-conceptual levels because, for example, he thinks that conceptual level phenomena of opacity have to be explained by reference to the syntactic form of the cognitive representations ('Propositional Attitudes', in Fodor 1981).

falsification). Conceptual structure is, indeed, precisely that structure which is required to model all of the logical inferences which are required for some analysis. So psychology is held to be explanatorily dependent on logic. What a concept is is fixed by its location in an inferential network, and its psychologically extrinsic connections to the world. By contrast, non-conceptualist theorizing exploits the non-conceptual structure *within* atomic concepts. Once the shape of individual concepts has been correctly modelled, the possibilities of inferential combination will follow (and be explicable) as a tessellation of these elements. (In the argument which follows, I shall focus on the systematicity (§7) of contents. If we can show that it is possible to capture systematicity at the level of non-conceptual content, then we will have shown that it is possible to capture inferential connections at the level of non-conceptual content.[73] What it would be to get right the shape of the elements so that the correct combinatorial tessellations follow is considered in §7. How one could achieve this is considered in §8.)

It is not correct that modelling the combinatorial structure of cognition *entails* modelling cognition with a representational system with a syntax and a semantics (Fodor and Pylyshyn 1988). For, if non-conceptualist ideas are correct, it is possible to capture combinatorial structure by means of modelling the non-syntactic/non-semantic representational structure within concepts. Instead of the nature of concepts being held to be explanatorily dependent on the nature of the inferential connections between concepts, the nature of the inferential connections will be explained as a consequence of the non-conceptual nature of the constituent concepts.

In summary, the following elements are intrinsic to conceptualist theorizing:

1. The presupposition of basic concepts;
2. The connection between concepts and the world is a 'peripheral' connection which is largely innate;
3. The presupposition of objectivity: the cognitive starting-point assumes a task-domain which is available to the organism;
4. The problem of the acquisition of a capacity to register the world as the objective world is not a psychological problem, but a problem for neurophysiology or the theory of natural selection.
5. The location of psychologically explanatory bedrock: the lowest level of psychological theorizing is the manipulation of basic concepts. Conceptual properties are merely implemented in non-conceptual properties.

[73] The claim is that inference is just a special case of the general phenomenon of systematicity of contents. Inference is the systematicity of *thought* contents.

6. Representational structure is intended, in the first instance, to be the structure of the inferential connections between concepts. The nature of a concept is explained in terms of the nature of its inferential connections, rather than vice versa. The explanatory centrality of inference, rather than learning.[74]

Each of these elements is a consequence of restricting psychological explanation exclusively to conceptual content and is, therefore, a consequence of the computational and psychological application of the S/S theory of representation, and hence a result of LOT. Suppose we want psychology to explain how physical organisms come to acquire basic concepts; that we want a psychological explanation for the possession by organisms of the elements of thought (that we want—for whatever reasons—psychological explanation that does not have the consequences (1)–(6)). Then we need an alternative to LOT. Given our assumptions that psychological explanation must be based on *some* notion of content, and that an alternative to LOT must still be a computational theory, then we require a theory based on the psychological and computational use of non-conceptual content.[75] But what *could* such psychological explanation be like? How could there be an alternative to conceptualist psychological explanation?

Armed with a notion of non-conceptual content (CTC), how can we do justice to the conceptual characteristics of cognition? If a cognitive life could be characterized only at the level of non-conceptual content, it would involve merely a very primitive registration of the world. But even a *primitive* registration of the world is a primitive *registration of the world* only if it is possible to exhibit it as a simple form of, or constituent in, a sophisticated, or fully conceptual, registration of the world. How can we do that?

6. NON-CONCEPTUALIST PSYCHOLOGICAL EXPLANATION: THE EMERGENCE OF OBJECTIVITY

Conceptualist theorizing presupposed a ready registered world—a task-domain—for its semantic theory. Because CTC is specified in terms of the properties of the s-domain rather than the t-domain, psychological

---

[74] (1) and (4) are recognized in Fodor (1976, 1980, 1981a), (2) is recognized in Fodor (1983), (3)—like the other elements—is implicit in the practice of much of AI, and (5) and (6) are exploited in Fodor and Pylyshyn (1988).

[75] It doesn't *follow* that we require a theory based on the psychological and computational use of CTC, since there may be other kinds of non-conceptual content. But my aim throughout the paper is to indicate how a certain kind of theory is possible, rather than necessary (even though I believe that too!).

theorizing based on CTC does not have to presuppose a ready registered, fully objective, world available to the subject. All that is presupposed are the basic, non-conceptual, organismal capacities of the s-domain. Non-conceptual content, in isolation from background conceptual capacities, presents the world (of course), but not yet objectively: as the world which is available from any perspective, and about which one may be mistaken.[76]

Is the Realm of Reference Explanatorily Prior to the Cognitive Realm?

Where the content of experience is *exclusively* conceptual content, some element of the objective world is given to the subject *as* (and only as) an element of the objective world (and thus as a task-domain element). For this reason, it is an adequate specification of an atomic conceptual content simply to use whatever linguistic item (name or predicate) stands in a primitive semantic relation to the objective element of the world which is the referent of the content. (And molecular contents can be specified as logical constructs out of atomic contents.) Thus, if I perceive a traffic light as a traffic light (and not as, say, an elongated, colour-coded candy), then an adequate specification of the content of this aspect of my experience will be the use of the phrase 'traffic light'. This is a *referential* specification of content. It picks out traffic lights as task-domain elements.[77] If the content of experience was exclusively conceptual then it would be possible to take any aspect of the content of experience, assume that the aspect presented an objective element of the world as an objective (i.e. task-domain) element, and then name that element as the referent of the content.

But it is not possible. Consider a particular colour-experience, divorced from a possible theoretical background.[78] Were we to attempt to specify

---

[76] I explain why this is so in §7.

[77] And thus distinguishes the content from an elongated colour-coded candy content, because, in a task-domain, elongated colour-coded candies are different objects from traffic lights (although they might occupy the same location). This is because a task-domain is a region of the world *under a given conceptualization*.

[78] By colour-experience which is divorced from a certain theoretical background, I mean colour-experience considered apart from a subject's knowledge that one might, for example, individuate colours by their matching conditions to all objects (Goodman 1951). Such basic colour-experience is governed by the two principles which I have adapted from Peacocke (1986). (Because these principles make reference to the causal conditions of colour perception, *basic* colour-experience is not *primitive* colour-experience which would be experience considered apart from a subject's knowledge that it is brought about causally in a certain way.) Considerable logical sophistication might give rise to a coherent referent for colour contents, but these colour contents will no longer be the colour contents of basic colour-experience, but rather the contents of super-sophisticated set-theoretical concepts. Basic colour contents are essentially observational: one can tell, normally, just by looking, whether they apply to the world.

the content of this experience referentially as an experience *of* a particular colour shade we would take our experience to be governed by the following two principles[79] (amongst others): where a subject's colour-experience of a surface A is not discriminably different from his or her colour-experience of a surface B, and the subject's perceptual faculties are functioning normally and correctly, the colour of surface A is identical to the colour of surface B. And, where a subject's colour-experience of a surface A is discriminably different from his or her colour-experience of a surface B, and the subject's perceptual faculties are functioning normally and correctly, the colour of surface A is different from the colour of surface B. But since, notoriously, basic colour-experience is a dimension along which non-discriminable difference is non-transitive, our attempt would quickly lead to contradiction, since we would obtain the result that the colour of A is identical to the colour of B, the colour of B is identical to the colour of C, but the colour of A differs from the colour of C. Therefore, the colour of A is identical to, and is not identical to, the colour of B. So basic colour-experience cannot be specified referentially in the familiar conceptualist fashion. There are no precise colour shades by means of reference to which we could specify the content of basic, observational, colour-experience.[80]

We have to be a little careful here, since there is (more or less) a concept *red*—we think of the world as objectively being coloured red—which can be a content of basic colour-experience. So we can specify a colour-experience as being as of a red surface. The question for the conceptualist, though, is whether we can presuppose an account of what it is for a surface to be coloured red, in order to characterize the content of basic colour-experience in the familiar conceptualist fashion by means of a semantic relation of reference to the presumed item, a red shade. But what the argument above shows is that there are no red shades that can play this role; in so far as there are red shades, they can't be task-domain elements. Hence the explanation of what it is for a surface to *be* red cannot be prior to the explanation of what it is to *experience* a surface as red. (Thus, the theory of the content of our colour-experience cannot *presuppose* a world of objective colour properties; it cannot presuppose its own task-domain. To specify a content by the concept *red* is to specify it

---

[79] This argument is given in Peacocke (1986), which makes use of the argument in Dummett (1978: 'Wang's Paradox').

[80] There are numerous attempts to define artificially a non-paradoxical colour referent. Some of these may be successful, but all of them involve some departure from our intuitive practice. The artificial referent will not be determined by the cognitive significance of our basic colour contents (as sense should determine reference). As mentioned in n. 78, I am concerned here only with basic, observational, colour contents of experience.

in a conceptual, but not a conceptual*ist*, fashion. It will then be for the non-conceptualist to give the proper account of concept ascription.)

Trying to get clear about the conclusion of this argument makes apparent an implicit consequence of conceptualist psychological theorizing: by presupposing the psychological possession of basic concepts, the conceptualist is also presupposing a psychologically independent account of what it is for there to exist in the world an objective referent as a suitable denotation for a basic concept. Perhaps it is for physics to tell us, but if so, psychology can play no explanatory role in this. By presupposing the possession of basic concepts, the conceptualist is assuming that all that can be said, from a psychological perspective, about what is essential to basic concepts (apart from their interaction with other concepts) is that their content consists in a semantic relation to an objective item of the world. So for a conceptualist, nothing can be said, by psychology, about what it is for the referents of basic concepts to exist. For psychological explanation, we presume the world, and try to explain the nature of mind in terms of it; that is, there is an explanatory priority of world to mind.

But it is just this explanatory priority which the argument from colour-experience casts doubt on. (And it is anyway highly dubious: do we really imagine that physics can tell us what chairs, or soccer matches, or university degrees, or crumpled shirts are, independently of a psychological account of what it is for organisms like us to recognize something as a chair, or . . . ?)

It would be equally implausible to adopt (with Berkeley) the reverse priority. The more satisfying alternative is to suppose that the explanation of cognition and the explanation of the world are inter-dependent. If we combine this idea with the idea that we should use a notion of non-conceptual content to provide a construction of our conceptual capacities, the result is a glimpse of non-conceptualist psychological explanation: the story of the capacity to think is the story of the non-conceptual *emergence of objectivity*. For the non-conceptualist, the elements of psychological explanation do not depend on the applicability of a mind/world distinction to the explananda 'subjects' or organisms. These non-conceptual elements are then employed in an account of how such a distinction comes to be applicable, and thus in an account of what it is for the explananda organisms to become subjects of experience and thought.

The Emergence of a Mind/World Distinction

Let me say a very little about what this means. Our ordinary talk about cognition embeds a mind/world distinction; we characterize our cognitive states as being about a mind-independent world, and so we characterize cognition in terms of an external relation between two independently existing entities: mind and world. I don't just mean that the world continues to exist even though nothing is perceiving it, or that there are more truths about the world than anyone knows, or that there are truths which we are incapable of recognizing, but that there is the kind of gap between mind and world that allows minds to be wrong about the world, to refer to the world (rather than merely being immersed—like a paramoecium—in the world), and therefore to be able to think thoughts about the world.

A manifestation of this is that we draw a sharp distinction between those predicates that can be appropriately combined with subject-terms which refer to the world, and those predicates that can be appropriately combined with subject-terms which refer to the mind. Thus we can say of a ball, but not of an experience, that it is round. We can say of a memory, but not of a football pitch, that it is veridical. And where a single predicate can be used in both contexts, we insist that, strictly speaking, it has a different meaning in each context, or is literal in one context and metaphorical in the other. Both memories and football pitches can be shrouded in mist, but . . .

One of the very few exceptions to this is the predicate ' . . . is objective'. Both an experience and a shape property can be objective. How can this predicate have a special status? Suppose that the mind/world distinction is a phylogenetic or ontogenetic *achievement*: paramoecia don't manifest it, but we do; very young human infants don't manifest it, but adult humans do. Natural selection has evolved creatures, by means of gradual and continuous genetic changes, which are capable of a mind/world distinction; and the processes of learning within the infant develop, through gradual and continuous neurophysiological changes, cognitive capacities which present an independent world to an independent mind. The question then arises, can we describe, explain, and make sense of the pre-objective stages of development before a mind/world distinction is applicable, by means of a theory of processes defined over non-conceptual content? Can we explain the transition from such a pre-objective stage where no concepts are possessed to an objective concept-exercising stage, by means of a theory of the computational manipulations of non-conceptual content? Non-conceptualist psychological explanation is the attempt to do just this. It is, therefore, the attempt to show how the mind/world distinction—objectivity—can emerge from a pre-objective stage at which there is merely an undifferentiated mind/world continuum. The mind is *embedded* (Cussins May 1987) rather than solipsistic or formal

(Fodor 1981a), in that a theory counts as a theory of the possession of a concept if, and only if, it is also a theory of what it is for the referent of the concept to exist.

All of our descriptions and explanations are, of course, from our conceptual perspective, so we can describe the undifferentiated mind/world continuum either from the perspective of the mind, or from the perspective of the world. Adopt, for a moment, the former perspective. Then the emergence of objectivity is the transition from mere experience to experience *of* the world.[81] Take the latter perspective. Then the emergence of objectivity is the transition from atoms swimming in the void, to ironing crumpled shirts and designing collaborative document-processing office environments.

Claim-Setting[82]: Objections from Supervenience and Superstructure

One might suppose that learning—the emergence of objectivity—is just a ladder to be kicked away once climbed. In other words, the conceptualist might grant that we need a non-conceptualist explanation of phylogenetic and ontogenetic development, but insist that once developed, adult human cognition is adequately treated in the conceptualist fashion. But this would be to get my argument back to front. I have not argued that we need to give a non-conceptualist account of learning and therefore we need to give a non-conceptualist account of adult cognition. Rather, I have discerned a psychological necessity to recognize non-conceptual content within adult, human cognition, and then seen that the notion so introduced is appropriate for explanation of learning.

There are two kinds of significance that the notion of learning can have, according to whether the notion is embedded in a conceptualist or non-conceptualist theory. If the former, then learning is just a ladder to be kicked away, because what is reached by climbing the ladder is something whose nature is explained by a theory (the theory of inference) which is independent of the theory of learning. But if the latter, then learning is essential to (mature) cognition because the nature of what is learned is not

---

[81] 'We can imagine a series of judgements "Warm now", "Buzzing now", made by a subject in response to changes in his sensory state, which have no objective significance at all. But we can imagine a similar series of judgements, prompted by the same changes in the subject's sensory state, which do have such a significance: "Now it's warm", "Now there's a buzzing sound"—comments upon a changing world. What is involved in this change of significance?' (Evans 1980). The exchange between Strawson (1959: ch. 2), Evans (1980), and Strawson's response to Evans (1980) is a classic discussion of adopting the mental perspective on the emergence of objectivity.

[82] Where a little stage-setting goes on by laying out some claims and counter-claims in a largely metaphorical fashion.

explanatorily independent of the mechanisms of learning. For the non-conceptualist, learning is the emergence of objectivity, and cognition (without learning) is the maintenance of objectivity. The explanatory notions in terms of which the maintenance of objectivity is to be explained are the same as those employed in the theory of learning. It would be illegitimate to assume the significance that the conceptualist attaches to the notion of learning, and then argue that C3 must be mistaken because it supposes that learning is essential to cognition.

My claim is that we are—as adult humans—still awash in a partially differentiated, partially objective, mind/world continuum, at which pain-experience lies at one end, various sorts of emotional experience a little further in, then colour-experience, then, perhaps, shape-experience, and the experience of democratic justice. Our concept possession is, at all moments of our cognitive lives, a matter of staying afloat (shape-experience), or only half drowning (the paradoxes of colour-experience), or being almost completely submerged (pain-experience). The exercise of a concept is the result, both literally and metaphorically, of our ability to find our way in the environment; to stay afloat. Our non-conceptual capacities *sustain* our conceptual capacity; they are not merely part of a transition phase.

The conceptualist's objection might take two forms. First, he might suppose that there is a counter-example to C3 in the form of an organism, which, by cosmic coincidence, comes into existence *de novo* as a fully formed adult human, perhaps molecule for molecule type-identical with a human who has come into existence in the ordinary way. Wouldn't C3 be committed to denying that such a creature possessed concepts, and wouldn't such a denial be incompatible with supervenience?[83] But, in asserting the centrality of learning, C3 does not hold that a person who came into existence *de novo* would not be a concept-possessor (as the objector supposed). Only that the *explanation* of the conceptual capacities which, we agree, he would possess, would be by means of a theory of learning. My claim is a claim about the direction of explanation, not about the ontology of supervenience. As is the conceptualist's who could grant the theoretical possibility of a concept-possessing system which had never engaged in inference, while also holding that the explanation of the possession of concepts would be by means of a theory of inference. So a non-conceptualist can claim that the explanation of what it is to possess concepts is to be given in terms of a theory of learning, while granting the possibility of a concept-possessing system that has never engaged in learning. Though in both cases, as soon as the concepts come to be *exercised*,

[83] An objection raised by Ned Block during a discussion at Birkbeck.

the proprietary processes (inference in the former case, learning in the latter) will also be exercised.

A second form that the conceptualist's objection might take has to do with cognitive superstructure, such as those cognitive capacities which involve language. The objector is willing to grant that linguistic capacites are *grounded* in non-conceptual capacities, but nevertheless insists that, once grounded, they build up on their own, yielding the higher aspects of cognition for which LOT is appropriate.[84] I consider, briefly, linguistic cognition less metaphorically towards the end of the paper, but, in the spirit of claim-setting, I note here the form of the C3 response.

We must be careful to keep in mind the distinction between vehicle and content. Linguistic and linguistically-infected cognition is crucial to the scientific psychology of human cognition, but we should not argue from the central role of linguistic vehicles in human cognition to a LOT-style theory which supposes that *conceptual content*, traditionally associated with language by the S/S theory, is basic to the scientific psychology of human cognition. Indeed, I have argued that linguistic vehicles often express *non-conceptual content*. The non-conceptualist, recognizing the important role of linguistic vehicles, maintains that it is the non-conceptual content of these vehicles, not their conceptual content, which is psychologically potent. Non-conceptual content doesn't so much form the foundation of cognition, as provide the building blocks out of which the cognitive superstructure is constructed. What is of concern is whether S/S theory is required for scientific psychology to respond to the problem of embodied cognition. The conceptualist's claim that it is, is not only the claim that language is the psychologically basic form of representational vehicle, but also the claim that S/S theory is the right theory of language. A proper recognition of the role of language in cognition entails neither of these claims.

Words do, often, express concepts, and that they do so is of great significance for our cognitive life. This is not denied by C3 which aims to show first, how the S/S theory of this phenomenon is unworkable because it cannot capture, in a theoretically adequate way, the cognitive significance of indexical and demonstrative contents and, therefore, is ill-placed to yield adequate theories of learning, perception, memory, and action (all of which are essentially indexical). And secondly, and ultimately more importantly, that there is a way of treating the conceptual phenomena of language which does not rest on S/S theory, but which is naturalistically acceptable. We need to identify the importance of concepts (§7) and then

---

[84] I have in mind, here, objections raised by Andy Clark and Martin Davies. Not that they are committed to conceptualism!

show how a computational theory of non-conceptual content can capture this importance (§§ 8 and 9).[85]

There are colour concepts, but because of the threat of paradox and problems of cognitive significance, basic colour cognition cannot be explained in the conceptualist fashion. The non-conceptualist supposes that all of our cognitive life is like basic colour-experience, only a bit better or a bit worse; we can usually treat it conceptually but never conceptualistically. Conceptual content is a valuable idealization, rather than the basis of psychological explanation. Obviously, it is more of an idealization with respect to colour cognition than with respect to number cognition, but the broad base of our cognition is not too dissimilar from the case of colour: the argument from the non-transitivity of non-discriminable difference of colour-of-surface experiences can be extended to cover any vague concept for which the Sorites paradox obtains: *bald*, *heap*, many shape concepts (e.g. where I have a concept of a 25-sided figure, but not *as* a 25-sided figure), the concept of the letter 'A' (which cannot be specified geometrically or topologically), the concept of the phoneme [p] (which cannot be specified acoustically), ethically and aesthetically evaluative concepts, the concept of democracy, my concept of macaroni . . . (If there are exceptions to this, they are mathematical or set-theoretical concepts.)

Conceptual contents are the availability in experience of (part of) a task-domain. CTC contents are the availability in experience of substrate-domain abilities. Non-conceptual contents, we saw, cannot be explained by means of t-domain specifications, so non-conceptual contents cannot be explained in the way distinctive of conceptual contents. *Non-conceptualist psychological explanation depends on the converse possibility*: that conceptual content *can* be explained in the way distinctive of non-conceptual content. The exercise of concepts—which we know to consist in the availability in experience of a t-domain—turns out to be a special case of the availability in experience of s-domain abilities. For the non-conceptualist, the notion of experiential availability of s-domain abilities is a generalization of the notion of the experiential availability of a t-domain. Scientific psychology should, therefore, dispense with the less general notion, and model cognition entirely in terms of the theoretical apparatus distinctive of CTC content. But we will then need to explain the conditions under which the experiential availability of an s-domain constitutes the experiential availability of a t-domain.

---

[85] Then, too, there are important cognitive phenomena which are dependent on linguistic—and other communicative—vehicles. Recognition of this is entirely within the spirit of C3 (§9).

7. OBJECTIVITY CONSTRAINTS

If non-conceptual content is to play the role in non-conceptualist psychological explanation that I have indicated—as the base for a progressive construction of objectivity—then we need to understand why certain non-conceptual cognitive states are not properly interpreted as presenting the objective world to a subject, and why certain other, more sophisticated, non-conceptual states do count as providing objective registrational capacities.

The conceptualist need say very little about objectivity[86] since an objective relation between mind and world is a presupposition of his psychological theorizing. But the non-conceptualist cannot afford this luxury; because, for him, psychological explanation is the explanation of the transition from a pre-objective state to an objective state. We need to understand how there can be a principled, if not sharp, distinction between creatures like paramoecia and creatures like us, between infantile and adult cognition, perhaps also between normal and demented cognition. With a fair grasp of the principle of this distinction, we can then address, in §§ 8 and 9, how best to model the computational processes that can transform a creature from lying on one side of the distinction to lying on the other side.

Some Intuitions about Objectivity

Our notion of the world is a notion of what is independent of an organism's idiosyncratic relations to the world, because it is a notion of what is common for all. If the idiosyncratic relations are specified informationally, then our notion of an element of the world is a notion of something which is independent of any particular informational relation to it. If the idiosyncratic relations are specified experientially, then our notion of an element of the world is a notion of something which is independent of any particular subjective experience of it. Different kinds of representational system may gain very different perspectives on the world, but we can talk of 'perspective' only because there is a common focus; a common something which may be glimpsed in very different ways, a common basis for agreement and disagreement. A theory of objectivity is a metaphysical theory of what it is for there to be a world in this sense. It is a theory of what it is for the significance of an organism's particular relations to the world to go beyond what is particular in those

---

[86] And we do hear very little about it from psychology. Certain areas of developmental psychology, esp. Piagetian psychology, are among the few exceptions.

relations, to go beyond the particular energy configuration, or the particular configuration of sense-data. If all there is to being in the relational state is having that energy configuration, then being in the relational state cannot 'present' anything which goes beyond the particular relation itself. So it cannot present something which is common for all. A theory of objectivity is, thus, a theory of the 'cognitive' separation between what is particular to an organism's relations to the world and the world itself; it is a theory of what it is for there to be a distinction between the cognitive processes of an individual and the common world.

One way to investigate what objectivity is and what it is to register the world objectively is to ask why we might think that this cognitive separation characterizes human relations to the world but not those of paramoecia or frogs. What are the criteria for the distinction between thermostats, voltmeters, paramoecia, and frogs, on the one hand, and cognizing persons on the other? Can we give a theoretically grounded distinction between conceptual response to the common-sense world and transducer response to nomic properties?[87]

A Coherence Test for Objectivity: The Case of Frogs and Automatic Screwdrivers

I propose a test for this distinction which is based on the insight that the explanation of mind and world is inter-dependent. The test, in essence, is this: take the capacities of the functioning system as non-conceptually described, and attempt to interpret these capacities as conceptual. The attempt will be successful, if and only if, the (putative) world which would be presented by means of the (putative) conceptual capacities is a coherent world. The distinction between concept-exercising and non-concept-exercising organisms is made to rest on a metaphysical theory of coherence conditions for the world, or what I call the objectivity constraints.[88]

Consider a simple example: is it correct to attribute concepts (and therefore a capacity to register the world objectively) to an automatic screwdriver equipped with some simple sensory apparatus which detects, as we loosely say, whether a screw is present, and if so whether it is screwed in or unscrewed. Must this talk of detecting screws be interpreted instrumentalistically[89] (Dennett 1987), or is it of essentially the same kind

---

[87] See Fodor (1986).

[88] This approach, like Evans's is influenced by ch. 2 of Strawson (1959).

[89] A way of talking which has utility but not truth. It may be useful to speak of the weather being fierce, even though the weather is not the kind of thing which *can* be fierce, strictly (i.e. truthfully) speaking. It is often held that the ascription of beliefs to thermostats is similarly instrumentalistic. Dennett (1978) holds that the ascription of beliefs even to us is instrumentalistic.

of talk as what we take to be the realist attribution of conceptually-based perceptual mechanisms to ourselves? In applying the test we would ask first, what would a world be like which was presented to a system with these non-conceptually characterized capacities? (Would it meet the constraints on objectivity?) We would attempt to interpret the capacities as conceptual, and would answer that it would have to be a world which just contained screws, and two properties, screwed or unscrewed. But such a world is not coherent: screws can only be part of a world in which there are factories which make them, properties of rigidity, length, and weight which they have, a location where they are, a direction in which they are screwed, . . . So the automatic screwdriver does not count as possessing any concepts at all.

This is what shows that talk of concepts of screws in connection with the screwdriver is extrinsic talk, in virtue of the artefact's functional location within the human world. We characterize instrumentalistically the abilities of the screwdriver in terms of elements of the human world, because we can happily presuppose the human world for the purposes of designing and evaluating the artefact. It is only because we can make this presupposition that we can talk of the artefact 'detecting screws'. So understanding the abilities of the screwdriver does not help at all with understanding how it can be that, for certain physical systems, a world is conceptually available to them.

Similar points apply to our descriptions of, say, frogs' behaviour, where our talk of frogs' detection of flies again depends on the extrinsic attribution of concepts, an attribution which we undertake in order to understand the evolutionary 'design' of the frog. Again we presuppose the human world because if we do so we get fairly good predictions of the frogs' behaviour, a fairly good understanding of the success of the frog design, without becoming submerged in the physiological details or the pure information-theoretic account. There is no intrinsic and non-instrumentalistic attribution of concepts to frogs because any attempt to interpret the non-conceptually characterized abilities of frogs as conceptually presenting a world yields an incoherent world. Nothing could[90] be a fly if it didn't have a size, but the success of the frogs' detection system depends on not discriminating flies in the near-distance from massive objects in the far distance. Nothing could be a fly if it couldn't be stationary, but the success of the frogs' detection system depends on the movement of flies. A frog's notion of a fly is a notion of an always-mobile

[90] For philosophers: the notions of possibility and necessity here are notions that are grounded in descriptive metaphysics (Strawson 1959).

something which has no size. So it is not a notion of a fly. So frogs don't have fly concepts.[91]

The Holism and Generality Constraints

These examples motivate a *holism constraint* on objectivity: nothing could count as a concept of an object or property unless it was a part of a complex, holistic web of concepts for the reason that nothing could be such an object or property unless it was a part of that complex, holistic web of objects and properties that is the referent of the conceptual system (and conversely).

The holism constraint adds power to Evans's generality constraint: an organism does not possess a concept *a* of an object unless it can think *a is F*, *a is G*, and so on for all of the concepts *F*, *G*, . . . of properties which it possesses (and which are not semantically anomalous in combination with *a*). And, similarly, an organism does not possess a concept *F* of a property unless it can think *a is F*, *b is F*, and so on for all of the concepts *a*, *b*, . . . of objects which it possesses. Thought is essentially structured[92] because the world is essentially structured, and conversely.

The generality constraint, in association with the holism constraint, which are grounded in the metaphysics of objectivity, are of enormous value for providing a principled basis for a separation of concept-exercising systems, for which there is a cognition/world distinction, from systems which cannot exercise concepts, for which there is no distinction. For example, consider the mechanism of auditory localization in humans. The information-processing mechanism carries out various computations which depend on the speed of sound and the distance between the two ears. We may speak of the system representing these quantities. How similar is this notion of representation of distance and speed to the representations which I employ when I plan to eat my lunch at Stanford?

---

[91] An objector might agree that frogs don't have OUR fly concepts, but still insist that they have FROG fly concepts; or that an automatic screwdriver doesn't have our concept of a screw, but it does have a 'black & decker' screw concept. But this is incoherent. Concepts just are our concepts; the way in which the human world is presented. Screws are part of the human world; they have their identity only in the context of the human world. Screws are necessarily the kind of thing that are presented to us by means of our screw concepts. Perhaps automatic screwdrivers live in a world which is incommensurable with ours? Well, if so, it makes no sense for us to try to characterize cognition for that world. We cannot make sense of a perspective from which frog or black-and-decker contents satisfy the objectivity constraints. If their world is incommensurable with ours, then it will be to us as if they have no world (see also Davidson 1974).

[92] For discussion of this notion, see Campbell (1986).

Are we justified in attributing concepts to the auditory mechanism, as we are justified in attributing concepts to the whole person?

By the generality constraint, the localization mechanism could only think *this sound is travelling at x m/s* (where x m/s is the speed of sound) and of thinking *the distance between the ears is y m* if it was also capable of thinking *this sound is travelling at y m/s* and *the distance between the ears is x m*. But there is no warrant of the kind that there is for attributing the content, *this sound is travelling at x m/s*, for attributing the content *this sound is travelling at y m/s*, or the content *the distance between the ears is x m*. But, by the generality constraint, nothing could be a warrant for the capacity to think *this sound is travelling at x m/s* or *the distance between the ears is y m*, unless it was also a warrant for the capacity to think *this sound is travelling at y m/s* and *the distance between the ears is x m*. Hence there was no proper warrant for the concept *x m/s* and *y m*. So, although there may be reasons for attributing representations, in the broad sense of 'representation', to the mechanism of auditory localization, it is not correct to attribute concepts to this mechanism. And, similarly, by the holism constraint, it makes no sense to suppose that a system possesses the concept of some number n, if it is not also capable of thinking of $(n-1)$. And again, . . .

Evans (1982) employs the generality constraint to show that possession of an information link[93] characteristic of a perceptual non-conceptual content is insufficient for possession of a concept. If my registration of my coffee cup consisted *solely* in the perceptual information link currently between me and the cup, then although we could imagine attributing to me the thought *that cup is grey* (cf. *that fly is to my right*) on the ground that my information link delivers information about the colour of the cup, there would be no ground for attributing to me the thought *that cup was manufactured in Stoke*, or *that cup won't move when the lights are out*, or *that cup will be smashed tomorrow* even though, we are to suppose, I have the concepts of these properties. The reason for this is that if my capacity to register the cup is *exhausted* by my capacity to discriminate colour and shape properties of the cup (on the basis of the information link), then I cannot register the cup as the kind of thing that exists in the dark, or was manufactured at some other location at some

---

[93] See Evans (1982: ch. 5). An information link is a link between an organism and an object by means of which the organism receives information about the object. One's judgements and movements may be responsive to changes in properties of an object on the basis of an information link. The content of the information which is picked up by the organism is not conceptual.

other time. Moreover, if my demonstrative thought about the cup consisted solely in my information link with it, there could be no basis for my ever being in error about the cup. A notion of cognition grounded solely in information links with the world would be a notion of *perfect* cognition of the world, and would therefore be a notion which provided for no distinction between cognition and the world. (The separation between me and the sensory deliverances from the cup would be no greater than, and no different from, the separation between me and the deliverances of my retina. But this latter separation is not a *cognitive* separation: there is no sense in which my cognition is *about* my retina.)

These objectivity constraints provide a metaphysically grounded way to assess physical systems for the possession of concepts, as the examples of the automatic screwdriver, the frog, the mechanism of auditory localization, and the information link with the cup demonstrate. But they also function as success criteria (a target) for the non-conceptualist psychological enterprise. Non-conceptualist explanation is explanation of how it is possible for physical systems to make the transition from being in states which are characterizable solely by means of specifications appropriate to non-conceptual content, to being in states which approximately satisfy the objectivity constraints, and which therefore are also specifiable in the way which is appropriate to conceptual content. And being in states which are specifiable in the way which is appropriate to conceptual content is to be such as to satisfy certain logical norms: for example that these states, or complexes formed from them, can enter into correct patterns of inferential connection, can be truth-value assessable, and so be the bearers of genuine, full-blooded intentionality. To show how to build non-conceptually characterized structures which approximately satisfy the objectivity constraints is to show how objectivity can emerge within the physical world.

8. PERSPECTIVE-DEPENDENCE

With a principled basis for objectivity now in play, we are in a position to explain why many types of CTC (§4) do not present the world to a subject as the objective world. And we are also in a position to understand what transformations must be applied to these types of CTC in order to yield types of CTC which *do* present the world to a subject as the objective world.

Perspective-Dependent Abilities v. Perspective-Independent Abilities

Conceptual contents are the availability in experience of a t-domain, and CTC contents are the availability in experience of an s-domain (§4). Non-conceptualist psychological explanation depends on showing that there is a spectrum of CTC contents, at one end of which the s-domain experiential availability entails t-domain experiential availability in virtue of the approximate satisfaction of the objectivity constraints. I want to suggest that this spectrum can be ordered as a dimension according to the degree of *perspective-dependence* of the s-domain abilities which are experientially available. The degree of perspective-dependence of the abilities by reference to which a CTC content is canonically specified is the degree to which the content fails the objectivity constraints. Conversely, the achievement of perspective-independence in these abilities is what is required for approximate satisfaction of the objectivity constraints, and therefore what is required in order for the content to present a t-domain. Therefore, the theory of the whole range of contents specified by means of abilities which are perspective-dependent to differing degrees is a generalization of the theory of conceptual content which concerns itself with only one end of this range.

One kind of perspective-dependent ability is the ability to find one's way through a city where the ability depends on starting at a certain location within the city and going to a particular location elsewhere in the city. The starting location will be identified by its appearance, and then the ability to follow the route from start to finish will depend on recognizing each landmark along the route partly in terms of its appearance, partly in terms of the fact of its being the nth landmark in the series of landmarks beginning with the starting location. Associated with each landmark will be an orienteering instruction: 'turn right and carry straight on', and so forth. The ability will not result in satisfactory performance if the project required not starting at the starting location, or not going from the start to the finish by the particular route, or going to a different finish location, or if the appearance or relative location of any of the landmarks alters. In all of these ways, the ability is heavily dependent on the perspective on the domain that the system must adopt. Certain *points of view* within the domain (from the starting location and from the landmarks) are privileged, in that the ability to find one's way through a city is dependent on occupying these particular, privileged points of view (perhaps in a certain order). The notion of point of view is an experiential notion. A perspective-dependent ability to find one's way around is an ability which depends upon the system's having certain particular kinds of experience: those it will enjoy from the privileged points of view.

A perspective-independent ability to find one's way through a city is such that were one to emerge within the city from a manhole, then wherever one emerged one would be able, barring external obstacles, to find one's way to any other point within the city. The ability to find one's way does not depend on a particular perspective (from the starting location) or a particular set of perspectives (a route). There are no privileged points of view, so the perspective-independent ability to find one's way through the city is an ability which yields satisfactory performance *whatever* one's point of view within the city. The ability is dependent on experience of course, but it is not dependent on the system's having particular kinds of experience (a landmark-1 type experience followed by a landmark-2 type experience, etc.). Performance is maintained whatever the experiential perspectives which the system happens to enjoy.

The general notion, here, is that of experience-based ways of finding one's way through a domain which depend, more or less, on having particular kinds of experience: those gained from privileged points of view. The literally spatial case provides the most direct example, but the contrast between perspective-dependent ways of finding one's way through a domain and perspective-independent ways of finding one's way through a domain can be drawn however abstract the domain. Recognition of properties, for example, may depend on the context in which the property is satisfied, whether the property is that of coffee, electromagnetism, or 'freedom fighter'.

From Perspective-Dependent Abilities to Perspective-Dependent Contents

Given a contrast between perspective-dependence and perspective-independence for abilities, we may apply the contrast to non-conceptual contents which are canonically specified by reference to these abilities. Thus, a *perspective-dependent content* is a content which is canonically specified by reference to perspective-dependent abilities. In virtue of the intrinsic nature of the content,[94] it can be entertained only from a particular perspective, or restricted set of perspectives.

This is a very strict notion of perspective-dependence of content, because it is a notion which infects the content itself, rather than merely the external conditions under which the content may be grasped, or the

---

[94] What is captured in the theory of contents' canonical specification of the cognitive significance of the content.

conditions under which it has been learned. Thus, my ability to register my mother is an ability that depends in many ways on having the perspective of a son. Yet I still register her as someone to whom many other kinds of relations are possible; *this* sort of perspective-dependence of a registrational ability (which is quite independent of the theory of the specification of β contents by means of concepts of the s-domain) does not infect the content of the ability. My registration of an elm tree is a registration from a non-botanical perspective (indeed, I can only grasp, in my present state, the content *elm tree* from a non-botanical perspective), yet it is a registration of the referent as the kind of thing which is also available to a botanical perspective.[95] I can only grasp the content *grass is green* if I am not asleep, or at least not dreaming. But the nature of the content itself (*what* I grasp) is quite unaffected by this content-extrinsic perspective-dependence.

Thus, perspective-dependent contents, which are introduced as contents which are canonically specified within the theory of content by means of perspective-dependent s-domain abilities, involve a quite different and much more radical notion of perspective-dependence than the usual one which is concerned only with dependence on conditions under which a content may be grasped. The less radical notion does not help at all with the explanation of why the content of certain states fails the objectivity constraints. But the more radical notion is the basis of such explanation.

It may help to consider an extreme example. Pain contents are specified in terms of the capacity to be in pain. But this capacity is heavily perspective-dependent: as a way of negotiating the world, it depends on having only one particular kind of experience: experiencing pain or remembering experiencing pain.[96] The capacity depends on a unique, privileged point of view. This perspective-dependence *does* infect the content; pain-experience does not present the world as being a way which is independent of how it is experienced; as being a certain way which is available to concept-exercising creatures who can't experience pain. By contrast, perspective-independent contents present the world as being potentially available to the perspective of any concept-exercising creature.[97] They do

---

[95] See Putnam (1975).

[96] It is interesting how inadequate pain memory is. Although one can remember, non-conceptually, being in pain, the memory is often highly deceptive with respect to the intensity of the pain. It is said that there would only be singleton children if this wasn't the case.

[97] Note that this is not to say that there are any ways of cognizing the world which are not from a particular perspective. Every way of cognizing is from a perspective, but only some ways of cognizing present the world as (approximately) being the kind of thing which could be available to any conceptual perspective.

so because they function successfully whatever the particular kinds of experience which, on an occasion, a creature employing such contents uses to negotiate its way through the domain.

Perspective-Independence, the Objectivity Constraints, and Task-Domains

It is by approximately being perspective-independent that certain contents approximately satisfy the generality constraint. Contents which present my coffee cup to me present it to me as being the kind of thing which could, for example, be available to the experience of a cup-manufacturer in Stoke. So I am able to entertain the thought that my coffee cup was manufactured in Stoke.

Where a content fails to meet the generality constraint, the content is not fully structured, so that it cannot, for example, enter into a full range of inferences. We might modify the usual notation for conceptual content by representing the loss of structure by hyphenation. So, for an unlikely example, I might have a registration of Jones which is dependent upon adopting an 'eating perspective' with respect to Jones (as the frog's registration of a fly is not a fully spatial registration, but rather a 'direction- and motion-dependent registration'). If it were correct to interpret my content as a concept, then I could think *Jones is eating*. But since my Jones-recognizing capacity, in terms of which my Jones-contents are canonically specified, is dependent on an eating perspective, I could not entertain the thought *Jones has his mouth taped shut*, or *Jones used to be a zygote back in Manchester*, even though I had the concepts *. . . has his mouth taped shut*, and *. . . used to be a zygote back in Manchester*. Therefore my content fails the generality constraint in virtue of being perspective-dependent; it is thus unstructured and should be represented as *Jones-is-eating*.

Similarly in the case where my ability to think about a distant location in a city as *there* is canonically specified in terms of a route-dependent capacity to find my way around the city. Being here, I can think *there*, lobbing a thought to one of my accessible goal-locations. But my capacity to entertain thoughts about 'there' depends on my adopting the perspective of 'here' (for a particular set of locations), and thus is perspective-dependent and so fails the generality constraint. My content is really *here-to-there*.

Because certain contents are perspective-dependent, they do not present their objects as being the kind of thing which could be available to any contentful perspective; hence, they do not satisfy the generality constraint. If we attempt to specify their content in a conceptualist fashion, then we shall fail because the content fails the generality constraint, and therefore cannot be canonically specified by means of a semantic relation to an objective item of the world (the generality constraint was a constraint on objectivity). Perspective-dependence entails failure of the objectivity constraints entails failure of canonical t-domain specification. It is *because* certain non-conceptual contents are perspective-dependent that they do not present the world as the objective world; and it is because certain non-conceptual contents are very little infected by perspective-dependence that they count successfully as presenting the world as the objective world. Perspective-*in*dependence entails satisfaction of the objectivity constraints entails canonical t-domain specification. This suggests what kind of psycho-computational modelling is required to yield non-conceptualist explanation of conceptual capacities: psycho-computational transformations defined over non-conceptual contents which have the effect of reducing the perspective-dependence of the contents. I will turn in a moment to how we might achieve this effect.

Task-Domain Independence and the Centrality of Learning

The general case of perspective-dependence can be easily stated: where the *psychological success* of a whole system or ability (rather than the evaluation of a particular exercise of an ability at a particular time) depends on being evaluated with respect to[98] a task-domain (in the sense defined in §4), the content which is canonically specified in terms of these abilities is perspective-dependent and so cannot ground the possession of concepts. The frogs' 'fly-thoughts' are not really fly thoughts because their success (and hence their content) depends on special features of the frog task-domain (the cost of tongue-swipes at massive distant objects is outweighed by the benefit of successful fly catches); frog 'cognition' is dependent on the perspective of a particular task-domain. It cannot generalize. My 'Jones cognition' depends on an eating task-domain, and my ability to find my way around Palo Alto depends on Palo Alto being a route-structured task-domain. (Redesign a few buildings at a major route intersection and my entire capacity could be wiped out.[99])

The way to make a system with contents which are doomed to

[98] Or Situated within, in the sense of Cussins (May 1987).

[99] When I first moved to Palo Alto, I used to recognize the street off El Camino on which I lived, from amongst the indiscriminable buildings and roads on my side of the highway, by means of the purple colour of a Taco stand on the corner of the street. One day I drove several miles past my street. The local housing association had objected to the colour, and required the owner of the Taco stand to repaint with a shade of grey.

perspective-dependence is to make it so that its success is task-domain dependent. So, conversely, in order to have a chance of building a system with perspective-independent contents, one must build it so that its success does not depend on the contingencies of some task-domain. Ironically, for a cognitive system to be an objective system whose contentful states are canonically specifiable by reference to a t-domain, the capacities of the whole system—as non-conceptually characterized—must be t-domain independent. It must operate at a level at which its abilities can modify themselves so as to transfer between t-domains; that is, at the level of those learning processes that give rise to the registration of task-domains. Hence this non-conceptualist theorizing[100] puts learning at the core of cognition, rather than at a peripheral transition stage through which we must pass in order to reach cognition proper.

In summary, contents are non-conceptual in virtue of being perspective-dependent,[101] so we can represent the transition from a non-conceptual content to a conceptual content as being the transition from perspective-dependence to perspective-independence. I need a registration of Jones which is available from any perspective that I could adopt towards him. For example, I need to be able to recognize him, or know what it would be to recognize him, not just when eating, but when swimming, reading, with his mouth taped shut, and as a zygote back then in Manchester. Thus my ability to recognize Jones has to be independent of any particular task-domain. Similarly, I need to have a registration of Palo Alto which is not route-dependent, so that I could, for example, pop up out of a man-hole anywhere in the city and find my way to any other point in the city which

---

[100] For examples of people who think just the reverse, i.e. that the route to intelligent cognition is by exploiting the features of particular task-domains, see Israel (1987), or Barwise (1987), or Rosenschein (forthcoming). This paragraph gives a reason why learning is the core of cognition, and why 'problem-solving' (in the GPS sense) is at the periphery. It is important to realize that perspective-independence, like the objectivity constraints, is an ideal to which no physical system can perfectly approximate; my Palo Alto abilities will always be route-contaminated. The point is that this ideal identifies the dimension along which different systems may be assessed for their conceptuality, and therefore for their intelligence. To have one's theory of intelligence (like Situated theory) make task-domain dependence a virtue (so that the way in which to design successful systems is to Situate them in a task-domain) is to abandon psychological theorizing, however good engineering it may be. A theory counts as a scientific psychological theory if, and only if, it shows what it would be to alter a physical system so that it is more nearly conceptual. My Palo Alto capacity, although route-contaminated, is a conceptual capacity because it is essentially a part of a flexible, learning system that, with experience, is moving my ability along the dimension of greater and greater perspective-independence. I was slow, but after the first couple of months I did manage to recognize my street without depending on a feature or landmark, like the colour of the Taco stand.

[101] More strictly: 'in virtue of being canonically specified by means of perspective-dependent abilities'. This expansion should be read into the shorter form of expression, whenever it is used.

might be my goal location. This is not a registration of Palo Alto which is from no point of view (whatever that might be), but a registration which does not depend on one's occupying some particular place within Palo Alto (or set of places, e.g. a route). My registration of Palo Alto is a view from anywhere in that I can find my way wherever I pop up (so with no prior route). Escape from perspective-dependence, and, hence, the achievement of objectivity, is gained not by somehow cutting one's cognitive life free of perspective altogether (a God's-eye view), but by making it a better and better approximation to being such as to work *whatever* perspective one adopts. Cognitive life couldn't be cut free from perspective, but it must be cut free from task-domains.

9. COMPUTATIONAL VEHICLES AND COGNITIVE MAPS

So far I have considered whole-system contents which are specified in terms of abilities of the whole system. The *representational* vehicles for these contents are whole-system abilities. I have indicated the form of a non-conceptual analysis of these contents. I have shown what must be achieved in order for this analysis to yield approximate satisfaction of the objectivity constraints, and thereby for the representational vehicles to carry conceptual contents, with their constitutive inferential roles. This has all been at the level of psychological explanation (Fig. 2). Finally, we must go subpersonal and examine the *computational* vehicles for these contents, in terms of which they will be modelled by cognitive science.

Map-Making and Map-Use

In order to pursue the C3 programme, we need to look at computational ways to reduce the perspective-dependence of the content of the representational states of the system. A good research strategy for this is to examine the transformations of external, communicative representations which have this effect. The importance of such consideration from the point of view of C3 lies not in the communicative representations which are constructed (maps or marks on a white-board), but in the process of representation *construction* and representation *use*. It will be analogues of these processes that a C3 model will need to implement computationally.

Consider how an ordinary map works. We may imagine that the map is constructed by a person walking around a territory, observing how the territory appears to him from each of his perspectives. The map-maker walks through the territory and thereby has an egocentric registration of the layout; that is, a registration of where things are which is in terms of how they are related spatially *to the map-maker*. This is a perspective-dependent ability, because were a map to reflect only this knowledge, it would yield a registration of the territory which depends on following one particular route through the territory: the route which the map-maker followed. And it would therefore be of no use to anyone who wished to follow a different route through the territory. We may say that such a 'map' represents only egocentric space, and not objective space; just as an organism which has accessible only non-conceptual contents can register only egocentrically, not objectively of the world.

Notice that at these early stages, there is no independence between the map-user's registration of where he is and the map-user's registration of what properties there are at the location. Places are identified simply by how they appear, or by their proximity to landmarks, so if one finds oneself at a place where there is no wood, then one could not be at a location marked on the simple graphical representation as having a wood. Such a simple map cannot be in error about what it is like at the places it represents, since the way in which it represents the locations which make up the space (the location contents) is wholly dependent on how it represents what it is like at those places. If one finds oneself at a place in the objective world where there is no wood then one is not at the location marked on the graphical representation as having a wood. The cognitive analogue of this simple spatial representation would be such as to yield a registration of a place about which one can only think that it has the properties that it appears to have. One could not think, for example, *they have chopped down the wood at this place*. There is no distinction between how things appear in one's cognition to be, and how they are.

With time, the map-maker would follow many different routes through the territory, and so would come to be able to represent a multiple route-based registration of the region. A preliminary graphical representation of this knowledge would represent particular routes, and features by means of which one could identify where on the route, or on which route, one was. (Many tourist maps are like this.) Were someone else to use such a map they would have a *route-dependent* knowledge of the territory. The more complex the map becomes, the more it becomes possible to identify a location by knowledge of where one has been and how one has travelled, rather than simply in terms of what properties the map represents as being true of the location. For the cognitive analogue, there is the beginnings of a distinction between how things appear and how they are; the beginnings of the possibility of error. The cognitive map cannot yet ground false judgements about represented locations, but other, less sophisticated norms can apply: the map can be misleading, for example.[102]

As he follows a number of routes through the space, and by using his tracking skills, he is able to determine how certain regions appear from different perspectives, how different perspectives are related to each other, and so he is able to start drawing what we think of as a (topological) map of the space. Such a map captures the space of the territory so that each place is represented in the same way as every other place[103] (rather than in terms of its relation to where the map-maker is, at stage one; or in terms of its relation to other places on the route, at stage 2). No place, or set of places, has a privileged representation. Once in possession of the map, one's registration of the space is not limited to a sense of how the territory looks when following certain routes through it, but rather is a registration of the space from no particular point of view (a view from anywhere); a registration which has utility *wherever* one finds oneself in the space. At this stage the identification of places is no more dependent on the identification of properties, than the identification of properties is dependent on the identification of places. Because of the holistic character of the representation in the map, it would be possible for the local quarry to remove a hillside, and one still identify where one was. Hence the map-user could tell that the map was now wrong about the hillside. A subject with a fully fledged cognitive map could therefore form true judgements about the location. He or she could think *there used to be a hillside here*, because the way of thinking of the location is no longer exhausted by being in receipt of information about how the place now looks.

In terms of the map-user, this story identifies a sequence of progressively less perspective-dependent abilities to find his way around, and in terms of the map-maker, it identifies a progressive construction of abilities that are less and less perspective-dependent. Moreover, the sequence appears to be of just the sort that C3 requires, because as a more sophisticated map is constructed, the kinds of error that are possible and there-

[102] The idea here is that correlative with the spectrum of increasingly sophisticated non-conceptual contents, there is a spectrum of increasingly sophisticated *norms* which can apply to the non-conceptual contents. True/False can apply only to conceptual contents (and thus to states whose non-conceptual content is sufficiently perspective-independent to approximately satisfy the objectivity constraints). But there are lesser norms which can apply to lesser non-conceptual contents. In Britain it is wrong to drive on the right, and right to drive on the left. But consider a time near the beginning of the century. There was no right or wrong with respect to which side of the road one drove on. As there came to be more and more traffic, a convention of driving on the left started to establish itself. During this intermediate phase, one would not be *wrong* to drive on the right (as one now is) but one would be a *menace*. Being a menace is being judged by a lesser norm than being wrong, but a norm nevertheless.

[103] See Evans (1982: ch. 6). The phrase 'a view from anywhere' was used by Brian Smith to capture a difference from Nagel's 'The View from Nowhere'.

fore the level of objectivity becomes more sophisticated. C3's task, then, is to show how to provide a computational analogue of the map-making and map-using story, for each cognitive domain.

The attainment of *cognitive* perspective-independence may be thought of as the computational construction of *cognitive* map-making and using abilities (O'Keefe and Nadel 1978). In the limit, possession of a cognitive map would entail that every place was thought about in the same way as every other place. Of course, no actual cognitive map could fully achieve this aim, since the map will always be bounded. But that is as it should be, because the objectivity constraints are idealizations to which, in principle, nothing could absolutely approximate. The point, though, is that the construction of a cognitive map on the basis of egocentric, perspective-dependent abilities is the right way to achieve a relative approximation to the objectivity constraints. If a subject can think of place A in the same way as he or she thinks of place B, then the ability to entertain any thought about place A (*A is F*) will entail the ability to think that thought about place B (*B is F*).

It may have worried the reader that the map-making story I have told was only in terms of an atemporal registration of locations. I helped myself to the registration of properties, and to the registration of time. But it does not seem too far-fetched to suppose that one could tell analogous stories for these too. We talk of the ways we have of locating ourselves in time, so the idea of a map of 'temporal space' seems quite natural. Likewise with properties: it seems natural to tell a map-making story of how our discriminatory capacities can come to be less and less perspective-dependent. There is a great need for lots and lots of detail here, but my purpose in this paper has been to see how C3 can provide a possible alternative to LOT, not to predict the results of years of detailed empirical research in a cognitive science informed by C3.

Connectionism: An Appropriate Computational Architecture for C3?

Can connectionism[104] provide a computational architecture which is appropriate for modelling cognition according to the C3 framework?

Paul Smolensky (1988a) has held that the Proper Treatment of Connectionism (PTC)—in order to provide an alternative way of modelling cognition—requires that connectionism model 'sub-conceptually' at a level between the 'symbolic' and the neural. If he is right, then given the discussion in this paper, it follows immediately that PTC must not aim to give *conceptualist* psychological explanation (§5). The conceptualist does not

[104] See Rumelhart, McClelland, and the PDP Research Group (1986).

recognize any representational level beneath the level of atomic concepts; any level of explanation beneath the symbolic conceptual level is implementation theory. But PTC must open up a cognitive explanatory gap between the conceptual level and any level of implementation theory. So PTC cannot be conceptualist.

Tracing this consequence down through the hierarchy of Fig. 2 (§1), it follows that PTC needs to make use of a notion of content which is different from a notion of conceptual content, and hence use a representational theory different from the S/S representational theory (§4). But we also saw (§3) that LOT was able to account for how computational psychology can respond to the problem of embodied cognition only if it employs S/S theory.

So PTC appears to be caught in a bind: on the basis of the only account we have, PTC must employ S/S theory in order to respond to the problem of embodied cognition. But, to provide the cognitive alternative that it aspires to, it cannot aim to give conceptualist psychological explanation. It can only find some alternative in so far as it doesn't employ conceptual content, and hence in so far as it doesn't employ S/S theory. Just as Fodor and Pylyshyn (1988) thought, S/S theory is PTC's nemesis.

But I hope to have shown (§§ 4, 6, 7, 8, and 9) how the CTC notion of non-conceptual content makes possible an alternative to S/S theory which can nevertheless speak to the problem of embodied cognition. C3 can do this because (1) CTC and the vehicle of CTC are not explanatorily independent of each other. Explanatory independence is essential to the relation between a semantic account of content, and syntax; (2) C3 provides a notion of non-conceptual content which cannot be specified by a semantic theory, and (3) there is a coherent, non-conceptualist notion of psychological explanation which can be built on top of the notion of CTC, and in terms of which we can account non-conceptually for the existence in the natural world of concept-exercising creatures. PTC and C3 belong together in a match as tight as GOFAI[105] and LOT.

But there is one point I am anticipating: that connectionist architectures are appropriate for modelling with CTC. Unlike most of the other connections between levels of Fig. 2's hierarchy, this one is largely empirical. But it is impossible to read much of the connectionist literature without having their need for a notion of non-conceptual content impressed forcefully upon one. I shall close with some simple, merely suggestive, reasons for thinking that connectionist architectures are appropriate for C3 modelling. These reasons can only be suggestive because this is an empirical issue, and, in any case, there is no reason why von Neumann computa-

---

[105] Haugeland's term for Good Old Fashioned Artificial Intelligence.

tional architectures cannot be used to model C3:[106] although it is natural to implement S/S theory in von Neumann architectures, the power of these architectures goes well beyond the use that S/S theory can make of them. It may be that PTC needs C3 more than C3 needs PTC.

Smolensky holds that in order for connectionism to model the productivity of thought, it must exploit the representational constituent structure that distributed representation makes available. Smolensky (1987, 1988a) considers (in a deliberately toy example) how connectionism represents *cup with coffee* in a structured way, without doing it as a syntactic relation between an item associated with *cup* and an item associated with *coffee*. His example is useful for our purposes because it indicates the first reason why connectionism is naturally appropriate for C3 modelling: that connectionist systems naturally adopt perspective-dependent representation. Figure 3 shows a representation of a connectionist representation of *cup with coffee*.

UNITS   MICROFEATURES

● Upright Container

● hot liquid

○ glass contacting wood

● porcelain curved surface

● burnt odour

● brown liquid contacting porcelain

○ oblong silver object

● finger-sized handle

● brown liquid with curved sides and bottom

Fig. 3.

While Figure 4 is a representation of a connectionist represention of *cup without coffee*.

---

[106] It would still be 'C3': The Computational Construction of Concepts.

```
UNITS      MICROFEATURES
```

●	Upright Container
○	hot liquid
○	glass contacting wood
●	porcelain curved surface
○	burnt odour
○	brown liquid contacting porcelain
○	oblong silver object
●	finger-sized handle
○	brown liquid with curved sides and bottom

Fig. 4.

In order to see what the representation of *coffee* might be in a connectionist system, we just have to subtract the representation of *cup without coffee* from the representation of *cup with coffee*. A representation of the result is shown in Figure 5.

The point of this is to see that the connectionist representation of *coffee* is heavily context dependent; it is a representation of *coffee-in-the-context-of-cup*. There will then also be representations of *coffee-in-the-context-of-instant-coffee-granules-in-a-jar*, *coffee-spilt-all-over-my-paper*, *coffee-in-the-context-of-someone-who-has-drunk-too-much-coffee-looking-rather-grey*, and so forth. Evidently, the inferential possibilities of combination of any one of these context-dependent representations of *coffee* will be limited to inferences that are appropriate in that context. All coffee helps you digest. But coffee-in-the-context-of-being-spilt-on-my-paper has the opposite effect. The concept *coffee* works very differently. For example, anything that can be said about coffee can be said with words that express the concept *coffee*. But an advertiser working for *Nestlé* cannot advertise *Nescafé* by expressing *coffee-in-the-context-of-someone-who-has-drunk-too-much-looking-rather-grey makes you feel wonderful*.

Connectionist representations, then, are naturally perspective-depen-

UNITS   MICROFEATURES

○ Upright Container

● hot liquid

○ glass contacting wood

○ porcelain curved surface

● burnt odour

● brown liquid contacting porcelain

○ oblong silver object

○ finger-sized handle

● brown liquid with curved sides and bottom

FIG. 5.

dent, and so fail to satisfy the generality constraint. But this need be so only for the representations on the input and output units. Much of the attention in developing connectionist algorithms has been directed to understanding how you can get useful behaviour out of systems whose input and output representations are perspective-dependent. If C3 is right, this is just what we want for modelling cognition. Weights between hidden units can be evolved so that processes which involve them are responsive to the connections between many different perspective-dependent coffee representations. And this is clearly encouraging in a context which has argued that our possession of the concept *coffee*, must—like all concepts—depend on a perspective reducing construction from numerous perspective-dependent representations of coffee.

A second reason for thinking that connectionism is appropriate for C3 modelling is that, unlike von Neumann architectures, 'learning' is central to connectionism. This makes it easier to think about how to implement C3 models in a connectionist architecture, because C3 takes learning to be central to, indeed, the essence of, cognition.

A third reason is that in any non-toy example, connectionist representations can only be interpreted by means of an analysis of the complex connections between patterns of input and patterns of output. It has often seemed to be a mystery as to what kind of analysis can be given of the significance of the activity of hidden units, especially in cases where the input units are connected directly to some sensory system, and the output units are connected directly to some effector system.[107] It was the failure to penetrate this mystery that led Fodor and Pylyshyn (1988) astray in their criticism of connectionism. They supposed that the 'sub-conceptual' space of microfeatures had to be thought of as an assignment of values to good, old-fashioned semantic features, thinly sliced:

Many connectionists hold that the mental representations that correspond to commonsense concepts (CHAIR, JOHN, CUP, etc.) are 'distributed' over galaxies of lower level units which themselves have representational content. To use common connectionist terminology, the higher or 'conceptual level' units correspond to vectors in a 'sub-conceptual' space of microfeatures. *The model here is something like the relation between a defined expression and its defining feature analysis*: thus, the concept BACHELOR might be thought to correspond to a vector in a space of features that includes ADULT, HUMAN, MALE, and MARRIED; i.e. as an assignment of the value + to the first three features and of − to the last. . . . Since microfeatures are frequently assumed to be derived automatically (i.e. via learning procedures) from the statistical properties of samples of stimuli, we can think of them as expressing the sorts of properties that are revealed by multivariate analysis of sets of stimuli. In particular, they need not correspond to English words; they can be finer-grained than, or otherwise atypical of, the terms for which a non-specialist needs to have a word. *Other than that, however, they are perfectly ordinary semantic features, much like those that lexicographers have traditionally used to represent the meanings of words* (my emphases).

But this is due simple to Fodor and Pylyshyn's blindspot. We may call it 'conceptualism's scotoma': the representational structure of a content can only be its conceptual structure, because the only kind of content there is is conceptual content. But I have not only shown that this is not so, I have shown that it can not be so. C3 provides a notion of non-conceptual content, and shows how to do justice to classical constraints on cognition by means of the perspective-dependence reducing transformations involved in the formation of a cognitive map. Connectionism may not be syntactically systematic, but PTC relies on this fact to show how connectionism can provide an alternative to classical ways of modelling cognition.

An objection which has been raised frequently against connectionism is of the form: a black box, even a very successful black box, adds nothing to

---

[107] It is sometimes thought that this difficulty is confined to the significance of individual hidden units, rather than to activity vectors over many units. The reason for this is that it is assumed that although the hidden units may well have no conceptual significance, the activity vectors will. I hope that C3 will show that this is not necessary either. The psychological analysis of cognition is non-conceptual through and through. The conceptual level of description merely provides psycho-computationally inert constraints on the psycho-computationally causally active processes.

our psychological understanding (see §1). Train up a connectionist network in some stimulus environment and it may come to perform very successfully in that environment, but since the network has programmed itself via its learning algorithm, we will understand nothing about how that performance was achieved unless we are able to examine the weights on its hidden units and *find an interpretation for those weights that allows us to understand why the system is successful.* The problem here is similar to a problem which classical artificial intelligence has also faced: how can a complex pattern of manipulations of LISP s-expressions constitute a psychological theory? LOT solved that problem, but the solution which it provided is not available to connectionism because, usually, no coherent semantics can be defined over the hidden units. It may be that all that can be determined about the functioning of a pattern of connectivity for the units is the way in which it mediates a connection between certain kinds of input activity vectors, and certain kinds of output activity vectors. How could that advance our psychological understanding?

CTC, like patterns of connectivity, is specified in terms of its powers of mediation between input and output. These powers are powers of intentionality if it can be shown how processes implemented in them can approximately satisfy the constraints on objectivity. Can theorists of connectionism show us how the powers of the hidden units can evolve under training to form a cognitive map? If so, we will be able to explore empirically a new way in which to understand how things in the world are capable of thinking about the world.

A Speculative Conclusion: Connectionist Vehicles for Cognition

In the first part of §9 we saw the kinds of computational processes of map-making and map-use which are required for increasingly perspective-independent abilities and therefore for CTC contents which come to approximate increasingly the objectivity constraints. In the second part of §9 we saw some reasons for thinking that connectionism provides an appropriate architecture in which to implement these computational processes, though of course this will remain an open question for a long while yet. But granting this, we can ask about the connectionist vehicles for CTC and cognitive maps.

The *representational* vehicles for CTC contents are whole-system abilities, but these contents also have *computational* vehicles which are the subpersonal, causal ground of particular exercises of the abilities of the whole system. It is natural to identify two kinds of connectionist vehicle: patterns of activity distributed over many processing units, and patterns of weighted connectivity between many processing units. The former patterns are more transitory than the latter patterns, but both are causally active: the processing state of the system at the next instant is a function of both the patterns of activity and the patterns of weighted connectivity. A pattern of activity will not, by itself, manifest any particular degree of perspective-dependence, since whether or not it is the partial ground of a perspective-dependent *ability* will depend on the context of the input which gave rise to the activity, the context of the output to which it will give rise, and the further dispositions of the system to which it belongs. Nevertheless, we can, in a derivative sense, speak of the perspective-dependence of a pattern of activity over some hidden units in virtue of its role in mediating between a context-dependent input representation and a context-dependent output representation (as in the coffee example). In this sense, the patterns of activity of very sophisticated and of every crude connectionist systems will be equally perspective-dependent. The differences between the contents implemented in sophisticated and crude networks comes in virtue of the other kind of connectionist vehicle: patterns of connectivity.

Tentatively, we may say that a pattern of connectivity implements a cognitive map if it mediates between context-dependent patterns of input and context-dependent patterns of output in a maplike way. The map-maker had always to employ egocentric, context-dependent presentations of the terrain as the basis for his map-making, and the map-user had always to generate from the map egocentric, context-dependent instructions for moving within the terrain. We can, therefore, think of a map as a function from egocentric, context-dependent representations to egocentric, context-dependent representations. A sophisticated map which represents the mapped space from no particular point of view will belong to the class of functions which will enable the map-user to find his or her way around the terrain in a perspective-independent manner. A pattern of connectivity within a connectionist network can implement a cognitive map if it implements a function from egocentric, context-dependent patterns of activity distributed over the input units to egocentric, context-dependent patterns of activity distributed over the output units. It will implement a sophisticated, objectivity-grounding cognitive map if this function belongs to the class of functions which will enable a perspective-independent ability to negotiate the mapped domain.

It is the possession of a cognitive map of the right class which is the causal ground of a perspective-independent capacity of the system to find its way around the mapped domain. Since we saw that it is the perspective-independence of capacities to find one's way around which is the basis of the satisfaction of the objectivity constraints by those contents which are canonically specified by reference to these capacities, it is natural to identify the cognitive map as the causal ground of the possession by the system of a concept. Hence we may say that concepts are implemented in a connectionist system as a pattern of weighted connectivity.

A given set of weighted connections may implement more than one cognitive map. Indeed we should expect this, given the holism of concepts. (The causal ground of the logical impossibility of possessing one concept without possessing a set of conceptually connected concepts might be the fact that it is the same set, or extensively overlapping sets, of connections which implement the cognitive maps corresponding to each of these concepts.) In an extreme case, every concept possessed by a system would be implemented in every connection of the entire system. Concept possession would be causally legitimated by the scientific levels of a C3 framework, but conceptual characterizations of content would play no role in the scientific psychological explanations of the behaviour of the system.[108]

[108] So if this account provides a sense in which concepts have scientific reality, it is a sense which is entirely compatible with considerable indeterminacy of *conceptual* content.

This paper is dedicated to the memory of my father, Manny Cussins (1905–1987).

I would like to thank for their help with the paper or with the development of its ideas: Dan Brotsky, John Campbell, David Charles, Bill Child, Ron Chrisley, Andy Clark, Michael Dummett, John Haugeland, Dimitri Kullmann, David Levy, Michael Martin, Geoff Nunberg, Gerard O'Brien, Christopher Peacocke, Jeff Schrager, Paul Skokowski, Scott Stornetta, Brian Smith, Susan Stucky, and Debbie Tatar. I have benefited from talks at Temple University, at the 1988 Society for Philosophy and Psychology Annual Meeting in North Carolina, at the Institute for Research on Learning, at the System Sciences Laboratory at Xerox PARC, at CSLI, Stanford, at David Charles's Oriel Discussion Group, and at a Birkbeck Discussion Group on Connectionism. I am grateful for support from a Junior Research Fellowship at New College, Oxford, a Post-doctoral Fellowship at CSLI, Stanford, resource support from the System Sciences Laboratory, Xerox PARC, and the remarkable research environment at PARC. And above all to Charis, for her inspiration and confidence.

REFERENCES

Armstrong, D. M. (1968). *A Materialist Theory of Mind*. London: Routledge & Kegan Paul.

Barwise, J. (1987). 'Unburdening the Language of Thought.' In *Two Replies*, CSLI Report No. CSLI-87-74.

Bennett, J. (1976). *Linguistic Behaviour*. Cambridge: Cambridge University Press.

Block, N. (ed.) (1980). *Readings in Philosophy of Psychology*, Vol. I. London: Methuen.

Brachman, R. J., and Levesque, H. J. (1985). *Readings in Knowledge Representation*. Los Altos: Morgan Kaufmann Publishers.

Campbell, J. (1986). 'Conceptual Structure.' In C. Travis (ed.), *Meaning and Interpretation*, pp. 159–74. Oxford: Basil Blackwell.

Churchland, P. M. (1979). 'Eliminative Materialism and the Propositional Attitudes.' *Journal of Philosophy* 78: 67–90.

Churchland, P. S. (1986). *Neurophilosophy: Toward a Unified Science of the Mind–Brain*. Cambridge, Mass.: MIT Press/Bradford Books.

Cussins, A. (Jan. 1987). 'Varieties of Psychologism.' *Synthese* 70: 123–54.

—— (May 1987). 'Being Situated Versus Being Embedded.' Stanford University, *CSLI Monthly* 2 (7):14–20.

—— (1988). 'Dennett's Realisation Theory of the Relation between Folk and Scientific Psychology', Commentary on Dennett: *The Intentional Stance*. *Behavioural and Brain Sciences* 11 (3): 508–9.

—— (1992). 'The Limitations of Pluralism.' In D. Charles and K. Lennon (eds.), *Reduction, Explanation and Realism*. Oxford: Oxford University Press.

—— (forthcoming). 'The Emergence of Objectivity: Why People are not just Complex Frogs.'

Davidson, D. (1967). 'Truth and Meaning.' In Davidson (1984).

—— (1974). 'On the Very Idea of a Conceptual Scheme.' *Proc. & Addresses of Amer. Philos. Assoc.* 47, and in Davidson (1984).

—— (1984). *Inquiries into Truth and Interpretation*. Oxford: Clarendon Press.

Dennett, D. C. (1969). *Content and Consciousness*. London: Routledge & Kegan Paul.

—— (1978). 'Toward a Cognitive Theory of Consciousness.' In D. C. Dennett, *Brainstorms: Philosophical Essays on Mind and Psychology*, pp. 149–73. Montgomery, Vt: Bradford Books.

—— (1987). *The Intentional Stance*. Cambridge, Mass.: MIT Press.

Dretske. F. (1981). *Knowledge and the Flow of Information*. Cambridge, Mass.: MIT Press.

Dummett, M. (1975). 'What is a Theory of Meaning.' In S. Guttenplan (ed.), *Mind and Language*, pp. 97–138. Oxford: Clarendon Press.

—— (1976). 'What is a Theory of Meaning (II).' In G. Evans and J. McDowell (eds.), *Truth and Meaning*, Oxford: Oxford University Press.

—— (1978). *Truth and Other Enigmas*. London: Duckworth.

Evans, G. (1982). *The Varieties of Reference*. Oxford: Oxford University Press.

—— (1980, 1985). 'Things Without the Mind – A Commentary Upon Chapter Two of Strawson's *Individuals*.' In Evans (1985), *Collected Papers*, pp. 249–90. Oxford: Clarendon Press.

Fodor, J. A. (1976). *The Language of Thought*. Sussex: Harvester Press.

—— (1980). 'On the Impossibility of Acquiring More Powerful Structures', and 'Reply to Putnam.' In M. Piatelli-Palmarini (ed.), *Language and Learning: The Debate between Jean Piaget and Noam Chomsky*, pp. 142–62 and 325–34. London: Routledge & Kegan Paul.

—— (1981a). 'Methodological Solipsism Considered as a Research Strategy in Cognitive Psychology.' In Fodor (1981b).

—— (1981b). *Representations*. Cambridge, Mass.: MIT Press.

—— (1983). *The Modularity of Mind: An Essay on Faculty Psychology.* Cambridge, Mass.: MIT Press/Bradford Books.

—— (1986). 'Why Paramoecia Don't Have Mental Representations.' *Midwest Studies in Philosophy* 10: 3–23.

—— (1987). *Psychosemantics.* Cambridge, Mass.: MIT Press.

—— and Pylyshyn, Z. W. (1988). 'Connectionism and Cognitive Architecture.' *Cognition* 28: 3–71.

Frege, G. (1891). 'On Sense and Meaning.' In P. Geach and M. Black (1960), *Translations from the Philosophical Writings of Gottlob Frege*, pp. 56–78 Oxford: Oxford University Press.

—— (1918, 1977). 'Thoughts.' In P. T. Geach (ed.), *Logical Investigations*, pp. 1–30. Oxford: Basil Blackwell.

Gibson, J. J. (1979). *An Ecological Approach to Visual Perception.* Boston: Houghton Mifflin.

Goodman, N. (1951). *The Structure of Appearance.* Cambridge, Mass.: Harvard University Press.

Hofstadter, D. R. (1985). 'Waking Up from the Boolean Dream; Or, Subcognition as Computation.' In D. R. Hofstadter, *Metamagical Themas: Questing for the Essence of Mind and Pattern*, pp. 631–65. New York: Basic Books.

Israel, D. (1987). 'The Role of Propositional Objects of Belief.' CSLI report No. CSLI-87-72.

Lenat, D. B., and Feigenbaum, E. A. (1987). 'On the Thresholds of Knowledge.' MCC-AI Non-Proprietary Technical Report.

Lewis, D. (1966). 'An Argument for the Identity Theory.' *Australasian Journal of Philosophy* 63: 17–25.

Marr, D. (1977). 'Artificial Intelligence – A Personal View.' *Artificial Intelligence* 9: 37–48, and reprinted in this volume.

—— (1982). *Vision: A Computational Investigation into the Human Representation and Processing of Visual Information.* San Francisco: W. H. Freeman.

McDowell, J. (1977). 'On the Sense and Reference of a Proper Name,' *Mind* 86: 151–85.

Millikan, R. (1984). *Language, Thought and Other Biological Categories.* Cambridge, Mass.: MIT Press/Bradford Books.

O'Keefe, J., and Nadel, L. (1978). *The Hippocampus as a Cognitive Map.* Oxford: Oxford University Press.

Peacocke, C. (1986). *Thoughts: An Essay on Content.* Oxford: Basil Blackwell.

—— (1989). Inaugural Lecture, given in the Examination Schools, University of Oxford, during Trinity Term.

—— (forthcoming). 'Perceptual Content.' In a volume in honour of D. Kaplan, edited by J. Perry, J. Almog, and H. Wettstein.

Pellionisz, A., and Llinas, R. (1979). 'Brain Modelling by Tensor Network Theory and Computer Simulation.' *Neuroscience* 4: 323–48.

—— (1980). 'Tensorial Approach to the Geometry of Brain Function.' *Neuroscience* 5: 1125–36.

—— (1982). 'Space-Time Representation in the Brain.' *Neuroscience* 7: 2949–70.

—— (1985). 'Tensor Network Theory of the Metaorganization of Functional Geometries in the Central Nervous System.' *Neuroscience* 16: 245–73.

Perry, J. (1979). 'The Problem of the Essential Indexical', *Nous* 13: 3–21.

Place, U. T. (1970). 'Is Consciousness a Brain Process?' In C. Borst (ed.), *The Mind Brain Identity Theory*, pp. 42–51. London: Macmillan.

Putnam, H. (1973). 'Reductionism and the Nature of Psychology.' *Cognition* 2: 131–46.

—— (1975). 'The Meaning of Meaning.' In H. Putnam, *Mind, Language and Reality*, pp. 215–71. Cambridge: Cambridge University Press.

Pylyshyn, Z. W. (1984). *Computation and Cognition: Toward a Foundation for Cognitive Science*. Cambridge, Mass.: MIT Press.

Quine, W. (1960). *Word and Object*. Cambridge, Mass.: MIT Press.

Reifel, S. (1987). 'The SRI Mobile Robot Testbed: A Preliminary Report.' technical note 413, SRI International, Menlo Park, Calif.

Rosenschein, S. (forthcoming). *An Introduction to Situated Automata*. Forthcoming in the CSLI Lecture Note Series, Chicago Press.

Rumelhart, D., McClelland, J., and the PDP Research Group (1986). *Parallel Distributed Processing: Explorations in the Microstructure of Cognition*, Vol. 1 and 2. Cambridge, Mass.: MIT Press/Bradford Books.

Schiffer, S. (1987). *Remnants of Meaning*. Cambridge, Mass.: MIT Press/Bradford Books.

Smart, J. (1970). 'Sensations and Brain Processes.' In C. Borst (ed.), *The Mind Brain Identity Theory*, pp. 52–66. Place: Macmillan.

Smith, B. (1987). *The Correspondence Continuum*. Center for the Study of Language and Information, Report No. CSLI-87-71.

Smolensky, P. (1987). 'The Constituent Structure of Connectionist Mental States: A Reply to Fodor and Pylyshyn.' *The Southern Journal of Philosophy* 26 Suppl.: 137–60.

—— (1988a). 'On the Proper Treatment of Connectionism.' *Behavioral and Brain Sciences* 11: 1–74.

—— (1988b). Department of Computer Science, University of Colorado at Boulder, Technical Report on Tensor Representation.

Stich, S. (1983). *From Folk Psychology to Cognitive Science: The Case Against Belief*. Cambridge, Mass.: MIT Press.

Strawson, P. (1959). *Individuals*. London: Methuen.

Swinburne, R. (1986). *The Evolution of the Soul*. Oxford: Clarendon Press.

Williams, S. G. 'Computers, Validity and the Runabout Inference Ticket', Worcester College, Oxford University unpublished paper.

Winograd, T. (1973). 'A Procedural Model of Language Understanding.' In R. C. Schank and K. M. Colby (eds.), *Computer Models of Thought and Language*, pp. 152–86. San Francisco: W. H. Freeman.

NOTES ON THE CONTRIBUTORS

MARGARET A. BODEN is Professor of Philosophy and Psychology in the School of Cognitive and Computing Sciences, University of Sussex.

PAUL M. CHURCHLAND is Professor of Philosophy in the Cognitive Science Faculty at the University of California at San Diego.

ANDY CLARK is Lecturer in Philosophy with Cognitive Studies in the School of Cognitive and Computing Sciences, University of Sussex.

ADRIAN CUSSINS is Assistant Professor of Philosophy at the University of California, San Diego.

DANIEL C. DENNETT is Professor of Philosophy and Director of the Center for Cognitive Studies at Tufts University.

HUBERT L. DREYFUS is Professor of Philosophy at the University of California at Berkeley.

STUART E. DREYFUS is Professor of Industrial Engineering and Operations Research at the University of California at Berkeley.

PATRICK J. HAYES is Senior Research Scientist at Xerox Palo Alto Research Center.

GEOFFREY E. HINTON is Professor of Computer Science at the University of Toronto.

JAMES L. MCCLELLAND is Professor of Psychology at Carnegie-Mellon University.

WARREN S. MCCULLOCH was a psychiatrist and neurophysiologist, and a member of the Research Laboratory of Electronics at the Massachusetts Institute of Technology.

DREW MCDERMOTT is Professor of Computer Science at Yale University.

DAVID C. MARR was Director of the Vision Research Laboratory at the Massachusetts Institute of Technology.

ALLEN NEWELL is Professor of Psychology and Computer Science at Carnegie-Mellon University.

WALTER H. PITTS was Lecturer in Mathematics at the Massachusetts Institute of Technology.

DAVID E. RUMELHART is Professor of Psychology at Stanford University.

JOHN R. SEARLE is Professor of Philosophy at the University of California at Berkeley.

HERBERT A. SIMON is Professor of Psychology and Computer Science at Carnegie-Mellon University.

AARON SLOMAN is Head of School of Computer Science, University of Birmingham.

ALAN M. TURING's biography (by Andrew Hodges) is: *Alan Turing: The Enigma.*

SELECTED BIBLIOGRAPHY

BOOKS

Items marked with an asterisk provide extensive references to primary sources in AI.

ARBIB, M. A. (1985). *In Search of the Person: Philosophical Explorations in Cognitive Science*. Amherst: University of Massachusetts Press.

BODEN, M. A. (1972). *Purposive Explanation in Psychology*. Cambridge, Mass.: Harvard University Press.

*—— (1987). *Artificial Intelligence and Natural Man*, 2nd edn. London: MIT Press; New York: Basic Books.

*—— (1988). *Computer Models of Mind: Computational Approaches in Theoretical Psychology*. Cambridge: Cambridge University Press.

—— (1990). *The Creative Mind: Myths and Mechanisms*. London: Weidenfeld & Nicholson.

BOLTER, J. D. (1984). *Turing's Man: Western Culture in the Computer Age*. London: Duckworth.

*CHARNIAK, E., and McDERMOTT, D. (1984). *Introduction to Artificial Intelligence*. Reading, Mass.: Addison-Wesley.

CHURCHLAND, P. S. (1986) *Neurophilosophy: Toward a Unified Science of the Mind–Brain*. Cambridge, Mass.: MIT Press/Bradford Books.

CLARK, A. J. (1989). *Microcognition: Philosophy, Cognitive Science, and Parallel Distributed Processing*. Cambridge, Mass.: MIT Press/Bradford Books.

—— (1993). *Associative Engines: Connectionism, Concepts, and Representational Change*. Cambridge, Mass.: MIT Press.

DENNETT, D. C. (1984). *Elbow Room: The Varieties of Free Will Worth Wanting*. Cambridge, Mass.: MIT Press/Bradford Books.

DREYFUS, H. L. (1979). *What Computers Can't Do: The Limits of Artificial Intelligence*, 2nd edn. New York: Harper & Row.

—— and DREYFUS, S. E. (1986). *Mind Over Machine: The Power of Human Intuition and Expertise in the Era of the Computer*. New York: Free Press, Macmillan.

FODOR, J. A. (1975). *The Language of Thought*. Hassocks, Sussex: Harvester Press.

—— (1983). *The Modularity of Mind: An Essay on Faculty Psychology*. Cambridge, Mass.: MIT Press/Bradford Books.

—— (1988). *Psychosemantics: The Problem of Meaning in the Philosophy of Mind*. Cambridge, Mass.: MIT Press/Bradford Books.

HAUGELAND, J. (1985). *Artificial Intelligence: The Very Idea*. Cambridge, Mass.: MIT Press/Bradford Books.

*HINTON, G. E., and ANDERSON, J. A. (eds.) (1981). *Parallel Models of Associative Memory*. Hillsdale, NJ: Erlbaum.

HOFSTADTER, D. R. (1979). *Godel, Escher, Bach: An Eternal Golden Braid*. New York: Basic Books.

HOLLAND, J. H., HOLYOAK, K. J., NISBETT, R. E., and THAGARD, P. R. (1986). *Induction: Processes of Inference, Learning, and Discovery*. Cambridge, Mass.: MIT Press/Bradford Books.

JACKENDOFF, R. (1987). *Consciousness and the Computational Mind*. Cambridge, Mass.: MIT Press/Bradford Books.

JOHNSON, M. L. (1988). *Mind, Language, Machine: Artificial Intelligence in the Poststructuralist Age.* London: Macmillan.

JOHNSON-LAIRD, P. N. (1983). *Mental Models: Towards a Cognitive Science of Language, Inference, and Consciousness.* Cambridge: Cambridge University Press.

LANGLEY, P., SIMON, H. A., BRADSHAW, G. L., and ZYTKOW, J. M. (1987). *Scientific Discovery: Computational Explorations of the Creative Processes.* Cambridge, Mass.: MIT Press.

MARR, D. C. (1982). *Vision: A Computational Investigation into the Human Representation and Processing of Visual Information.* San Francisco: Freeman.

PUTNAM, H. (in press). *Representation and Reality.* Cambridge, Mass.: MIT Press/ Bradford Books.

PYLYSHYN, Z. W. (1984) *Computation and Cognition: Toward a Foundation for Cognitive Science.* Cambridge, Mass.: MIT Press/Bradford Books.

*RUMELHART, D. E., and McCLELLAND, J. L. (eds.) (1986). *Parallel Distributed Processing: Explorations in the Microstructure of Cognition.* Vol. 1: *Foundations*; Vol. 2: *Psychological and Biological Models.* Cambridge, Mass.: MIT Press/ Bradford Books.

SEARLE, J. R. (1984). *Minds, Brains, and Science.* London: BBC Publications.

SLOMAN, A. (1978). *The Computer Revolution in Philosophy: Philosophy, Science, and Models of Mind.* Brighton: Harvester Press.

—— (forthcoming). *Artificial Intelligence: The Philosophical Foundations.* To be published by Ellis Horwood.

STICH, S. C. (1983). *From Folk Psychology to Cognitive Science: The Case Against Belief.* Cambridge, Mass.: MIT Press/Bradford Books.

SUSSKIND, R. (1987). *Expert Systems in Law: A Jurisprudential Inquiry.* Oxford: Clarendon Press.

THAGARD, P. (1988). *Computational Philosophy of Science.* Cambridge, Mass.: MIT Press/Bradford Books.

WINOGRAD, T., and FLORES, F. (1986). *Understanding Computers and Cognition: A New Foundation for Design.* Norwood, NJ: Ablex.

COLLECTIONS AND ANTHOLOGIES

ANDERSON, A. R. (ed.) (1964). *Minds and Machines.* Englewood-Cliffs, NJ: Prentice-Hall.

*ANDERSON, J. A., and ROSENFELD, E. (1988). *Neurocomputing: A Reader.* Cambridge, Mass.: MIT Press/Bradford Books.

BODEN, M. A. (1981). *Minds and Mechanisms: Philosophical Psychology and Computational Models.* Ithaca: Cornell University Press.

—— (1989). *Artificial Intelligence in Psychology: Interdisciplinary Essays.* Cambridge, Mass.: MIT Press/Bradford Books.

*BRACHMAN, R. J., and LEVESQUE, H. J. (eds.) (1985). *Readings in Knowledge Representation.* Los Altos, Calif.: Morgan Kaufmann.

CHURCHLAND, P. M. (in press). *The Neurocomputational Perspective.* Cambridge, Mass.: MIT Press/Bradford Books.

DENNETT, D. C. (1978). *Brainstorms: Philosophical Essays on Mind and Psychology.* Cambridge, Mass.: MIT Press/Bradford Books.

—— (1987). *The Intentional Stance*. Cambridge, Mass.: MIT Press/Bradford Books.

FEIGENBAUM., E. A., and FELDMAN, J. (eds.) (1963). *Computers and Thought*. New York: McGraw-Hill.

FODOR, J. A. (1981). *Representations: Philosophical Essays on the Foundations of Cognitive Science*. Brighton: Harvester Press.

GRAUBARD, S. R. (ed.) (1988). *The Artificial Intelligence Debate: False Starts, Real Foundations*. Cambridge, Mass.: MIT Press.

HAUGELAND, J. (ed.) (1981). *Mind Design: Philosophy, Psychology, Artificial Intelligence*. Cambridge, Mass.: MIT Press/Bradford Books.

HOFSTADTER, D. R., and DENNETT, D. C. (eds.) (1981). *The Mind's I: Fantasies and Reflections on Mind and Soul*. New York: Basic Books.

HOOK, S. (ed.) (1960). *Minds and Machines: A Symposium*. New York: New York University Press.

MCCULLOCH, W. S. (1965). *Embodiments of Mind*. Cambridge, Mass.: MIT Press.

MACHLUP, F., and MANSFIELD, U. (eds.) (1983). *The Study of Information: Interdisciplinary Messages*. New York: Wiley.

MOHYELDIN SAID, K., and WILKES, K. V. (eds.) (in press). *Seminars on Cognitive Science*. Oxford: Oxford University Press.

PUTNAM, H. (1975). *Mind, Language and Reality* (Philosophical Papers, Vol. 2). Cambridge: Cambridge University Press.

PYLYSHYN, Z. W. (ed.) (1987). *The Robot's Dilemma: The Frame Problem in Artificial Intelligence*. Norwood, NJ: Ablex.

RINGLE, M. (ed.) (1979). *Philosophical Perspectives in Artificial Intelligence*. Brighton: Harvester Press.

STICH, S. P., RUMELHART, D. E., and RAMSEY, W. (eds.) (forthcoming). *Philosophy and Connectionism*. Hillsdale, NJ: Erlbaum.

TORRANCE, S., and VIENNE, J.-M. (eds.) (1984). *The Mind and the Machine*. Chichester: Ellis Horwood.

—— and SPENCER-SMITH, R. (eds.) (in press). *Philosophy and Computation*. Norwood, NJ: Ablex.

ARTICLES

Articles included in a collection or anthology listed above are not itemized below.

ABELSON, R. P. (1973). 'The Structure of Belief Systems.' In R. C. Schank and K. M. Colby (eds.), *Computer Models of Thought and Language*, pp. 287–340. San Francisco: Freeman.

BECHTEL, W. (1988). 'Connectionism and Rules and Representation Systems: Are They Compatible?' *Philosophical Psychology* 1: 5–16.

BROADBENT, D. (1985). 'A Question of Levels: Comments on McClelland and Rumelhart.' *J. Experimental Psychol: General* 114: 189–92.

CHURCHLAND, P. M., and CHURCHLAND, P. S. (1982), 'Functionalism, Qualia, and Intentionality.' In J. I. Biro and R. W. Shehan (eds.), *Mind, Brain, and Function: Essays in the Philosophy of Mind*, pp. 121–45. Norman: University of Oklahoma Press.

CLARK, A. J. (1987a). 'Connectionism and Cognitive Science.' In J. Hallam and C. Mellish (eds.), *Advances in Artificial Intelligence*, pp. 3–15. Chichester: Wiley.

CLARK, A. J. (1987*b*). 'The Kludge in the Machine.' *Mind and Language* 2: 277–300.

COHEN, L. J. (1981). 'Can Human Irrationality be Experimentally Demonstrated?' *Behavioral and Brain Sciences* 4: 317–31.

DAVIES, M. (1987). 'Tacit Knowledge and Semantic Theory.' Mind (NS) 96: 441–62.

—— (in press). 'Connectionism, Modularity, and Tacit Knowledge.' *Brit. J. Phil. Science.*

DENNETT, D. C., and HAUGELAND, J. (1987). 'Intentionality.' In R. L. Gregory (ed.), *The Oxford Companion to Mind*, pp. 383–6. Oxford: Oxford University Press.

DRESHER, E., and HORNSTEIN, N. (1976). 'On Some Supposed Contributions of Artificial Intelligence to the Scientific Study of Language.' *Cognition* 4: 321–78.

DRETSKE, F. (1985), 'Machines and the Mental.' *Proc. & Addresses of Amer. Philos. Assoc.* 59: 23–33.

FODOR, J. A., and PYLYSHYN, Z. W. (1988). 'Connectionism and Cognitive Architecture: A Critical Analysis.' *Cognition* 28: 3–71.

HARRE, R. (1988). 'Wittgenstein and Artificial Intelligence.' *Philosophical Psychology* 1: 105–16.

HOFSTADTER, D. R. (1985). 'Waking Up from the Boolean Dream; Or, Subcognition as Computation.' In D. R. Hofstadter, *Metamagical Themas: Questing for the Essence of Mind and Pattern*, pp. 631–65. New York: Viking.

ISRAEL, D. (1985). 'A Short Companion to the Naive Physics Manifesto.' In J. C. Hobbs and R. C. Moore (eds.), *Formal Theories of the Commonsense World*, pp. 427–47. New York: Ablex.

McCARTHY, J., and HAYES, P. J. (1969). 'Some Philosophical Problems from the Standpoint of Artificial Intelligence.' In B. Meltzer and D. Michie (eds.), *Machine Intelligence 4*, pp. 463–502. Edinburgh: Edinburgh University Press.

McCLELLAND, J. L., and RUMELHART, D. E. (1985). 'Levels Indeed!— A Response to Broadbent.' *J. Experimental Psychol: General* 114: 193–7.

MALCOLM, N. (1971). 'The Myth of Cognitive Processes and Structures.' In T. Mischel (ed.), *Cognitive Development and Epistemology*, pp. 385–92. New York: Academic Press.

MARTIN, M. (1973). 'Are Cognitive Processes and Structure a Myth?' *Analysis* 33: 83–8.

NEWELL, A. (1980). 'Physical Symbol Systems.' Cognitive Science 4: 135–83.

PEACOCKE, C. (1986). 'Explanation in Computational Psychology: Language, Perception, and Level 1.5.' *Mind and Language* 1: 101–23.

PINKER, S., and PRINCE, A. (1988). 'On Language and Connectionism: Analysis of a Parallel Distributed Processing Model of Language Acquisition.' *Cognition* 28: 73–193.

PUTNAM, H. (1983). 'Computational Psychology and Interpretation Theory.' In H. Putnam, *Realism and Reason* (Philosophical Papers, Vol. 3). Cambridge: Cambridge University Press.

ROBINSON, G. (1972). 'How to Tell Your Friends from Machines.' *Mind* (NS) 81: 504–18.

SAYRE, K. (1986). 'Intentionality and Information Processing: An Alternative Model for Cognitive Science.' *Behavioral and Brain Sciences* 9: 121–60.

SEARLE, J. R. Peer-commentary on J. R. Searle's 'Minds, Brains, and Programs.' *Behavioral and Brain Sciences* 3 (1980): 424–57; 5 (1982): 338–48.

SIMON, H. A. (1981). 'Cognitive Science: The Newest Science of the Artificial.' In D. A. Norman (ed.), *Perspectives on Cognitive Science*, pp. 13–26.

SLOMAN, A. (1979). 'The Primacy of Non-Communicative Language.' In M. McCafferty and K. Gray (eds.), *The Analysis of Meaning*, pp. 271–88. London: ASLIB and Brit. Comp. Soc.

—— (1986). 'Did Searle Attack Strong Strong or Weak Strong AI?' In A. G. Cohn and J. R. Thomas (eds.), *Artificial Intelligence and Its Applications*. London: Wiley.

—— and COHEN, L. J. (1986). 'Symposium: What Sorts of Machines Can Understand the Symbols They Use?' *Proc. Aristotelian Soc.* Supp. 60: 61–95.

SMOLENSKY, P. (1987). 'Connectionist AI, Symbolic AI, and the Brain.' *AI Review* 1: 95–110.

*—— (1988). 'On the Proper Treatment of Connectionism' (with peer-commentary and author's reply). *Behavioral and Brain Sciences* 11: 1–74.

THORPE, S., and IMBERT, M. (in press). 'Neuroscientific Constraints on Connectionist Modelling.' In R. Pfeiffer, Z. Schreber, F. Fogelman, and T. Bernold (eds.), *Connectionism in Perspective*. Amsterdam: North-Holland.

TURING, A. M. (1936). 'On Computable Numbers, with an Application to the *Entscheidungsproblem*.' *Proc. London Math. Society*, Series 2, 42: 230–65.

INDEX